MOLECULAR
MEDICINE

For Churchill Livingstone

Publisher: Timothy Horne
Project Manager: Ninette Premdas
Copy Editor: Tim Jackson
Project Controller: Kay Hunston

MOLECULAR MEDICINE

An Introductory Text

Second Edition

R. J. Trent

PhD BSc (Med) MBBS (Syd) DPhil (Oxon) FRACP FRCPA
Professor of Molecular Genetics, Department of Molecular Genetics,
Royal Prince Alfred Hospital, New South Wales,
Australia

**CHURCHILL
LIVINGSTONE**

EDINBURGH LONDON NEW YORK PHILADELPHIA ST LOUIS SYDNEY TORONTO 1997

Churchill Livingstone
An imprint of Harcourt Publishing Limited

First edition 1993
Second edition 1997
   Reprinted 1999

ISBN 0 443 053669

British Library of Cataloguing in Publication Data
A catalogue record for this book is available from the British Library.

Library of Congress Cataloging in Publication Data
A catalog record for this book is available from the Library of Congress.

Printed in China
NPCC/02

Medical knowledge is constantly changing. As new information becomes available, changes
in treatment, procedures, equipment and the use of drugs become necessary. The author
and the publishers have, as far as it is possible, taken care to ensure that the information
given in this text is accurate and up to date. However, readers are strongly advised to
confirm that the information, especially with regard to drug usage, complies with current
legislation and standards of practice.

The
publisher's
policy is to use
**paper manufactured
from sustainable forests**

CONTENTS

PREFACE

The years since the 1st edition of *Molecular Medicine* have been accompanied by many developments. Key genes identified included *BRCA1* (breast cancer), *OB* (obese) gene and those involved in DNA repair. With this wealth of information it is not surprising that the media have had a lot to say and the public, more than ever before, is aware of the importance of genes. Another noteworthy change has been the increase in gene therapy trials with the emphasis shifting from genetic disorders to the cancers and HIV infection. The rapidly evolving implications of DNA in medicine (molecular medicine) continue to stimulate and at the same time place additional pressures on health practitioners in terms of their education.

The biggest problem when writing the 2nd edition was to keep the size down since there were so many concepts,

facts or examples that needed to be included. In this endeavour there was only partial success. The helpful comments made by reviewers of the 1st edition have been incorporated into the text. Members of my laboratory who contributed DNA testing examples include Ms H. Le, Ms Cheryl Ryce and Dr Con Tomaras, I would like to thank Sr Regis Dunne for her helpful comments on Chapter 9. Ms Carol Yeung worked enthusiastically and skillfully in helping with the preparation of the text and diagrams just as she did for the 1st edition.

I dedicate this edition of *Molecular Medicine* to my mother and sister, Lynette, whose support over the years has been much appreciated.

Camperdown, 1996
R J T

1

HISTORICAL DEVELOPMENTS

A CHRONOLOGY OF MOLECULAR MEDICINE

THE EARLY DAYS: 1871 TO THE MID-1980s

DNA and molecular medicine

In 1871, Miescher isolated nucleoprotein. It was established to be an acidic material and was called 'nuclein'. From this came *nucleic acid*. In 1944, Avery and colleagues showed that the genetic information in the bacterium *Pneumococcus* was found within its DNA. Six years later, Chargaff demonstrated that there were equal numbers of the nucleotide bases adenine and thymine as well as the bases guanine and cytosine in DNA. This, as well as the X-ray crystallographic studies by Franklin and Wilkins and work by Watson and Crick in the 1950s, led to the description of DNA as a *double-stranded* structure. Subsequently, it was shown that the complementary strands which made up the DNA helix separated during replication. In 1956, a new enzyme was discovered by Kornberg. It was called *DNA polymerase* and enabled small segments of double-stranded DNA to be synthesised (see Ch. 2).

Other discoveries during the 1960s included the finding of *mRNA* (messenger RNA) which provided the link between the nucleus and the site of protein synthesis in the cytoplasm and identification of autonomously replicating, extrachromosomal DNA elements called *plasmids*. These were shown to carry genes such as those coding for antibiotic resistance in bacteria. Plasmids would later be used extensively by the genetic engineers. A landmark in this decade was the definition of the full *genetic code* which showed that each amino acid was encoded in DNA by a nucleotide triplet (see Ch. 2). In 1961, Lyon proposed that one of the two X chromosomes in female mammals was normally inactivated. The process of *X-inactivation* enabled males and females to have equivalent DNA content despite differing numbers of X chromosomes. In 1966, McKusick

published *Mendelian Inheritance in Man* (MIM), a catalogue of genetic disorders in humans. This became a forerunner to the many databases which would subsequently be created to store and transfer information on DNA from humans and many other species (see Chs 3, 10 and the *Human Genome Project* on p.6).

Technological developments in molecular medicine

The dogma that DNA → RNA → protein was shown to be incorrect when Temin and Baltimore showed in 1970 that *reverse transcriptase*, an enzyme found in the RNA retroviruses, allowed RNA to be copied into DNA. This enzyme would later provide the genetic engineer with a means to produce DNA copies (known as complementary DNA or cDNA) from RNA templates. Reverse transcriptase also explained how some viruses could integrate their own genetic information into the host's genome (see Chs 2, 3, 5–7).

Enzymes called *restriction endonucleases* were isolated from bacteria by Smith, Nathans, Arber and colleagues during the late 1960s and early 1970s. Restriction endonucleases were shown to digest DNA at specific sites determined by the underlying nucleotide base sequences. A method now existed to produce DNA fragments of known sizes (Box 1.1). At about this time an enzyme called *DNA ligase* was described. It allowed DNA fragments to be joined. The first *recombinant DNA* molecules comprising segments that had been 'stitched together' were produced in 1972. Berg was later awarded a Nobel Prize for his contribution to the construction of recombinant DNA molecules. This was one of many Nobel Prizes that resulted from work which had or would have a direct impact on molecular medicine (Table 1.1). Also in 1972,

Box 1.1 DNA mapping with restriction endonuclease enzymes

Restriction endonucleases are enzymes which have been isolated from bacteria. The unique features of the restriction enzymes are that they recognise specific nucleotide sequences in double-stranded DNA and cleave both strands of the duplex. For example, the restriction enzyme *Eco*RI (derived from the bacterium *E. coli* strain *R*YI) will recognise the sequence G▼AATTC and digest at the site marked▼. In the cell of origin each restriction enzyme is part of a restriction–modification system consisting of the *restriction* enzyme and a matched *modification* enzyme which recognises and modifies (usually by methylation) the same nucleotide sequence in DNA recognised by the restriction enzyme. Modification protects cellular DNA from cleavage whilst foreign (unmodified) DNA is cleaved. The restriction enzyme system is widespread in bacteria and the names of the enzymes are derived from the source bacteria, e.g. *Hpa*II, *Haemophilus parainfluenzae*; *Bam*HI, *Bacillus amyloliquifaciens*; *Not*I - *Nocardia ottidis*, etc. The usefulness of restriction enzymes in the analysis of DNA lies in their specificity for nucleotide sequences which are usually four to six base pairs in length. Restriction enzymes are analogous to specific proteolytic enzymes and are extensively used to digest DNA into fragments of known length. In bacteria, the restriction enzymes may function as a primitive immune system to protect from foreign DNA, e.g. viral DNA (a further description of these enzymes is given in Ch. 2).

Table 1.1 Molecular medicine and some Nobel Prize winners.

Year	Recipients	Subject
1959	S Ochoa, A Kornberg	In vitro synthesis of nucleic acids
1962	J D Watson, F Crick, M H F Wilkins	Structure of DNA
1975	D Baltimore, H Temin, R Dulbecco	Reverse transcriptase and oncogenic viruses
1978	W Arber, D Nathans, H D Smith	Restriction endonucleases
1980	P Berg, W Gilbert, F Sanger	Creation of first recombinant DNA molecule and DNA sequencing
1989	J M Bishop, H E Varmus	Oncogenes
1989	S Altman, T R Cech	RNA and 'ribozymes'
1993	R Roberts, P Sharp	Gene splicing
1993	K Mullis, M Smith	Polymerase chain reaction (PCR) and site-directed mutagenesis
1995	E Lewis, C Nusslein-Volhard, E Wieschaus	Genetic mechanisms in early embryonic development

Cohen and colleagues showed that DNA could be inserted into plasmids which were then able to be reintroduced into bacteria. Replication of the bacteria containing the foreign DNA enabled unlimited amounts of a single fragment to be produced, i.e. DNA could be *cloned*. The first eukaryotic gene to be cloned was the rabbit β globin gene in 1976 (see Chs 2, 3, 7 for further discussion on cloning).

The development of *DNA probes* followed from a 1960 observation that the two strands of the DNA double helix could be separated and then reannealed. Probes comprised small segments of DNA which were labelled with a radioactive marker such as ^{32}P. DNA probes were able to identify specific regions in DNA through their annealing (or *hybridisation*) to complementary nucleotide sequences. The specificity of the hybridisation reaction relied on the predictability of base pairing, i.e. the nucleotide base adenine (usually abbreviated to A) would always anneal to the base thymine (T) and guanine (G) would anneal to cytosine (C). Thus, because of nucleotide base pairing, a single-stranded DNA probe would hybridise in solution to a predetermined segment of single-stranded DNA (see Ch. 2 and the section on thalassaemia, p. 7).

Solution hybridisation gave way in 1975 to hybridisation on solid support membranes when DNA digested with restriction endonucleases could be transferred to these membranes by *Southern blotting*, a process named after its discoverer E M Southern. The ability of radiolabelled DNA probes to identify specific restriction endonuclease fragments enabled *DNA maps* to be constructed. This was the forerunner to DNA mutation analysis, which is discussed further in this chapter as well as in Chapters 2–6.

In the 1970s, the impressive developments in molecular medicine were matched by growing concern in both the public and scientific communities that *genetic engineering*, as molecular medicine was popularly called, was a perversion of nature and potential source of much harm. In 1975, a conference was convened at Asilomar in California to discuss these issues. Subsequently, regulatory and funding bodies issued *guidelines* for the conduct of recombinant DNA work. These guidelines dealt with the type of experiments allowable and the necessity to use both vectors (e.g. plasmids) and hosts (e.g. bacteria such as *E. coli* which carried the vectors) that were safe and could be contained within laboratories certified to undertake recombinant DNA work. Guidelines began to be relaxed during the late 1970s and early 1980s when it became apparent that recombinant DNA technology was safe if carried out responsibly. However, government and private funding bodies insisted that a form of monitoring be maintained which has continued to this day (see Ch. 9).

Gene structure and function

The structure of the gene became better defined with descriptions in 1975 and 1977 of two methods to sequence individual nucleotide bases in DNA. The significance of this achievement was acknowledged with a Nobel Prize to Sanger and Gilbert. In 1977, an unexpected observation revealed that eukaryotic genes were discontinuous, i.e. coding regions were split by intervening segments of DNA. To distinguish these two components in the gene, the terms *exons* and *introns* were first used by Gilbert in 1978. *Splicing*, the mechanism by which genes were able to remove introns to allow the appropriate exons to join, was described by Roberts and Sharp, for which they were awarded a Nobel Prize in 1993 (see Chs 2, 3). In the 1970s and 1980s three scientists (Lewis, Nusslein-Volhard and Wieschaus) laid the foundations for what would be exciting revelations in respect to genes which were critical for development in the fruit fly *Drosophila melanogaster*. The same genes were found in animals, including humans, and characterisation of them would subsequently show the molecular basis for normal and abnormal development in vertebrates. For their pioneering work the three were awarded a Nobel Prize in 1995.

Variations in the lengths of DNA segments between individuals (called *polymorphisms*) were reported in the mid-1970s, although their full potentials were not realised until the early 1980s when Botstein and colleagues described how it might be possible to use DNA polymorphisms as markers to construct a map of the human genome. Subsequently, DNA polymorphisms would form a component of many studies involving DNA (see Chs 2–6, 8). These markers also became useful in comparing different populations or species to identify evolutionary affinities and origins. In 1987, Cann and colleagues proposed, on the basis of mitochondrial DNA polymorphisms and sequence data, that *Homo sapiens* evolved from a common African female ancestor. Although evolutionary data based on DNA markers have evoked a number of controversies, many of which remain unresolved, the contribution of DNA polymorphisms in comparative studies has remained very significant (see Chs 3, 8).

Biotechnology

The ability to take DNA in vitro and produce a protein from it became an important step in the commercialisation of molecular medicine. The latter half of the 1970s saw the development of the *biotechnology industry* based on this type of DNA manipulation. The first genetic engineering company, called Genentech Inc., was formed in 1976 in California. Human insulin became the first genetically engineered drug to be marketed in 1982 following the successful in vitro synthesis of recombinant human growth hormone a few years earlier (see Ch. 7).

During the 1980s, *transgenic mice* were produced by microinjecting foreign DNA into the pronucleus of fertilised oocytes. Injected DNA became integrated into the mouse's own genome and the transgenic animal expressed both the endogenous mouse genes and the foreign gene. A 'supermouse' was made when a rat growth hormone gene was microinjected into a mouse pronucleus. In 1988, the first US patent was issued for a genetically altered animal. In the years to follow, particularly the 1990s, the debate about patents and genes would provoke considerable controversy (see Chs 9, 10).

RECENT ADVANCES

Functional and positional cloning

Until the mid-1980s, conventional approaches to understanding genetic disease relied entirely on the identification and then characterisation of an abnormal protein. This could be taken one step further with molecular medicine, since it became possible to use the protein to clone the relevant gene. From the cloned gene more information could then be obtained about the underlying genetic disorder. This was called *functional cloning* and is illustrated below (see p. 7) by reference to the thalassaemia syndromes. However, the identification of an abnormal protein was not always easy or indeed possible. For example, the genetic disorder Huntington disease was first described by Huntington in 1872 and more than 100 years later no abnormal protein had been found.

In the late 1980s, an alternative approach to the study of genetic disease became available through *positional cloning*. This method bypassed the protein and enabled direct isolation of genes on the basis of their chromosomal location and certain characteristics which identified a segment of DNA as 'gene-like'. Identification of the mutant gene as well as knowledge of the genetic disorder could then be inferred from the DNA sequence. The strategy was initially called 'reverse genetics'. Subsequently, the name was changed to the more appropriate one of 'positional cloning'. The first success stories involving positional cloning for human genes came in 1986 with the isolation of the gene for chronic granulomatous disease by Orkin and colleagues and in 1987 with the Duchenne muscular dystrophy gene by Kunkel and colleagues. Successes were slow to follow initially, but by the mid-1990s, it became difficult to keep up with the output from positional cloning! Genes to be found included those which caused:

- Chronic granulomatous disease
- Duchenne muscular dystrophy
- Retinoblastoma
- Wilms tumour

Table 1.2   Genes which have been cloned and characterised by the candidate gene approach.
A candidate or 'likely' gene is used to narrow the field when searching DNA clones isolated on the basis of their chromosomal location by positional cloning.

Genetic disorder	Candidate gene(s) and relevance
Retinitis pigmentosa (disease affecting the retina and producing blindness)	Rhodopsin, the gene for human rhodopsin (visual) pigment was isolated first. Subsequently, retinitis pigmentosa was localised to the same chromosome. Thus, rhodopsin became an obvious 'candidate' gene for retinitis pigmentosa
Familial hypertrophic cardiomyopathy (disease of heart muscle)	β myosin heavy chain, α tropomyosin, troponin T (all genes associated with the muscle sarcomere)
Marfan syndrome (disease of connective tissue)	Fibrillin (the fibrillin protein is a constituent of elastin-associated myofibrils)
Alzheimer disease (early onset presenile dementia	APP = gene for amyloid precursor protein) (amyloid protein described in plaques associated with Alzheimer disease)
Long QT syndrome (a cardiac disorder associated with arrhythmias)	HERG = gene predicted to have potassium channel-like activity on its DNA sequence was located in the same chromosomal position as the Long QT syndrome

- Cystic fibrosis
- Neurofibromatosis types 1, 2
- Testis determining factor
- Fragile X mental retardation
- Familial adenomatous polyposis
- Myotonic dystrophy
- Huntington disease
- DNA repair defects (ataxia telangiectasia, colon cancer)
- Bloom syndrome
- Breast cancer.

A variation of positional cloning enabled genes to be identified on the basis that they were potential 'candidates' for genetic disorders. This was a short-cut but required prior knowledge about the likely genes which might be involved (Table 1.2) (see Chs 2, 3, 6, 9, 10).

Polymerase chain reaction (PCR)

In 1985, work by Mullis, Saiki and colleagues in the Cetus Corporation, California, made it possible to target segments of DNA with oligonucleotide primers and then amplify these segments with the *polymerase chain reaction* or *PCR* as it is now best known. PCR soon became a routine procedure in the molecular biology laboratory.

In a short period of time, this technology has had a profound and immediate effect in both diagnostic and research areas. Reports of DNA patterns which were obtained from single cells by PCR started to appear. Even the dead were not allowed to rest as it soon became possible to study DNA patterns from ancient Egyptian mummies, old bones and preserved material of human origin. The extraordinary and rapid contribution made by PCR in medical, industrial and research projects was recognised by the award of a Nobel Prize to Mullis in 1993.

The availability of automation meant that DNA amplification had unlimited potential for mutation analysis in genetic disorders as well as the identification of specific DNA sequences from infectious agents. A patent was obtained to cover the use of PCR and illustrated the growing importance of commercialisation in recombinant DNA technology (see Chs 2, 3, 9).

Genes and cancer

During the 1980s and 1990s, the application of molecular biology techniques to medicine ('molecular medicine') has proved to be critical for our understanding of cancer. The first breakthrough came in 1910, when Rous implicated viruses in the aetiology of cancer by showing that a filterable agent (virus) was capable of inducing cancers in chickens. However, it was not until the early 1980s, when a DNA sequence from a bladder cancer cell line was cloned and shown to have the capacity to induce cancerous transformation in other cells, that real progress began. The cause of the neoplastic change in both the above examples was soon demonstrated to be dominantly acting cancer genes which were called *oncogenes*. These have assumed increasing importance in our understanding of how cancers arise and progress. For their work on oncogenes during the early 1980s, Bishop and Varmus were awarded a Nobel Prize in 1989.

More recently, the identification of cellular sequences which normally repress or control cellular growth led to the discovery of *tumour suppressor genes*. Loss or mutation of tumour suppressor gene DNA through genetic and/or acquired events was shown to be associated with unregulated cellular proliferation and hence neoplasm. Just as occurred with positional cloning, the list of oncogenes and tumour suppressor genes has expanded rapidly during the past decade. Information from both classes of genes has provided insight into the mechanisms which can lead to cancer and, perhaps more importantly, how the normal cell functions (see Ch. 6).

DNA changes found in cancers subsequently provided evidence for Knudson's *two-hit* hypothesis which was proposed to explain the development of certain tumours. Thus, earlier epidemiological data on which Knudson devised his model in 1971 were now confirmed (Box 1.2).

Box 1.2 Knudson's two-hit hypothesis for cancer

A genetic (familial) predisposition to cancer is sometimes found. In retinoblastoma (a tumour of the retinal cells in the eye) the genetic component was confusing, e.g. in some cases there was an obvious genetic factor because a number of family members were affected but in other cases the tumour was sporadic, i.e. there was no family history. Knudson explained the retinoblastoma dilemma by suggesting two genetic hits were required before retinoblastoma would develop. In the familial case, one of the two hits was present in the germ cells (egg or sperm) so that children of an affected parent had a 50% chance of inheriting this abnormality and would become predisposed to retinoblastoma. When the second hit occurred in the actual retinal (somatic) cells, a retinoblastoma would develop. In contrast, those with a sporadic form of the tumour did not have a family history because both hits occurred in somatic cells, i.e. the retinal cells. In this situation, the germ cells were normal. Knudson's theory was proven correct when the retinoblastoma (Rb) tumour suppressor gene (TSG) was identified. For cancer to develop, a mutation affecting *both* Rb TSGs was necessary. At the molecular level, familial retinoblastoma was associated with one Rb TSG mutation in germ cells and a second mutation in the tumour tissue. However, in sporadic cases, the Rb TSG in germ cells was normal but both Rb TSGs in the tumour tissue were mutated (see Ch. 6 for further discussion of retinoblastoma, tumour suppressor genes and cancer).

designing methods which allowed large DNA fragments to be *screened* for potential changes in their DNA sequence. From the screening results, more specific PCR-based mutation studies became possible. Methods such as SSCP (**s**ingle-**s**tranded **c**onformation **p**olymorphism), DGGE (**de**naturing **g**radient **g**el **e**lectrophoresis) and CCM (**c**hemical **c**leavage of **m**ismatch) were reported. However, none was ideal in all situations. Today, considerable technological development continues in this area of molecular medicine (see Chs 2, 3, 6, 10).

Another aspect of DNA mutation analysis became evident in the 1990s and this involved commercialisation through the availability of DNA kits. In the first instance these were provided to facilitate the implementation of DNA technology in less experienced diagnostic laboratories. Ultimately, it was suggested that the family practitioner or local health centres would utilise DNA kits to identify specific genetic parameters. Perhaps 'home kits' would become available. The implications of this development in molecular medicine will need careful consideration. It is interesting to note that the community response to these potential developments was less vocal than on previous occasions, e.g. when genetic engineering was first started or the use of gene therapy was considered. However, important questions will need to be asked, e.g. what DNA testing is necessary and what will cause more harm than good? How will the costs be met? The advantages and disadvantages of DNA mutation analysis will require forward and clear thinking by both health professionals and the community (see Ch. 9).

Mutation analysis

Mutation analysis, i.e. identifying defects in DNA, assumed a higher profile in the 1990s. This followed from the increasingly larger and more complex genes to be isolated by positional cloning as well as the realisation that DNA diagnosis could provide useful information for the clinical management of patients with genetic disorders or infectious diseases. The profile for mutation analysis increased further in the mid-1990s as genes for relatively common cancers, e.g. bowel, breast, began to be described.

The 'gold standard' in terms of DNA mutation analysis is *sequencing* so that an exact mutation can be defined. However, the size of genes involved as well as the very broad range of mutations observed in genetic disorders makes this impractical as a routine means to detect mutations. Although automated DNA sequencing became more accessible in the 1990s, it remains expensive. PCR quickly made a major impact on mutation analysis, but the very focused nature of PCR, i.e. a specific region of the genome is targeted and then amplified, was its weakness when it came to mutations which could be found anywhere in a very large gene, e.g. the 250 kb cystic fibrosis gene has in excess of 600 reported defects over this segment of DNA. Hence, the 1990s became a decade for

Therapeutic options

In 1987, the first *recombinant DNA vaccine* against the hepatitis B virus was produced by inserting a segment of viral DNA into a yeast expression vector. Successes with the expression of DNA in vitro (e.g. the production of recombinant DNA-derived drugs) and in vivo (e.g. transgenic mice), highlighted the potential for transferring DNA directly into cells for therapeutic benefits. However, concern about manipulation of the human genome through a *gene therapy* approach was followed by a moratorium and extensive public debate until guidelines for the conduct of this type of recombinant DNA work became firmly established. Permission for the first human gene therapy trial to proceed was only obtained in 1990. A child with the fatal adenosine deaminase (ADA) deficiency disorder was given somatic gene therapy using her own lymphocytes which had been genetically engineered by retroviral insertion of a normal adenosine deaminase gene. Lymphocytes with the normal ADA gene were then reinfused into the individual. Initial clinical and laboratory responses were gratifying and long-term follow-up will determine the effectiveness of this form of treatment which by the mid-1990s had been used in 10 other children with ADA deficiency (see Ch. 7).

During the late 1980s to the mid-1990s, novel approaches to gene therapy which avoided the use of retroviruses began to be discussed as potential ways to transfer DNA into cells or interfere with foreign RNA or DNA (e.g. a virus) present in a cell. One method was based on observations from 1981 that RNA from the protozoan *Tetrahymena thermophila* had enzyme-like activity (called *ribozyme*). The discoverers, Altman and Cech were subsequently awarded a Nobel Prize for this work. Ribozymes provided a new approach to gene therapy since they made use of naturally occurring RNA sequences which could cleave RNA targets at specific sites. Diseases for which ribozymes had particular appeal included cancer (to inhibit the mRNA from oncogenes) and AIDS (to inhibit mRNA from the human immunodeficiency virus (HIV)).

By the mid-1990s, an unexpected change in the indications for gene therapy became apparent as the number of proposals involving cancer and HIV infection surpassed those for genetic disorders. Methodologies which enabled genes to be targeted to their normal genetic locus through a process called *homologous recombination* were also described. This was attractive since once developed it would enable therapeutic manipulations to be conducted with greater accuracy (see Ch. 7).

The cloning of cattle by nuclear transplantation in 1987 highlighted ethical and social issues which could arise from irresponsible use of recombinant DNA technology in the human. Legislative prohibitions relating to certain types of human embryo experimentation and human gene therapy involving the germline were enacted in many countries (see Chs 4, 7, 9).

DNA polymorphisms

The number and types of *DNA polymorphisms* first described in the 1970s and early 1980s rapidly expanded. The inherent variability in DNA polymorphisms led to the concept of DNA *fingerprinting* in 1985 when Jeffreys and colleagues described how more complex DNA polymorphisms *(minisatellites)* were able to produce unique DNA profiles of individuals. DNA testing for minisatellites has subsequently had an increasingly important role to play in forensic practice. The courts of law became involved in molecular medicine when the potential for identification of individuals on the basis of their minisatellite DNA patterns was realised. In 1987, DNA 'fingerprints' were allowed as evidence in the first court case. Just as was mentioned earlier when describing the use of DNA polymorphisms in studies to identify the origins and affinities of peoples and species, DNA fingerprinting has also produced its share of controversies which continue to be debated in the courts (see Chs 2, 8). Another type of DNA polymorphism, known as a *microsatellite*, was shown in the 1980s to be dispersed throughout the human genome and so is very useful when it comes to studies

involving positional cloning (see Chs 2, 3–6, 8–10 and the following section).

Human Genome Project

The Human Genome Project, a multidisciplinary, multinational study initiated by the US Department of Energy and National Institutes of Health, was given the go-ahead in 1988. The ultimate aim of the project is to clone and sequence the entire human genome and some model organisms by the year 2005. Information resulting from this project will have scientific and social consequences as yet unrealised (see Chs 9, 10 for further discussion of the Human Genome Project).

Two essential components of the Human Genome Project (*genetic* and *physical mapping* strategies) were facilitated by key technological developments in the 1980s and 1990s. Genetic mapping is an indirect measure of distance along a chromosome and relies on family studies and DNA polymorphisms. The availability of microsatellite polymorphic markers enabled genetic mapping to expand rapidly. Better methods for measuring actual *distances* (physical mapping) included **p**ulsed **f**ield **g**el **e**lectrophoresis (PFGE), **y**east **a**rtificial **c**hromosomes (YACs) and **f**luorescence **i**n situ **h**ybridisation (FISH). These physical mapping strategies enabled longer segments of DNA to be identified and mapped in relation to each other.

Pulsed field gel electrophoresis is a variation of DNA mapping first described by Schwartz and Cantor in 1984. It enables megabases rather than kilobases of DNA (1 megabase or 1 Mb is equivalent to 1×10^6 base pairs; 1 kilobase or 1 kb is equivalent to 1×10^3 base pairs) to be measured. Similarly, DNA fragments up to 1 Mb in size could be cloned with the YAC vectors (conventional cloning vectors were restricted to DNA fragments ~10–40 kb in size). The studies which initiated YAC cloning were described by Burke and Olson in 1987. *Fluorescence in situ hybridisation* emerged from work carried out in the 1980s which assisted with the characterisation of chromosomes by using DNA probes labelled with radioisotopes or chemical markers. Subsequently, the labelling with fluorescent dyes provided a bridge between the conventional cytogenetic analysis for gross chromosomal changes and the discrete defects detectable by DNA mapping. These techniques quickly became established as key components in positional cloning strategies (see Chs 2–4).

Another component of the Human Genome Project is *informatics*. This is the means by which data can be stored and then transmitted so that it becomes rapidly accessible to a broad scientific community. The key elements in informatics are computer networks and databases. The Internet and local servers provide the former and the development of resource centres and databases act as a focus into which information can be fed and processed. The information explosion which has come as the Human Genome Project gains momentum has made the development of informatics a priority (see Chs 3, 9, 10).

THALASSAEMIA – A MODEL FOR MOLECULAR MEDICINE

Introduction

Many of the developments which have occurred in molecular medicine can be illustrated by reference to a group of clinical disorders known as the thalassaemia syndromes. Haemoglobin, the pigment in red blood cells, comprises iron and a protein called globin. Four polypeptide chains make up globin – two α globin chains and two β globin chains. A genetic defect which impairs globin synthesis produces thalassaemia and this manifests as an anaemia with red blood cells which are smaller and paler than normal. The clinical picture in the thalassaemia syndromes is diverse and ranges from an asymptomatic disorder which is detected fortuitously, to a life-long blood transfusion-dependent anaemia or an anaemia which is fatal in utero or soon after birth.

The first accurate clinical description of thalassaemia was given by Cooley in 1925. Cooley's anaemia, as it was then called, was shown to be genetic in origin during the late 1930s and early 1940s when relatives of severely affected individuals were observed to have similar but milder changes in their red blood cells. The word thalassaemia was coined in 1936 and comes from the Greek θαλασσα which means 'the sea'. The name arose since it was initially considered that thalassaemia was a disease which affected those who lived near the Mediterranean sea.

Functional cloning

Thalassaemia illustrates how human genetic disorders were studied by traditional approaches. Initially, clinical observations based on pedigrees failed to explain the transmission or variable severity observed with this disorder. During the 1950s and 1960s, extensive protein analyses were undertaken to characterise the globin proteins involved. These studies showed that there were at least two different proteins. In 1959, Ingram and Stretton proposed that the thalassaemias could be subdivided into α and β types corresponding to α globin and β globin proteins. During the 1960s, the biochemical defect in the thalassaemias was identified as an *imbalance* in the number of α and β globin chains, with the normal α/β ratio being 1. Failure to produce α globin gave α thalassaemia, which is fatal in its most severe form. Failure to produce any β globin (β thalassaemia) was associated with a life-long blood transfusion-dependent anaemia. Carriers of either thalassaemia defect were clinically asymptomatic although their blood counts ranged from normal to mildly abnormal. Despite the very elegant biochemical studies, an unexplained feature of the thalassaemias remained, i.e. the considerable variation in phenotypes involving both the clinical and laboratory pictures.

Following the discovery of reverse transcriptase, it became possible to take immature red blood cells from patients with homozygous β thalassaemia and isolate from them α globin mRNA which could then be converted to complementary DNA (cDNA) (Fig. 1.1). In this way DNA probes specific for the α globin genes became available. Using these probes and solution hybridisation techniques it was possible to show that there were distinct abnormalities affecting mRNA in the α thalassaemias. DNA probes

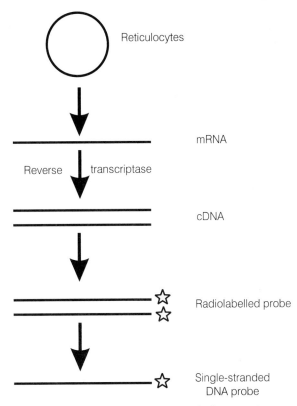

Fig. 1.1 Making an α globin gene DNA probe from mRNA. Reticulocytes are immature red blood cells that are transcriptionally very active. Therefore, there will be a lot of mRNA present. If the reticulocytes are derived from an individual with homozygous β thalassaemia there will be little, if any, β globin mRNA because the β globin genes in this individual are not functional. Thus, the cells' globin mRNA will have originated from the α genes. α globin mRNA can be converted to cDNA (complementary or copy DNA) with the enzyme reverse transcriptase. The cDNA functions as a DNA probe which can be labelled with radioactivity (★). The radiolabelled probe is made single-stranded which will enable it to bind (hybridise) with its corresponding single-stranded sequence in the genome.

Box 1.3   Genetic abnormalities in thalassaemia

In the α thalassaemias, it was possible to utilise α globin gene probes to quantitate gene dosage, i.e. how many of the four normal genes were present. For example, a form of α thalassaemia called haemoglobin H (HbH) disease was shown to have 25% of the normal gene-specific activity. This was subsequently confirmed when gene mapping demonstrated that in HbH disease there is only one of the four normal α globin genes remaining. Gene dosage in the β thalassaemias was normal but different defects at the mRNA level were found. DNA mapping subsequently established that the β globin genes were intact and so the mRNA defects had resulted from point mutations in the gene rather than deletions. Thus, the thalassaemias showed that a molecular classification of genetic disorders could be proposed on the basis of (1) deletional or (2) non-deletional mutations in DNA.

abnormal, indicating that the underlying gene defects involved loss, i.e. *deletions*, of DNA. On the other hand, the maps in the β thalassaemias were normal. Therefore, the molecular (DNA) abnormalities in the β thalassaemias were either *point mutations* (changes involving a single nucleotide base) or *very small deletions*. Once DNA probes became available the next stop was to clone the human α and β globin genes. This was achieved in the late 1970s. From this 'functional' cloning approach to the thalassaemia syndromes, it became possible to understand further the evolution of thalassaemia and its clinical consequences.

Mutation analysis

The first human genetic disorders to be diagnosed by identifying the mutant gene in the fetus whilst still in utero, i.e. prenatal diagnosis, were α thalassaemia (1976) using a liquid hybridisation technique and sickle-cell anaemia (1978) using a DNA polymorphism approach by Kan and colleagues. Subsequently, mutation analysis in the thalassaemia syndromes has developed along two lines, i.e. the detection of deletions associated with the α thalassaemias and the identification of point mutations (the predominant defect) in the β thalassaemias. Until PCR became routinely available, the many different point mutations in the β thalassaemias were unable to be distinguished individually and so indirect DNA diagnosis using DNA polymorphisms was used (Fig. 1.2, Chs 2, 3). Today, DNA diagnosis has become more sophisticated with individual mutations being sought by PCR. In terms of mutation analysis, the

specific for the β globin genes were isolated next. This was assisted by the cloning of the rabbit β globin gene since it had considerable homology (similarity) to its human equivalent. A number of different abnormalities in mRNA were also found in the β thalassaemias. Differences observed clinically were now beginning to be seen at the molecular level (Box 1.3).

With the availability of *restriction endonuclease* enzymes it became possible to construct DNA maps for the globin genes. DNA maps for the α thalassaemias were quite

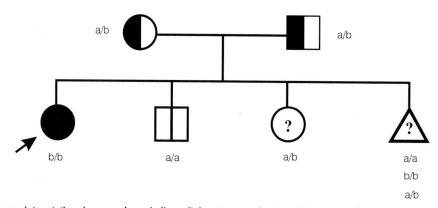

Fig. 1.2   Detecting point mutations in a gene by an indirect (linkage) approach using DNA polymorphisms.
Understanding how polymorphisms are used to follow a disease within a family (called linkage analysis) is a difficult concept to grasp and will be discussed further in Chapter 3. Essentially, a polymorphism is a marker for a chromosomal location or gene. In the case of the β globin gene, each individual has two genes (one on each chromosome 11 – see below) and so two polymorphic markers should be detectable. To undertake linkage analysis the first step involves identifying polymorphic markers in a family which will distinguish the two β globin genes. The polymorphisms are not mutations but simply DNA sequence changes which allow the two genes to be distinguished or marked. Once the polymorphisms are identified, they are traced in a family and compared to the clinical phenotypes. In the pedigree the two parents are β thalassaemia carriers. They have a female child who has homozygous β thalassaemia (indicated by →) and a normal male. The thalassaemia status for a third (female) child is unknown (?). The mother is also pregnant and the fetus is indicated by a triangle with unknown thalassaemia status. The two polymorphisms which distinguish the two β globin genes in this family are defined by the letters 'a' and 'b'. Both the parents are carriers and have the a/b polymorphic markers. This information alone is not enough for diagnosis. However, the homozygous-affected child is b/b and so the polymorphic marker 'b' identifies the mutant β thalassaemia gene in this family. It can also be assumed that the marker 'a' defines the normal gene. This is confirmed because the normal child is a/a. The child with the unknown status is a/b and so she must be a carrier. The fetus can have three combinations: a/a (= normal), b/b (= homozygous-affected) or a/b (= carrier).

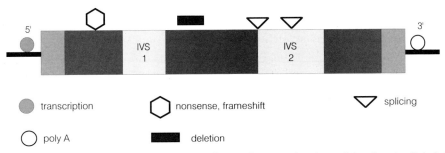

Fig. 1.3 Mutations within the β globin gene can have a variable effect on the gene's function and therefore the clinical phenotypes which will result.
The β globin gene is shown as a rectangle with light and dark areas. The IVS1 and IVS2 segments correspond to the two introns in the gene (IVS stands for **i**ntervening **s**equence). The three dark areas are the coding regions (known as exons). The beginning of the gene is the 5' part and the end is defined by 3'. Various symbols identify mutations in the gene. Starting at the 5' end the mutations are: transcription defect (a mild mutation because the gene can still function reasonably effectively); nonsense; frameshift defect or deletion (both are severe mutations because the globin protein will not be able to form); two splicing defects are next depicted (these are usually associated with mild clinical phenotypes); and at the 3' end is a poly A mutation. The poly A segment of the gene is important for mRNA stability but a mutation in this region often leads to a mild disorder.

thalassaemia syndromes have provided a relatively simple but effective model to develop this aspect of molecular medicine.

Mutation analysis in the thalassaemia syndromes has shown that the word *heterogeneity* is the most appropriate way to describe the underlying DNA defects. This heterogeneity emerges from different mutations which affect (1) the α or β globin genes, (2) other related globin genes, and (3) various combinations of (1) and (2). Because of the relatively small size of the globin genes, i.e. ~1 kb compared to 250 kb for cystic fibrosis, they have been extensively characterised and so remain a model for the increasingly complex genetic disorders which would follow.

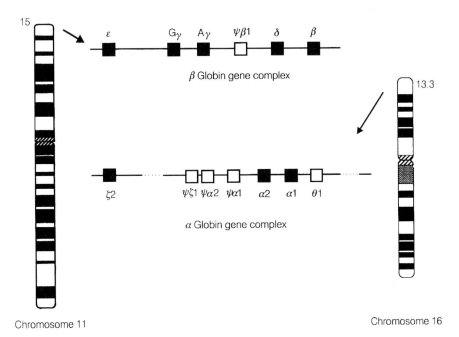

Fig. 1.4 The α and β globin gene clusters.
Functional genes are indicated as ■ and non-functioning genes (called pseudogenes) as □. On the short arm of chromosome 11 at band position 15 is found the β globin gene complex. There is one gene which is active during embryonic life (ε), two which are fetal specific (Gγ, Aγ) and two are expressed in adult life (δ, β). The α globin complex is on the short arm of chromosome 16 at band 13.3. There are a lot more genes situated in this complex but many are non-functional. The embryonic/fetal gene is ζ₂ and the two adult genes are α2 and α1. The evolution of the globin gene clusters from a common ancestral gene is seen by the similarity in structure and sequence which the above genes share even though they are on different chromosomes. The .... on the α globin complex marks the position of a DNA polymorphism (see Chs 2, 3, 8 for further discussion of polymorphisms).

Table 1.3   Globin gene switching in development.
Different genes are operational during specific periods of development resulting in sequential changes in the globin subunits of haemoglobin.

	Organ		
	Yolk sac	Fetal liver	Adult marrow
α Genes	α, ζ	α	α
β Genes	ε	γ	β,δ
Haemoglobins	$\xi_2\varepsilon_2$ $\alpha_2\varepsilon_2$	$\alpha_2\gamma_2$	$\alpha_2\beta_2$ $\alpha_2\delta_2$

Another contribution from the thalassaemias comes in the correlation of phenotypes (clinical and laboratory features) with genotypes (DNA changes). In this way the functional *significance of mutations* has been studied. Two examples of this type of correlation are (1) the co-inheritance of two different gene mutations such as α and β thalassaemia in the same individual can be predicted to be relatively benign since there is minimal globin chain imbalance and (2) there are nearly 200 point mutations which can lead to the β thalassaemia syndromes. Some of these can be predicted to produce a milder form of this disorder (described as β⁺ thalassaemia) while others will be more severe (β⁰ thalassaemia) (Fig. 1.3, Ch. 3). Phenotype/genotype correlations are now being attempted in other genetic disorders but with variable success, e.g. cystic fibrosis, familial hypertrophic cardiomyopathy (see Ch. 3).

DNA, evolution and society

Following the cloning of the human α and β globin genes in the late 1970s, it became apparent that the globins represented a gene family with a larger number of genes than initially considered. Cell fusion studies localised a cluster of such genes on chromosome 16 (α globin cluster) and a second cluster on chromosome 11 (β globin cluster) (Fig. 1.4). DNA studies also confirmed that the globin genes had evolved by duplication from a single ancestral gene.

Once a definitive (DNA) test had become available for thalassaemia, particularly for the α thalassaemia carrier states which were otherwise difficult to detect with conventional laboratory tests, the question as to why thalassaemia was so common could be asked. The original observation that thalassaemia was a disease of the Mediterranean region had by now been shown to be incorrect since it was present in many regions of the world, e.g. South East Asia, the Middle East and parts of the Pacific. A further observation was that malaria was also to be found in the same locations. Molecular 'epidemiology' studies confirmed the close relationship between thalassaemia and malaria with the mechanism being an evolutionary survival advantage for red blood cells which carried the thalassaemia trait, i.e. pale, small

Box 1.4   Identification of regulatory elements by mutation analysis

The thalassaemia syndromes illustrate how knowledge of spontaneously occurring genetic mutants can lead to a greater understanding of more fundamental biological issues. For example, what are important regulatory elements which control the activity of a eukaryotic gene? Information about this has come from the study of a rare thalassaemia.

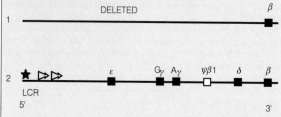

(1) Characterisation of DNA from this thalassaemia showed an extensive deletion starting 5′ of the β globin gene and extending a distance beyond the ε, γ and δ genes. The β gene was intact but did not function, which was puzzling since sequence analysis failed to disclose any mutation in the gene. Furthermore, the β gene was able to transcribe using in vitro assays.

(2) Molecular characterisation identified a segment of DNA 10.5 kb from the ε globin gene which was called the locus control region (LCR) because it had changes which are normally found in regions of DNA carrying out regulatory function. The LCR has now been shown to exert a powerful effect on the globin genes, enabling high levels of gene expression which is specific to red blood cells. Deletion of the LCR in the above thalassaemia prevented the remaining (normal) β globin gene from functioning. The finding of the LCR has identified a key regulatory element in the eukaryote genome. Because this sequence stimulates the gene's output it is called an enhancer.

red blood cells provided a poor environment for the growth of malarial parasites. A similar positive selective advantage against malaria was also shown with the sickle-cell defect, a mutation affecting the β globin chain. In contrast to the thalassaemias, the basis why some other genetic disorders occur relatively frequently, e.g. the cystic fibrosis mutation (1 in 22 in some populations) and the haemochromatosis defect (1 in 10 in some populations), remains to be adequately explained (see Chs 3, 10).

Another question which has often been asked is when did thalassaemia first occur in humans, since changes found in ancient skeletons and references to thalassaemia-like conditions in works by Hippocrates and Pythagoras would be difficult to distinguish from other causes of chronic anaemia. Molecular medicine has yet to provide an answer but an interesting PCR study on skeletal remains in an archaeological site has shown that one of the buried children had β thalassaemia. Coins found in some graves suggested that they may have dated from the Ottoman period (16th to 19th century). A further observation was that despite being homozygous for a severe β thalassaemia defect, the child survived until an estimated age of 8 years which would be unexpected without modern-day treatment options. However, molecular analysis identified a β globin gene DNA polymorphism which is associated with an increase in fetal haemoglobin (abbreviated to HbF). It is now known that one way to reduce the severity of β thalassaemia is to increase the level of HbF. Thus, the co-inheritance of two genetic defects in the same individual has provided an explanation for the relatively mild β thalassaemia phenotype in this child (see Filon et al 1995, Chs 3, 7).

Therapeutic options

One disappointment in the thalassaemia story is its failure in the area of gene therapy. Initially one of the first genetic disorders identified as suitable for this type of treatment, the thalassaemia syndromes are now a long way down in the priority list mainly because of the amount of globin which would need to be produced by an inserted gene. However, thalassaemia has been a valuable model to study how the activity of genes is *regulated* and this knowledge has had broad relevance to all eukaryotic genes.

The globin genes are also interesting because they display a developmental switch. This switch is present in the form of separate embryonic, fetal and adult-specific genes which are operational during different developmental periods (Table 1.3). As will be shown in Chapters 3 and 7, information gained from characterising mutations in the globin genes has now provided insight into the mechanisms which control gene expression and possible ways in which these could be overcome or utilised in future gene therapy strategies (Box 1.4).

FURTHER READING

A chronology of molecular medicine

Botstein D 1990 1989 Allan Award Address: The American Society of Human Genetics, Annual Meeting, Baltimore. American Journal of Human Genetics 47: 887–891 (an address on polymorphisms)

Collins F S 1995 Positional cloning moves from perditional to traditional. Nature Genetics 9: 347–350

Guyer M S, Collins F S 1993 The Human Genome Project and the future of medicine. American Journal of Diseases of Children 147: 1145–1152

Harper R 1995 World wide web resources for the biologist. Trends in Genetics 11: 223–228

King R C, Stansfield W D 1990 A dictionary of genetics. Oxford University Press, Oxford (for historical information)

McKusick V A 1992 Presidential Address, Eighth International Congress of Human Genetics: Human Genetics: The last 35 years, the present, and the future. American Journal of Human Genetics 50: 663–670

Mullis K B 1990 The unusual origin of the polymerase chain reaction. Scientific American 262: 56–65

Wallace D C 1995 1994 William Allan Award Address: Mitochondrial DNA variation in human evolution, degenerative disease and aging. American Journal of Human Genetics 57: 201–223

Watson J D, Gilman M, Witkowski J, Zoller M 1992 Recombinant DNA, 2nd edn. W H Freeman, New York

Thalassaemia – a model for molecular medicine

Filon D, Faerman M, Smith P, Oppenheim A 1995 Sequence analysis reveals a β-thalassaemia mutation in the DNA of skeletal remains from the archaeological site of Akhziv, Israel. Nature Genetics 9: 365–368

Weatherall D J 1980 Of some common inherited anemias: the story of thalassemia. In: Wintrobe M M (ed) Blood, pure and eloquent. McGraw–Hill, New York, p 373–414

Weatherall D J 1991 The new genetics and clinical practice, 3rd edn. Oxford University Press, Oxford

2

MOLECULAR TECHNOLOGY

DNA

DNA is deoxyribonucleic acid, a double-stranded macro-molecule which contains the organism's genetic information. Isolation of DNA requires: (1) nucleated cells, (2) enzymes to break up cell membranes and proteins, and (3) chemicals to separate proteins from nucleic acids.

Sources

DNA has a number of properties which are exploited in the laboratory. DNA in all cells of an organism is identical in its sequence. Therefore, obtaining a tissue specimen for DNA studies is relatively simple since 10 ml of blood will usually suffice. This yields ~250 µg of DNA. Isolation of DNA is straightforward. Nuclei are first separated from cellular debris by enzymatic means. DNA is then separated from proteins by using chemicals such as phenol and chloroform. Other convenient sources of DNA used in routine genetic diagnosis include the exfoliated cells in mouth washes and hair roots.

Structure and function

Two properties of DNA are of particular relevance in molecular technology. The first is the genetic code which is present in the form of nucleotide triplets called *codons* (Table 2.1). This means that the signal for individual polypeptides is coded by different triplet combinations. For example, the codons for a polypeptide such as glycine-serine-valine-alanine-alanine-tryptophan will read: –GGT–TCT–GTT–GCT–GCT–TGG–. Similarly, the positions where to start and where to end a polypeptide are clearly defined by the triplets ATG (start) and TAA or TAG or TGA (stop).

Table 2.1   The genetic code.
Nucleotides code in sets of three, or triplets, for individual amino acids. The triplets or codons are shown as they appear in DNA (T = thymine, C = cytosine, A = adenine and G = guanine). In mRNA, T is replaced by U (uracil). The code is degenerate, i.e. there can be more than one codon per amino acid. The genetic code is read from left to right, i.e. TTC = Phe (phenylalanine); TCC = Ser (serine); CCT = Pro (proline).

First nucleotide [5']	Second nucleotide				Third nucleotide [3']
	T	C	A	G	
T	Phe	Ser	Tyr	Cys	T
T	Phe	Ser	Tyr	Cys	C
T	Leu	Ser	STOP	STOP	A
T	Leu	Ser	STOP	Trp	G
C	Leu	Pro	His	Arg	T
C	Leu	Pro	His	Arg	C
C	Leu	Pro	Gln	Arg	A
C	Leu	Pro	Gln	Arg	G
A	Ile	Thr	Asn	Ser	T
A	Ile	Thr	Asn	Ser	C
A	Ile	Thr	Lys	Arg	A
A	Met	Thr	Lys	Arg	G
G	Val	Ala	Asp	Gly	T
G	Val	Ala	Asp	Gly	C
G	Val	Ala	Glu	Gly	A
G	Val	Ala	Glu	Gly	G

Amino acids are Phe = phenylalanine; Ser = serine; Tyr = tyrosine; Cys = cysteine; Trp = tryptophan;
Leu = leucine; Pro = proline; His = histidine; Gln = glutamine;
Arg = arginine; Ile = isoleucine; Met = methionine;
Thr = threonine; Asn = asparagine; Lys = lysine; Val = valine;
Ala = alanine; Asp = aspartic acid; Glu = glutamic acid;
Gly = glycine.

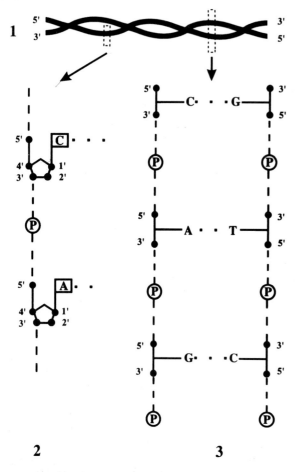

Fig. 2.1   The structure of DNA.
(**1**) A schematic drawing of the DNA double helix. There are two complementary strands (i.e. the base adenine always pairs with thymine; cytosine with guanine) which run in opposite directions: the sense strand 5′ → 3′ and the antisense 3′ → 5′. (**2**) An expanded view of a single strand showing the three basic components: the bases (C – cytosine, A – adenine, not shown are T – thymine and G – guanine). Nitrogenous bases are of two types: purines (A, G) and pyrimidines (T, C). The deoxyribose sugar has the position of its 5 carbons identified as 1′ to 5′. Finally, the phosphodiester (P) linkage between the deoxyribose sugars is shown. (**3**) An expanded view of the two strands. The two strands are held together by hydrogen bonds between the bases (two hydrogen bonds between A/T and three between G/C). The direction for transcription is 5′ → 3′.

Point mutations (single changes in the nucleotide bases) in any of the above codons can lead to genetic disease (see Detecting point mutations, p. 28).

The second significant property of DNA is that it is made up of *two strands* (Fig. 2.1). Each strand has a sugar phosphate backbone linked from the 5′ and 3′ carbon atoms of deoxyribose. At the end of a strand is either a 5′ phosphate group, the 5′ end, or a 3′ phosphate group, the 3′ end. One strand of DNA (sense strand) contains the genetic information in a 5′ to 3′ direction in the form of the four nucleotide bases adenine (A), thymine (T), guanine (G) and

cytosine (C). Its partner strand (antisense strand) has the complementary sequence, i.e. A pairs with T, G with C and vice versa. For example, the double-stranded DNA sequence for the polypeptide described above will be:

Sense   strand: 5′-ATG(start)–GGT–TCT–GTT–GCT–GCT–TGG–TAA(stop)-3′
Antisense   strand: 3′-TAC–CCA–AGA–CAA–CGA–CGA–ACC–ATT-5′

In biological terms, the double-stranded DNA structure is essential for replication to ensure that each dividing cell receives an identical DNA copy. From the above example, it can be seen that the genetic code needs to be read from the sense strand. Hence, transcription to give the appropriate mRNA sequence is taken from the antisense strand so that the single-stranded mRNA will have the sense sequence (antisense RNA is discussed further in Ch. 7).

RNA

Structural differences between RNA and DNA relevant to molecular technology include the single-stranded nature of RNA and its utilisation of uracil in place of thymine. The RNA which is sought in most instances is mRNA. In contrast to DNA, RNA is *less robust* and isolation techniques require the addition of chemicals to ensure that any RNase (also written RNAase) enzymes which may be present are inactivated to avoid degradation of RNA.

Another important difference between RNA and DNA lies in the former's *tissue-specificity*. Thus, the relevant mRNA can only be isolated from a tissue which is transcriptionally active in terms of the target protein. For example, reticulocytes, the red blood cell precursors, would contain predominantly erythroid-specific mRNAs (i.e. mRNAs for the α and β globin genes). The reticulocyte would be inappropriate as a source of neuronal-specific mRNA. The tissue-specificity requirement limited the use of mRNA until fairly recently. Now it has been observed that mRNA production in cells which are easy to access, such as peripheral blood lymphocytes, can be 'leaky', i.e. there is transcription of mRNA species which are not directly relevant to the lymphocytes' function. These 'ectopic' or 'illegitimate' mRNAs are found in minute amounts but the amplification potential of the **p**olymerase **c**hain **r**eaction (PCR) (see p. 19) can be utilised to isolate rare species or minute amounts of either DNA or RNA.

In terms of recombinant DNA technology, mRNA has one important advantage over DNA in that it contains only the essential genetic data resident in exons without the superfluous information found in introns, i.e. mRNA is much smaller than its corresponding DNA.

cDNA

cDNA refers to complementary (or sometimes called copy) DNA. The usual progression in the cell of DNA to RNA to

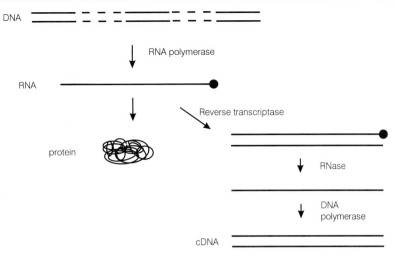

Fig. 2.2 Making cDNA.
Double-stranded DNA (exon represented as a solid line; introns as broken lines) is transcribed into RNA (solid line with ●). In the normal course of events, the RNA is then translated into protein. However, reverse transcriptase allows a copy (cDNA) of the RNA to be made (solid line). Once this occurs the RNA component of the cDNA is removed with an enzyme such as RNase. A DNA polymerase enzyme will then allow the second DNA strand of the cDNA to be formed. From the initial DNA template, a synthetic double-stranded segment containing only exon(s) has now been made. The type of PCR approach described above is often abbreviated to RT-PCR (reverse transcriptase-PCR).

protein can be perturbed in vitro and in vivo (see Chs 1, 6 and 7) with the enzyme reverse transcriptase. Now it is possible to take an mRNA template and produce from this a second strand which is the complement of the mRNA. The double-stranded structure formed from this is called cDNA. Unlike the starting or native DNA (known as genomic DNA), the cDNA does not have introns but contains only coding (exon) sequences (Fig. 2.2).

DNA probes

The double-stranded structure of DNA is exploited in making and utilising probes. These comprise single-stranded segments of DNA which have the complementary nucleotide sequences to bind a segment of DNA which is also single-stranded. For example, if the single-stranded target has the sequence 5'-GGTTACTACGT-3' the single-stranded DNA probe will be 3'-CCAATGATGCA-5'. The *specificity* of a probe thus resides in its *nucleotide sequence*. Since double-stranded DNA is held together by hydrogen bonds, it is relatively easy to make both DNA probe and target DNA single-stranded, e.g. by boiling or treating with sodium hydroxide. Once cooling occurs or the pH is neutralised, the complementary DNA strands will reanneal, i.e. reform into double, base-paired strands. Reannealing will occur between the following combina-

tions: DNA probe + DNA probe; target DNA + target DNA; DNA probe + target DNA. If the DNA probe is radiolabelled with ^{32}P, then the DNA probe + target DNA hybrids can be detected by autoradiography after a procedure such as DNA mapping (Fig. 2.3; see also p. 17). Probes can also be labelled with chemicals which allow the probe to be detected without the potential hazards associated with radioactivity. This approach is becoming increasingly more popular as the detection methods improve and DNA template is provided by means of DNA amplification, which will be described in detail below (Fig. 2.4).

DNA probes are of three types: cDNA, genomic and oligonucleotide (Box 2.1). Plasmid vectors are now available which allow the production of *RNA probes*. These can be used in a similar way as described for DNA probes or in the identification of RNA species (see also Expression vectors, p. 33). DNA probes have a variety of names, which can lead to confusion. To create uniformity, a nomenclature for DNA loci against which a number of related DNA probes can hybridise has been devised. For example, the locus D19S51 means human chromosome 19, segment 51. Thus, probe pTD3-21 (p = plasmid; TD = T Donlon – the scientist who prepared the probe) is more usefully identified by its official name of D15S10 (human chromosome 15, segment 10).

DNA MAPPING

DNA or gene mapping refers to a technique which identifies the pattern of restriction endonuclease recogni-

tion sites in relation to a particular gene or locus. This pattern will be disturbed by a gene or DNA rearrangement.

DNA mapping requires: (1) DNA, (2) restriction endonucleases, (3) electrophoresis, (4) Southern transfer, (5) DNA probes, and (6) autoradiography.

Restriction endonucleases

The unique property of restriction endonucleases (also called restriction enzymes), i.e. their ability to recognise specific base sequences, usually four to six nucleotides in length, is a key element in DNA mapping. How restriction

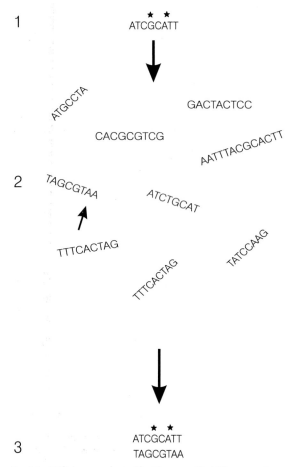

Fig. 2.3 Utilising a probe to identify a specific DNA segment. Probes are single-stranded segments of DNA or RNA which will bind to a complementary sequence in DNA which is also single stranded. The specificity of a probe resides in its nucleotide sequence. Nucleotides (e.g. cytosine) in the DNA probe are radiolabelled with ^{32}P (denoted by ★). **(1)** The probe is then made single-stranded. **(2)** Target DNA is digested into small fragments with a restriction endonuclease and then made single-stranded. In the mixture will be a fragment (indicated by →) whose sequence is complementary to the probe, i.e. the base pairing of adenine with thymine and guanine with cytosine is matched between target and probe. **(3)** Radiolabelled probe and target DNA bind together or anneal. Non-specific binding of the probe to other areas of the genome can be inhibited by the type of hybridisation (annealing) conditions used and a washing step following hybridisation. The washing uses high temperature and low salt to break non-specific joining of the probe with other DNA segments.

enzymes do this has been discussed in Chapter 1. At any DNA locus, there will be a number of sites which are recognised by restriction enzymes. DNA fragments produced following digestion with restriction enzymes will make up a map for that region. The restriction map may be represented by one or more restriction fragments or a composite of many. Disruption of the map will indicate an alteration in DNA sequence. A change in only one of many restriction sites occurs when there is a discrete modification such as a point mutation affecting a single nucleotide base. This may indicate a genetic disorder or more likely it is a neutral mutation which has given rise to a DNA polymorphism. An alteration in more than one restriction site usually indicates a structural rearrangement, e.g. deletion, has taken place (Fig. 2.5).

Restriction enzymes which recognise a four base pair sequence will digest DNA frequently whilst the rare cutting enzymes will give larger fragments (Table 2.2). Restriction enzymes can produce overlapping (sticky) ends following digestion or clean-cut (blunt) ends. The former are particularly useful if a fragment of DNA is to be cloned. Some

---

Box 2.1 Type of DNA probes

A *cDNA* probe is derived from mRNA by using reverse transcriptase. Thus, it represents DNA sequences which are only present in exons (Fig. 2.2). On the other hand, *genomic* probes can be derived from any segment of DNA, i.e. they comprise any combination of flanking sequences, exons and introns. A genomic probe may contain 'anonymous' DNA sequences, i.e. segments of DNA that are not genes and may not even have a known chromosomal location. The above two types of probes vary in size from a few hundred base pairs to a number of kilobases. *Oligonucleotide* probes are much smaller (e.g. 20–30 base pairs) and are synthesised by automated means according to the required DNA sequence. Probes are usually labelled with ^{32}P using techniques such as random primer labelling, nick translation or end-labelling. To date non-radiolabelled probes have been of limited value because sensitivity with these probes is less than that obtained with the corresponding ^{32}P-labelled probes. This has now changed with the ever increasing utilisation of amplified DNA (obtained through the polymerase chain reaction – PCR) and the substitution of colorimetric methods with chemiluminescent substrates (see Fig. 2.4). Oligonucleotides are unsatisfactory as probes if total genomic DNA is used because the small size of the probes produces non-specific background hybridisation. However, oligonucleotides are very effective if hybridised against amplified DNA. DNA probes can be purchased commercially or obtained through central processing facilities for a nominal fee, e.g. ATCC, the American Type Culture Collection in Maryland, lists an extensive catalogue of DNA probes. In the UK, the Human Genome Mapping Project Resource Centre in Hinxton Hall, Cambridge, offers a similar service.

restriction enzymes will not digest DNA if cytosine residues are methylated. This property is useful in distinguishing a segment of DNA which is *hypomethylated* compared to DNA which has been methylated. The former finding may provide a clue that the region is transcriptionally active (see also Box 3.12).

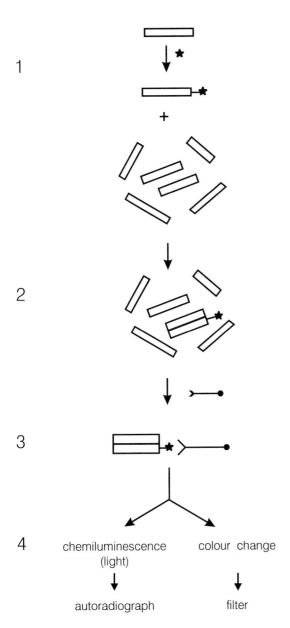

Fig. 2.4 Labelling a DNA probe without using radioactivity.
This follows the same steps as depicted in Figure 2.3 except that
(1) a chemical substance (★) is now attached to the DNA probe.
(2) The probe and its complementary sequence join, i.e. hybridise.
(3) The chemical substance is made detectable by using an antibody against it. **(4)** This antibody is also tagged with another chemical (●) which can either induce a colour change or give off light after combining with a third substrate.

Southern analysis

DNA fragments generated by restriction endonucleases are separated into their different sizes by electrophoresis in agarose gels. DNA fragments are then made single-stranded by treatment with sodium hydroxide. Single-stranded DNA fragments are transferred from agarose to a more robust medium such as a nylon membrane. The transfer step is called Southern blotting. DNA is now ready for hybridisation to a DNA probe which is labelled with ^{32}P or, less commonly, with a non-radioactive marker.

Non-specific hybridisation of the probe to the membrane or to sequences which might have imperfect homology to that found in the probe is prevented in a number of ways: (1) prehybridising the membrane with

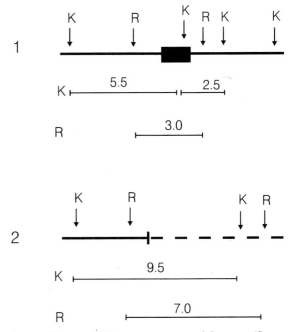

Fig. 2.5 How a restriction enzyme map can define a specific DNA locus.
Restriction enzymes recognise a specific base sequence at which they will cleave DNA. A change in the DNA sequence will disrupt the site(s) of cleavage, altering the map of fragments. A small change, such as point mutation, will only affect one site; a larger change such as a deletion will affect more than one restriction enzyme cleavage site. A DNA probe is selected to hybridise against a gene (■). Two restriction endonucleases (designated K and R) are used to digest DNA in the vicinity of this gene. **(1)** The restriction enzyme digests give 5.5 kb and 2.5 kb bands with enzyme K and a single 3.0 kb band with R. This combination of restriction fragments in association with the probe used provides a 'gene map' for this particular locus. Note that to the right of the 2.5 kb fragment there is another potential fragment generated by K. However, since this fragment does not overlap the probe region, it is not seen in the map. **(2)** Any rearrangement around this locus, such as a deletion (– – – –), will disrupt the restriction enzyme pattern since the novel DNA sequence found in association with the deletion will have different restriction endonuclease recognition sites, i.e. enzyme K now gives a 9.5 kb fragment and enzyme R a 7.0 kb fragment.

DNA from another species (heterologous DNA, e.g. salmon sperm DNA), and (2) following the hybridisation step, the membrane is washed under stringent conditions (e.g. high temperature, such as 65°C, and a low salt solution). The stringent wash breaks hydrogen bonds which may have formed by chance between random complementary base pairs. The end result will be binding of the radiolabelled probe to its specific target DNA fragment which can be detected by autoradiography (Fig. 2.6).

Controls for DNA mapping include normal DNA which has been processed under identical conditions and DNA markers which enable the size of the hybridised fragment to be determined. A DNA map may require a number of different restriction endonuclease digestions before interpretation can be made. The upper limit of resolution for

DNA mapping is ~30-40 kb. Differences in restriction fragments 100–200 base pairs in size can be detected by conventional DNA mapping.

RNA may also be 'mapped', although restriction endonucleases are not used since these enzymes digest *double-stranded* nucleic acid. RNA 'mapping' involves an estimation of the RNA size. Another difference between RNA and DNA noted earlier is the requirement that RNA must be prepared from transcriptionally active tissues. Electrophoresis of RNA is also undertaken in a denaturing gel to prevent secondary structures from forming during electrophoresis. Transfer of RNA from an agarose denaturing gel to nitrocellulose membranes is known as 'northern' blotting. Both northern, and also western, are usually

Table 2.2 Cleavage sites and some properties of restriction endonucleases.
Restriction enzymes recognise palindromic sequences, i.e. the DNA sequence is the same when one strand is read from left to right and the second strand from right to left. Fragments produced by digestion with restriction enzymes which have sticky ends, i.e. there is an overlap because a few bases are not paired, are particularly useful in cloning since they anneal with greater efficiency compared with blunt-ended fragments.

Enzymes	Cleavage site	Properties
*Eco*RI	5' G AATTC 3' 3' CTTAA G 5' ↓ G AATTC + CTTAA G	Sticky ends; 6 base pair recognition
*Hpa*II	5' CCGG 3' 3' GGCC 5' ↓ C CGG + GGC C	Sticky ends; methylation sensitive; 4 base pair recognition
*Rsa*I	5' GTAC 3' 3' CATG 5' ↓ GT AC + CA TG	Blunt ends; 4 base pair recognition
*Not*I	5' GCGGCCGC 3' 3' CGCCGGCG 5' ↓ GC GGCCGC + CGCCGG CG	Sticky ends; recognises 8 base pair sequence (i.e. a rare cutter enzyme and so is useful in PFGE)

▼ ▲ indicate the site of digestion; PFGE = pulsed field gel electrophoresis

α Globin Gene Probe Restriction Enzyme *Bam*HI

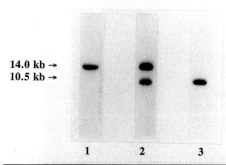

14.0 kb →
10.5 kb →

1   2   3

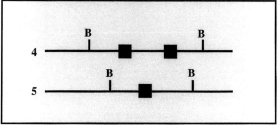

Fig. 2.6 Analysis of α thalassaemia by using a DNA probe: autoradiograph illustrating a DNA restriction map.
The DNA probe is specific for the α globin gene locus on chromosome 16 and the restriction enzyme used to digest DNA is *Bam*HI. The normal (also known as wild-type) α globin gene structure at this locus is represented by a 14 kb band on a gene map and a form of α thalassaemia associated with a single gene deletion is identified by a 10.5 kb band. Track (1) on the gene map has only the 14 kb band, i.e. normal. Track (2) has both 14 and 10.5 kb fragments and so is heterozygous for the α thalassaemia deletion. Track (3) has only the lower (10.5 kb) fragment and so is homozygous for this form of α thalassaemia. The box below the gene map depicts the α globin gene as a ■ and the position of the *Bam*HI recognition sites as B. In (4) is found the normal structure, i.e. 2 α globin genes which have on either side a *Bam*HI recognition site to give a single 14 kb fragment. In (5) there is loss through deletion of one of the two α globin genes (such as occurs in the form of α thalassaemia mentioned above). This brings the two *Bam*HI restriction sites closer together and the 14 kb band becomes 10.5 kb. There are other possible explanations for the interpretation of the DNA map in lanes 1 and 3, but these can be excluded by taking into consideration both the haematological parameters and the gene map.

written in lower case because they do not refer to an individual, cf. Southern blotting.

Specific RNAs can be identified by using radiolabelled DNA or RNA probes. Detection and quantitation of low abundance mRNAs is possible through a technique called S1 mapping or RNase protection. In essence this means that RNA probes which will anneal with a corresponding mRNA species are less liable to breakdown by single-strand specific enzymes such as SI nuclease or RNase. The 'protection' resulting from formation of a double-stranded structure, i.e. the RNA probe + mRNA hybrid, provides indirect evidence for the presence of mRNA.

DNA AMPLIFICATION

The ability to target a segment of DNA (and also RNA) and then produce multiple amounts of that targeted region is known as DNA amplification. The usual method to achieve this is by PCR. DNA amplification requires: (1) DNA (or RNA), (2) oligonucleotide primers, (3) *Taq* polymerase, (4) a mixture of the four nucleotide bases, and (5) temperature cycling apparatus.

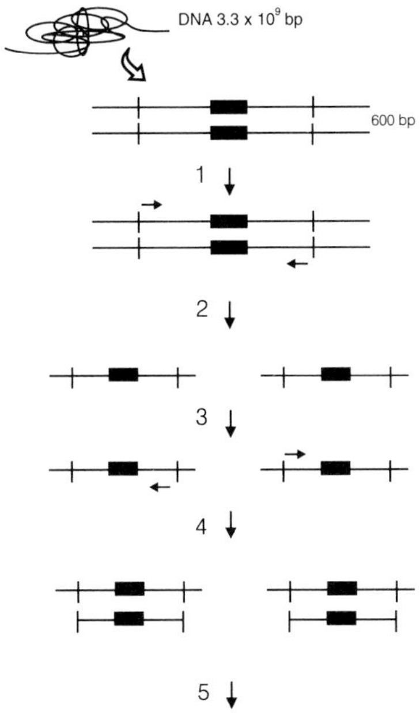

Polymerase chain reaction (PCR)

PCR utilises a DNA extension enzyme (polymerase) which can add nucleotide bases once a template is provided (Fig. 2.7). There are three basic steps in PCR. (1) Denaturation of double-stranded DNA into its single-stranded form. (2) Joining of oligonucleotide primers to both ends of a target sequence. The oligonucleotides are constructed so that they are complementary to the target DNA or RNA sequence and this gives PCR its specificity. Oligonucleotide primers are present in excess and so will not be limiting in the subsequent amplification steps. (3) Addition of the four nucleotide bases and a polymerase. *Taq* polymerase is used since it is relatively heat resistant and so the denaturation step can be incorporated into the overall cycle without interfering with the polymerase activity. One PCR cycle comprises steps 1–3 described. After a cycle, each of the single-stranded DNA target segments has become double-stranded through the polymerase's activities. The cycle is then repeated and each time a new target segment of DNA is synthesised. Theoretically, the number of templates produced equals 2^n, i.e. after 20 cycles of amplification there should be $\sim 1 \times 10^6$ templates.

Amplified DNA products can be visualised in a number of ways. The simplest makes use of electrophoresis and staining of DNA with ethidium bromide which intercalates between the nucleotide bases and fluoresces under ultra-violet light (Fig. 2.8). Radiolabelled nucleotides can also be incorporated into the PCR steps and the amplified products visualised by autoradiography. Amplified DNA products can be transferred to nylon membranes by Southern blotting and detected by their probe hybridisation patterns. Colorimetric assays have been used with PCR. These depend on oligonucleotide primers which have been modified to allow them to be labelled with chemicals such as biotin or fluorescein.

A feature of PCR is its exquisite *sensitivity* so that even DNA from a single cell can be amplified. The ability of PCR to amplify small numbers of target molecules has been put to use in detecting 'illegitimate transcription'. As described earlier, mRNA is tissue-specific except for some

Fig. 2.7 The polymerase chain reaction (PCR).
This technique allows amplification of a targeted DNA sequence by using a DNA extension enzyme (polymerase) to make new copies of the sequence. Oligonucleotide primers give PCR its specificity. In this example a sequence of DNA 600 bp in size from the total 3.3×10^9 bp human genome is required. **(1)** The first step is to design primers which span either side of the 600 bp sequence. The primers $(\rightarrow, \leftarrow)$ are in fact single-stranded DNA sequences complementary to the ends of the targeted sequence. **(2)** Double-stranded DNA is now made single stranded by heating. **(3)** Primers are added to the single-stranded DNA and they will anneal on either end. **(4)** DNA polymerase and a mixture of the four nucleotide bases are added. The combination of primers and reagents in step **(4)** will lead to a copying of the single-stranded segment. The final product is double-stranded DNA which comes from the region defined by the primers. At this stage of the PCR, an initial single template of DNA has now been formed into two copies, i.e. duplicated. **(5)** The steps **(2)**–**(4)** are now repeated to produce (in theory if the process is 100% efficient) 2^n times the amount of template DNA (where n = number of cycles), e.g. 20 cycles should amplify the original segment about 1×10^6 times.

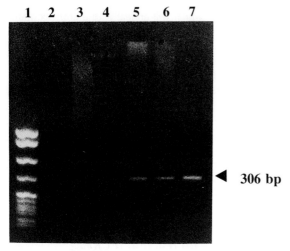

◀ **306 bp**

Fig. 2.8 Amplified DNA can be visualised directly by staining with ethidium bromide.
After electrophoresis, a gel is submerged in a solution of ethidium bromide. Ethidium bromide intercalates between the DNA bases and fluoresces under ultraviolet light and so will identify the presence of DNA. Lane 1 has DNA size markers. Lanes 2–7 show increasing concentrations of DNA 306 base pairs in size which have been amplified from exon 5 of the human aldolase B gene.

'leakiness' which occurs in cells such as the lymphocyte. Thus, mRNA specific for muscle tissue in disorders such as Duchenne muscular dystrophy and the hereditary cardiomyopathies has been characterised by amplification of mRNA from lymphocytes (see Box 3.7 for further discussion). PCR is rapid and can be automated with a 30-cycle procedure taking ~2–3 hours to complete. Applications of this technique are extensive with new modifications or innovations being constantly described (Box 2.2).

Problems

PCR has two major disadvantages. The first is the requirement to know a DNA sequence before oligonucleotide primers can be synthesised and the second reflects the relative ease with which *contamination* can occur. The latter may come from extraneous DNA such as other samples or the operator. The commonest source for contamination is the previously amplified products. Many strategies have been described to avoid contamination, which is one reason why PCR has not been utilised in more of the routine diagnostic laboratories.

The sequence fidelity of amplified products is an additional consideration when assessing the usefulness of DNA amplification since in vitro DNA synthesis is an error-prone process. The error rate associated with *Taq* DNA polymerase activity is relatively low and has been estimated to be about 0.25%, i.e. one misincorporation per 400 bases, over 30 cycles. The more recent commercially produced *Taq* polymerases are claimed to have less reading errors.

Box 2.2 Applications of the polymerase chain reaction (PCR)

DNA amplification by the polymerase chain reaction (PCR) is used in the diagnosis of many genetic disorders. PCR enables a segment of DNA to be screened for polymorphisms or point mutations. In the infectious diseases, PCR can detect DNA from microorganisms which are present in too few numbers to be visualised or whose growth characteristics are such that there will be a delay in diagnosis. The sensitivity of PCR is useful in forensic pathology where tissue left at the scene of the crime is small in amount or degraded. Research applications utilising PCR are numerous. Amplification has enabled cloning, direct sequencing and the creation of mutations in DNA segments. Investigation of DNA/protein interactions by a process called 'footprinting' is possible. Quantitation of DNA by PCR can be undertaken although it requires a number of inbuilt control reactions to allow for variability in amplification. Non-radioactive labelling becomes feasible if used in conjunction with PCR. Novel automated procedures for amplifying DNA on a large scale are described and will enable rapid and sensitive screening for a number of defects or polymorphisms. Multiplex PCR refers to the simultaneous amplification of a number of DNA segments using a combination of multiple oligonucleotide primer sets. RNA may also be studied along the same lines as described for DNA. An additional step, which incorporates the enzyme reverse transcriptase, enables RNA to be first converted to cDNA from which amplification is then able to proceed (see also Fig. 2.2).

Variations of PCR

A comment that PCR is limited only by the ingenuity of the scientist is very relevant. *Multiplex PCR* refers to the mixing of various primer combinations so that simultaneous amplifications are occurring. *Nested* PCR describes the use of two sets of primers, the second of which is 'inside' the first. This approach is taken to increase the sensitivity and specificity of PCR. *In situ PCR* allows genes, or more commonly mRNA, to be identified in tissue sections. *RT-PCR* (reverse transcriptase PCR) is used to amplify RNA. *Long PCR*, as its name implies, has been marketed so that large segments of amplified DNA are able to be produced. This is an achievement since conventional PCR products are usually relatively small fragments measuring in the hundreds of base pairs. With long PCR it is claimed that DNA sizes ranging from 5 to 40 kb can be amplified. See Harper (1994) for further descriptions of PCR variations.

Attempts have also been made to find alternative methods to amplify DNA. One of these is the *ligase chain reaction*. Here the amplification of DNA comes from the 'stitching' together of synthetic oligonucleotides onto a DNA template. Because there is no adding of individual nucleotide bases, the ligase chain reaction does not have the potential problem of misincorporation errors.

CLONING DNA

Cloning DNA involves the production of an unlimited and pure amount of the same DNA segment using a vector and host system. Cloning of DNA requires: (1) DNA fragments prepared to a specific size range, (2) a vector which will accept DNA to be cloned, (3) a host to allow propagation of vector and its inserted DNA, and (4) a method to screen a DNA library for the clone of interest.

Strategies

The principles behind cloning are summarised in Figure 2.9 and a summary of cloning vectors is given in Table 2.3. The first step is to decide on the size of DNA to be cloned. Thus, a limited segment of DNA needs to be cloned if the gene being sought is small (for example, the globin genes). On the other hand, trying to clone a large gene (for example, cystic fibrosis) or finding a gene somewhere on a chromosome will mean a different approach is necessary, i.e. the largest possible DNA segments will need to be cloned. Once size is known, the appropriate vector system required to carry the cloned DNA fragment can be selected. Random shearing by physical methods or restriction endonucleases will break DNA into fragments. Target DNA will be found within one or more of these fragments.

Vectors

Cloning with plasmid vectors

The simplest cloning system involves a plasmid vector. In this situation, genomic DNA and plasmid are both digested with the same restriction endonuclease creating, if possible, fragments with unpaired bases at each end (sticky ends). Mixing the two together in the presence of DNA ligase will allow ligation to occur so that each individual plasmid will have 'stitched' to it one fragment of genomic DNA. Using electrical or calcium chloride shock, it is possible to get the plasmids plus their cloned inserts to be taken up (in a process called *transformation*) by E. coli. The host E. coli divides and is then plated out as a lawn on an agarose plate. If all goes well there should be a broad representation of DNA from most of the genome (including the target DNA) inserted into the plasmids. Thus, a *library* of DNA fragments has been produced. The next step involves screening of the library to find the target DNA. For screening, a replica of the colonies on the agar plate is obtained by placing the plate against a nylon membrane. DNA which has been transferred to the nylon by this contact is then made single-stranded and can be screened for target DNA with the appropriate probe. When a positive colony is found it is traced back to the relevant colony on the agar plate which is then isolated and used as a source of the cloned DNA (Fig. 2.10).

Table 2.3 Vectors available for DNA cloning.
Vectors will carry segments of DNA into a host cell where it can be produced in larger amounts. The vector chosen will depend on the size of the DNA to be cloned.

Vector	DNA insert size	Features
Plasmid	<10 kb	Technically simple but relatively inefficient
Phage	~20 kb	Most conventional approach; size of inserts is limited
Cosmid	~40 kb	Larger inserts possible
P1 phage	<100 kb	Recent vector; in between cosmids and yeast artificial chromosomes
YAC	~300–1000 kb	Complex to make and difficult to screen; large inserts possible

Cloning with bacteriophage vectors

The next level of complexity for cloning is the bacteriophage ('phage') vector. This has the advantage over plasmids in that it can accept a larger DNA insert and is overall a more efficient way to clone. The same steps as described for the plasmids are undertaken except that phage plus its DNA insert need to be prepared into an infectious unit (called 'packaging') for insertion into E. coli. Phage vector cloning will produce plaques (areas of lysis) wherever a phage infects an E. coli. These plaques on the agar plate represent individual clones. A disadvantage of using phage as a vector is poor yield of target DNA once it has been isolated. This is usually overcome by finding the relevant DNA in a phage library and then subcloning that DNA into a plasmid vector. There are now many commercial kits available which allow cloning into a variety of phage vectors.

Cloning with yeast artificial chromosomes (YACs)

To clone larger DNA fragments requires more sophisticated vectors. These include cosmids, P1 phage or yeast artificial chromosomes (YACs). The last are autonomously replicating stable chromosomes with selectable markers suitable for yeast. They are particularly valuable since large DNA fragments, e.g. 300–500 kb in size, can be cloned. DNA plus vector are inserted into a eukaryotic (yeast) host rather than the prokaryotic bacterial host described above. One major advantage in cloning with YACs is the inserts (DNA fragments) are so large that it is very likely target DNA will be cloned within a single fragment.

The larger cloned fragments also make it easier for directional movement along a chromosome ('chromosome walking') to occur. This will be discussed further under Positional cloning (p. 42). To isolate a gene, it is necessary

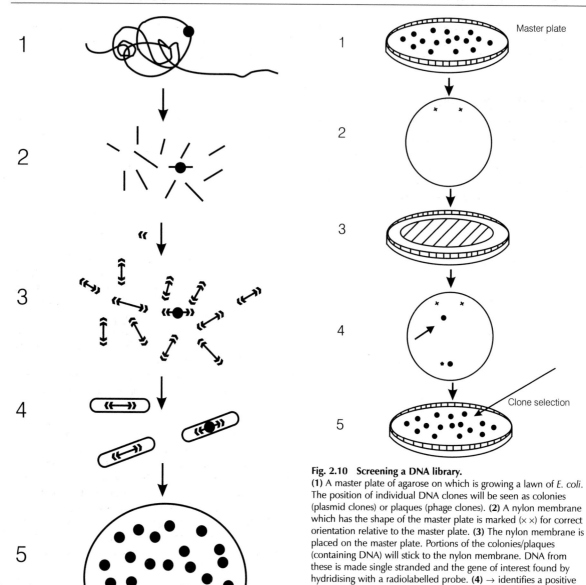

Fig. 2.9 Steps involved in cloning.
(1) Genomic DNA with target DNA is depicted as (●). (2) DNA is randomly fragmented, by physical methods or restriction enzymes, into many pieces which are then size-selected. (3) The correct-sized DNA is then joined, using the enzyme DNA ligase, to the DNA vector (<< and >>) such as a plasmid or a phage. There will be many ligated DNA segments. Hopefully, the target gene will be included in this 'library'. (4) Vector plus inserts are taken up into a host such as *E. coli* in a process known as transformation. (5) Vector, inserts and host are plated out onto an agarose plate. Here each *E. coli* forms a discrete colony. The colonies can be screened to identify which one contains the DNA segment of interest (see Fig. 2.10 for a description of screening). The relevant colony can then be used to grow large quantities of bacteria containing the DNA of interest. The latter is then purified from bacterial products.

Fig. 2.10 Screening a DNA library.
(1) A master plate of agarose on which is growing a lawn of *E. coli*. The position of individual DNA clones will be seen as colonies (plasmid clones) or plaques (phage clones). (2) A nylon membrane which has the shape of the master plate is marked (× ×) for correct orientation relative to the master plate. (3) The nylon membrane is placed on the master plate. Portions of the colonies/plaques (containing DNA) will stick to the nylon membrane. DNA from these is made single stranded and the gene of interest found by hydridising with a radiolabelled probe. (4) → identifies a positive clone/plaque on the membrane, ★ is a positive control. (5) Using the position of the positive clone relative to the x x markers, it becomes possible to go back to the original master plate and select the right colony/plaque.

to have overlapping, contiguous segments of DNA known as 'contigs' to assist in chromosome walking (discussed further below and in Ch. 3 under Cystic fibrosis). The larger the contigs the easier is 'the walk'. It should also be noted that some segments of the genome are difficult to clone and so phage or cosmid libraries may be inadequate to provide contigs over those regions. The ability to clone into YACs is very useful in these circumstances.

Not surprisingly, there are technical problems associated with YACs. The libraries are difficult to make. To get around this a number of laboratories have functioned as core facilities. YAC libraries with inserts which are much larger than previously described are now available through CEPH

(Centre d'Etude du Polymorphisme Humain), one of the international resource centres. Screening of YAC libraries can be difficult, but this has improved with the availability

of PCR (see p. 19). Finally, DNA inserts in YACs, particularly the larger ones, are frequently unstable, i.e. they can recombine or delete.

SEQUENCING DNA

Sequencing of DNA refers to the enumeration of individual nucleotide base pairs along a linear segement of DNA. Sequencing requires: (1) cloned or amplified DNA, (2) DNA primers (also known as oligonucleotide primers), (3) DNA polymerase, (4) the four nucleotide bases (A, T, G,

C), (5) the four nucleotide bases in modified form as dideoxy bases, (6) radiolabelled or fluorochrome-labelled bases, and (7) a method to separate DNA, e.g. electrophoresis.

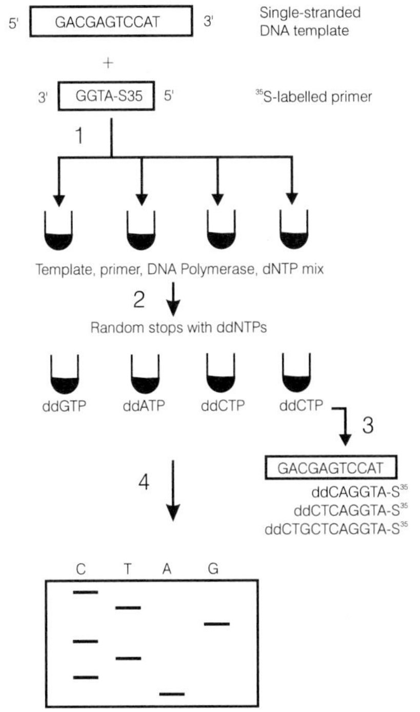

Fig. 2.11 DNA sequencing with the dideoxy chain termination method.
Single-stranded DNA template and radiolabelled (^{35}S) single-stranded primer (GGTA) are prepared. The two are allowed to anneal and are then aliquoted into four tubes. (1) In the presence of DNA polymerase and a mixture of the four deoxynucleotides (dGTP, dATP, dTTP, dCTP) there is extension from the primer/template double-stranded site. (2) Random stops in the extension are then produced by adding to each tube one of the dideoxynucleotides (ddNTP). (3) As illustrated in one tube, ddCTP will produce random stops wherever there is a cytosine nucleotide and so a number of double-stranded DNA products are formed of varying sizes. Note that the ddCTP-containing fragments are complementary to the original template sequence. The remaining three dideoxynucleotides will do likewise in their individual reactions. The end result is a mixture containing variable lengths of extended DNA segments. (4) Each mixture is electrophoresed in the gel track corresponding to the dideoxy nucleotide added, e.g. G = ddGTP, etc. The DNA sequence is read from bottom to top. In this example the sequence reads: ACTCGTC which represents DNA sequence 3' to 5' following annealing between GGTA (primer) and its complementary sequence in the template (CCAT). See also Figure 2.12 which will give you practice at reading an actual sequencing gel.

Manual methods

The majority of genetic disorders are caused by point mutations and, less frequently, small discrete deletions. In these circumstances, DNA mapping alone is unlikely to detect an abnormality unless, by chance, the mutation deletes or creates a recognition site for a restriction endonuclease. The latter occurs infrequently and so sequencing of individual nucleotide bases may be required. The methodology for sequencing DNA has evolved rapidly since the first descriptions in the mid-1970s by Sanger, Maxam and Gilbert. Two procedures were initially developed. One utilises chemical cleavage of DNA and the second is the more popular enzymatic or dideoxy chain termination method (Fig. 2.11).

Since first described, the dideoxy chain termination method has undergone a number of modifications which have improved resolution, e.g. utilisation of the isotope ^{35}S rather than ^{32}P for radiolabelling; the isolation of a more efficient polymerase enzyme (Sequenase^{TM}) and better gel electrophoresis techniques to increase the length of readable DNA sequence. Generation of cloned single-stranded DNA required for sequencing has been facilitated with the development of a special phage vector (M13) or plasmids containing M13 replication origins. An alternative to cloned DNA for sequencing is DNA which has been amplified by PCR. Single-stranded or double-stranded DNA generated by PCR can be used directly for DNA sequencing and is now the preferred method (Fig. 2.12). DNA sequencing can now be obtained as part of the actual PCR in a process known as 'cycle sequencing'.

Automated methods

While a satisfactory manual sequencing run can produce 200–400 base pairs of readable DNA sequence, it is now possible with automated sequencing to obtain over twice that amount in a single run. The instruments in use at present rely on variations of fluorescence labelling, PCR and gel electrophoresis (Fig. 2.13). As the Human Genome Project develops it is expected that more effective and cheaper sequencing protocols will become available (see Ch. 10).

POSITIONAL CLONING

Positional cloning refers to the isolation and cloning of a gene on the basis of its chromosomal position rather than its functional properties. Positional cloning requires: (1) chromosomal location, (2) candidate gene, (3) DNA polymorphisms, (4) genetic map, (5) physical map, and (6) gene 'decoding'.

Functional cloning

Haemophilia A, a genetic bleeding disorder, provides an example of what is meant by *functional* cloning. In this case, a disease has been sufficiently characterised to enable identification of an abnormal end-product (the factor VIII coagulation protein). From this the relevant gene can be cloned and mapped at the DNA level (Fig 2.14). Cloning has resulted in a wider range of diagnostic tests, an increased knowledge of the disorder's pathogenesis and alternative therapeutic approaches (discussed further in Chs 3 and 7). However, the situation found in most genetic disorders, e.g. Duchenne muscular dystrophy, is quite different. In these circumstances, the underlying defect remains unknown and there can be no further progress with understanding of the disease or even the possibility of diagnosis. The molecular strategy known as *positional* cloning (formerly called *reverse genetics* but positional cloning is a more appropriate description) can now be adopted.

Chromosomal location

The first step in positional cloning is to obtain, if possible, a clue as to the likely locus or chromosome involved. This usually comes from case reports or observations in which chromosomal rearrangements have been noted to occur in association with the clinical picture. An example of this is illustrated by the familial colon cancer called FAP (familial adenomatous polyposis). The clue that this disorder was associated with the long arm of chromosome 5 came from the chance observation of a deletion involving this chromosome and FAP in one family. Positional cloning was then started at the chromosome 5q locus and eventually it led to the underlying FAP gene being found (see Ch. 6 for further discussion).

Candidate genes

Another approach, if a disease locus has not been identified, is to look with DNA markers derived from 'candidate genes'. For example, a good candidate gene for a heart muscle disorder would be myosin, a component of muscle (see Table 1.2 and Ch. 3, Familial hypertrophic cardiomyopathy, for further discussion). Whether a locus is known or a candidate gene is available, the most important tool available to the molecular biologist to confirm that a

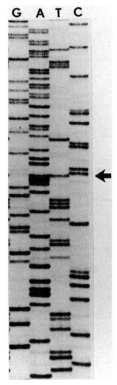

Fig. 2.12 Direct sequencing of double-stranded DNA amplified with PCR.
The four gel lanes marked GATC indicate the relevant nucleotide in the sequence. Reading from the bottom of the gel the DNA sequence in this example is: ATCTTGACTGTTGA, etc. An arrow marks where two lanes (A and T) have a band present in the same position. Therefore, because this is double-stranded DNA sequencing, the simultaneous appearance of bands in the A and T tracks means one allele has adenine and the second has thymine. This may represent a DNA polymorphism or a point mutation producing a genetic defect.

disease is associated with that locus/gene is linkage analysis with DNA polymorphisms. This is the first step in making a genetic map.

DNA polymorphisms

These refer to variations in DNA fragment lengths caused by the following mechanisms.

1. *Point mutations* which fortuitously delete or add restriction endonuclease recognition sites. This polymorphism is called an RFLP (**r**estriction **f**ragment **l**ength **p**olymorphism). An RFLP is biallelic, i.e. the polymorphism comprises either a large or small fragment.

2. *Insertions* made up of tandem repeats which are found between two restriction endonuclease recognition sites in a segment of DNA. The repeats can be of any number and so the end result is a multiallelic polymor-

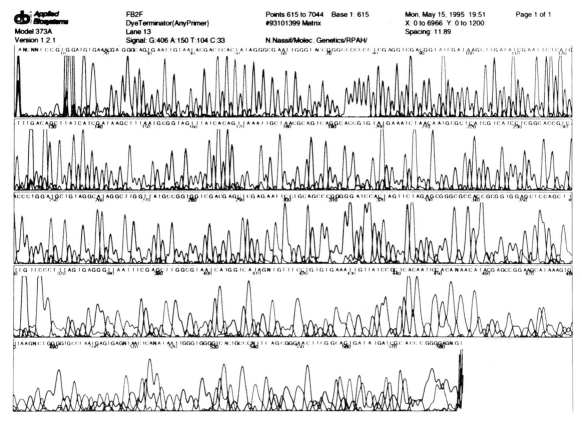

Fig. 2.13 Trace of an automated sequencing run.
The automated DNA sequencer has made life easier for the molecular biologist since it enables relatively large segments of DNA to be sequenced and read automatically. A black and white picture will not allow the reader to practise reading from the automated printout which comes in four different colours to represent the four nucleotide bases. The DNA sequence is seen above the various peaks. N in the DNA sequence means the automated sequencer cannot decide which base is correct.

phism which is of variable length (called minisatellites or VNTR – **v**ariable **n**umber of **t**andem **r**epeats).

3. *Variations* in the number of repeats associated with a simple repeating motif composed of 2 to 4 nucleotide bases (called microsatellites or SSRs – **s**imple **s**equence **r**epeats) (Fig. 2.15).

In a family, a DNA polymorphism is inherited along Mendelian lines. For convenience, a single-base change is called a DNA polymorphism if it occurs at a frequency of 1% or more in the population. DNA polymorphisms have proven to be invaluable markers to distinguish between two alleles at a locus (for example, wild-type versus mutant). DNA polymorphisms enable a gene locus to be detected *indirectly* by allowing co-segregation between a phenotype (which can be normal or abnormal) and a particular DNA polymorphism to be followed within the context of a family study. It is not even necessary to have identified a gene or its underlying defects to utilise DNA polymorphisms. Predictive estimations concerning disease status of individuals on the basis of their DNA polymorphism patterns are now frequently undertaken (examples may

be found in Chs 3 and 4). DNA polymorphisms also provide patterns which are useful in forensic analysis (see Ch. 8). There are few diagnostic or research studies which do not utilise these invaluable markers.

Genetic maps

Once a DNA marker has been linked to a genetic defect it is necessary to follow this with the construction of genetic and physical maps for that locus. The genetic map is made by looking at DNA polymorphisms within affected families. The closer the polymorphism is to the gene, the fewer will be the recombinations (breaking and rejoining of the DNA) that are observed (see Ch. 3 for a more detailed description of recombination). Eventually, a polymorphism associated with the gene itself will produce no recombination events.

The steps described earlier involving the identification of a chromosomal locus or candidate genes to initiate a positional cloning strategy is very helpful in narrowing a search, but is not essential. In many circumstances, positional cloning starts off with random testing of DNA probes

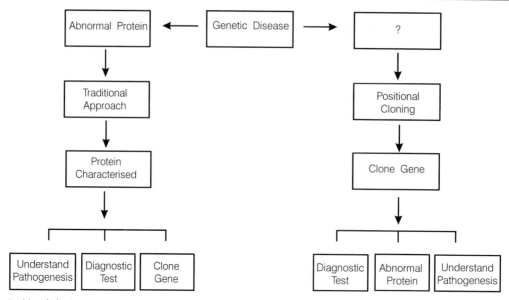

Fig. 2.14 Positional cloning.
The traditional way to study a genetic disorder is illustrated by haemophilia. First the underlying protein defect is identified and from this it becomes possible to understand the pathogenesis, develop diagnostic tests and go on to clone the gene ('functional cloning'), which will then provide additional knowledge about the disorder and perhaps DNA testing options. However, in most cases, e.g. Huntington disease, it is not always possible to isolate the underlying defective product or, more likely, it is unknown. Now positional cloning comes into its own. In this modern genetic approach, the protein is bypassed completely and the relevant gene is cloned on the basis of its position in the genome. From this, DNA tests become available, the protein is produced by expression systems (see Expression of rDNA, below). Finally, information coming from the positional cloning of the gene allows an understanding of pathogenesis.

for linkage to a disease locus. This is tedious and risky but has been successful on a number of occasions as illustrated by the Huntington disease and adult polycystic kidney disease examples (see Ch. 3). The increasing availability of SSR-type polymorphisms generated particularly from the Human Genome Project will facilitate positional cloning. There are now a number of international laboratories with the expertise and resources to conduct *whole* human genome searches in a relatively short and efficient time frame.

Physical maps

Unlike genetic maps, physical maps are based on actual measurements, e.g. kb or Mb. The availability of **p**ulsed **f**ield **g**el **e**lectrophoresis (PFGE), **f**luorescence **in** **s**itu **h**ybridisation (FISH) and YACs has helped in the construction of physical maps and chromosome walking. These new techniques make positional cloning a more realistic strategy for isolating genes (see Chs 1 and 3 for examples of how positional cloning has been utilised). From clones contained within YACs, genes can be identified and these are then further characterised until the correct one is found. Information derived from this enables a diagnostic test to be developed and the function of the underlying genetic disorder to be determined. The many successes of the positional cloning approach have already been mentioned in Chapter 1.

Pulsed field gel electrophoresis (PFGE)

PFGE, a modification of gene mapping, was described by Schwartz and Cantor in 1984. This technique has increased the upper limit of resolution for DNA mapping from kilobases to megabases. Technology developments which have made this possible include: (1) the isolation of restriction endonucleases which cleave DNA into much larger fragments than is possible with the more frequently used restriction enzymes and (2) the application of a non-homogeneous and interrupted electric field.

DNA for PFGE needs to be prepared in a special way since it is essential to avoid random shearing which will break DNA into fragments which are relatively small. Rare cutting restriction enzymes then digest DNA into fragments ranging in size from 100 to 2000 kb. Application of a constant and homogeneous electric current, such as that used for conventional gene mapping, will not separate large DNA fragments (e.g. 100–500 kb) since the forward motion brought about by the electric charge and the drag due to friction as the DNA passes through agarose pores does not allow differential migration of large molecules. In PFGE, a differential separation of DNA is obtained by periodically altering the orientation of the electric field. Each time this happens, large molecules need to reorientate and then find a new path through the gel matrix. The reorientation time will be dependent on the size of the fragment (Fig. 2.16). Restriction fragments up to 10 Mb in size can be distinguished by PFGE. This technique is

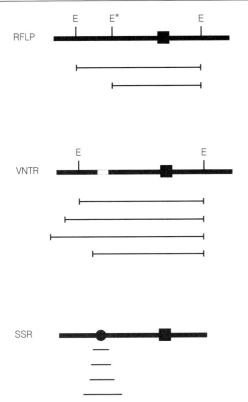

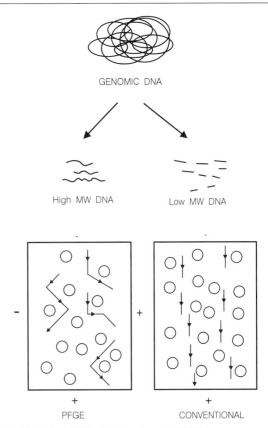

Fig. 2.15 Three types of DNA polymorphisms: restriction fragment length polymorphisms (RFLPs), variable number of tandem repeats (VNTRs) and simple sequence repeats (SSRs).
DNA polymorphisms are produced by changes in the nucleotide sequence. These result in variations in the fragment length pattern produced after digestion of DNA with restriction enzymes or PCR amplification of a specific DNA segment.

RFLP: A segment of DNA is digested with a restriction enzyme 'E'. This segment can be identified in Southern analysis by using a DNA probe which will hybridise against the segment marked (■) or PCR can be used to amplify this specific region. RFLPs are caused by point mutations affecting a single restriction enzyme recognition site (E*) which will either be absent or present. If absent, enzyme E will digest the DNA at the two end E sites; if present, enzyme E will digest DNA at E* and the right-hand side E. Therefore, RFLPs are biallelic, i.e. they give two options (large or small) depending on whether the polymorphic restriction fragment site (*) is absent or present, respectively.

VNTR: In contrast, the multiallelic VNTR has the potential to be more polymorphic (and so more informative) since the changes in the E-specific restriction fragment are brought about by the insertion of a variable number of repeat units at the polymorphic site (hatched area). Thus, the number of polymorphic DNA fragments generated is potentially much greater (e.g. the four different sized fragments illustrated). Because of their greater intrinsic variability, VNTRs are usually more informative in polymorphism studies since there is a greater chance that heterozygous patterns will be detected at any one locus.

SSR: In contrast to RFLPs or VNTRs which can be identified by Southern analysis and PCR, SSRs are much smaller in size and so are detectable only if PCR is used. SSRs are polymorphic because of repeats in simple sequences (●) such as $(AC)_n$ where *n* is any number. Amplification of DNA containing a SSR will produce fragments of variable size. See Figures 1.2, 3.6, 3.14, 3.21, 4.3 which provide actual examples of how DNA polymorphisms are used in genetic diagnosis.

Fig. 2.16 Pulsed field gel electrophoresis (PFGE).
This technique enables large fragments of DNA (which would co-migrate if conventional electrophoresis conditions were used) to be separated. PFGE has made an important contribution to the development of positional cloning since it has enabled megabase fragments of DNA to be mapped, isolated and cloned. Requirements for PFGE include: (1) DNA of high molecular weight, (this usually comes from cell lines); (2) the use of restriction enzymes which recognise unusual DNA sequences, i.e. they digest DNA very rarely; and (3) a type of zig-zag electrophoresis. DNA is negatively charged and so will move towards the positive electrode through the agarose particles (o). Large fragments are separated from each other by altering the orientation of the electric field. The snake-like movement of the large particles, because there is a periodic change in polarity of the electric field, separates them. Their reorientation time to a new electric field is dependent upon the size of the fragment. In contrast to PFGE, the smaller DNA fragments in conventional electrophoresis move in one direction. +, − indicate the electrical charge through the agarose gel.

increasingly being used to construct physical maps of DNA loci. Large fragments generated by PFGE can also be isolated from gels and cloned. Fluorescence in situ hybridisation is described on p. 32.

Gene decoding

Gene 'decoding', a term used by Collins, describes the situation in which the molecular biologist finds him/herself when a positional cloning strategy has been successful, i.e. a gene has been found. At this stage the most important questions concern what has been found and whether it is the right gene. To answer this question the gene (usually cDNA since this represents only the exons and so is smaller than the total or genomic structure) is sequenced. The sequence is entered via the Internet into the various DNA and protein databases available, e.g. GenBank – one of the main repositories for DNA information. Computer programs then enable a search to be made which compares sequences in the databases with the recently discovered gene. Three scenarios are possible in this process of decoding. (1) A perfect match is found in the database, i.e. the gene has already been described and the scientist is probably too late in his/her discovery! (2) No match is found. This means the gene is novel but there is no clue as to what it might do. Often positional cloning stops at this point until some future time when additional information throws light on the identity of the gene. If the scientist is lucky, it might be possible to show that the gene and a disease are associated by demonstrating mutations in affected individuals. (3) There is some homology found to another entry in the database. This is the best result since the gene is still novel and a clue to its function might come from the gene in the database with which it shares some DNA sequence (see Box 3.3).

MUTATION ANALYSIS

Mutation analysis refers to the identification of changes in DNA which produce disease or dysfunction. Detecting mutations in DNA (or RNA) requires various combinations of: (1) physical mapping, (2) cloning or amplifying DNA, (3) screening methods to identify DNA changes involving one to a few nucleotide bases, (4) electrophoretic separation of DNA fragments, and (5) sequencing.

Detecting DNA deletions

Very small deletions in DNA can be detected by PCR (e.g. cystic fibrosis, Ch. 3). Larger deletions (e.g. α thalassaemia, Ch. 1) may require Southern blotting and the largest deletions (e.g. contiguous gene syndromes, see Ch. 3) are detected by PFGE or FISH.

Detecting point mutations

In terms of causing genetic disorders, point mutations occur more frequently than deletions. In contrast to deletions, point mutations are usually more difficult to identify because they are *small* (often it is a single base pair which is mutated) and *heterogeneous* (it is not unusual for each family to have its own specific point mutation). PCR is a particularly helpful technique when it comes to detection of point mutations. The significance of the subject as well as the considerable expansion which has occurred in the area are demonstrated by the fact that there is now a new international scientific journal which deals solely with mutation analysis.

Screening DNA

In the past few years, the necessity to 'screen' amplified DNA products to identify which one is likely to contain a mutation has assumed increasing importance. This will be illustrated in Chapter 3 by reference to Duchenne muscular dystrophy, haemophilia and familial hypertrophic cardiomyopathy. In these examples, there are a number of point mutations which cause the underlying disease but unfortunately the genes are too large to make sequencing a practical diagnostic approach. One solution to this problem involves the amplification of cDNA in a number of segments. For example, the cDNA itself might be 2 kb in size, but it is amplified in 10 overlapping segments each of which is ~0.2 kb in size. The 10 segments are next screened to see which has a point mutation. That amplified fragment alone is sequenced. The end result is sequencing of 0.2 kb rather than 2.0 kb. cDNA is used in preference to genomic DNA because the former is much smaller in size since it is only comprised of exons. These are also more likely to have mutations compared to introns (see Ch. 3 for examples).

A number of methods have been developed to screen DNA for point mutations. They include: **d**enaturing **g**radient **g**el **e**lectrophoresis (commonly abbreviated to DGGE); **s**ingle-**s**tranded **c**onformation **p**olymorphism (SSCP) and the **c**hemical **c**leavage of **m**ismatch method (CCM). The first two depend on the mobility shifts of single-stranded DNA which has one or more nucleotide base mismatches. The last relies on chemical reactivity if there is a mismatch between nucleotide bases. The above methods do not identify the actual changes in DNA sequence but they provide a screening approach by which differences (e.g. mutations, polymorphisms, etc.) can be detected rapidly.

DGGE as a method to distinguish a segment of DNA containing a single base pair mutation from its corresponding wild-type segment was first described in 1983 by Fischer and Lerman. Target DNA (genomic, cloned or amplified) is annealed to a radiolabelled single-stranded DNA probe. If target DNA contains a point mutation, heteroduplexes (two strands of DNA in which there is a base mismatch) between the target and probe will form. DNA is then electrophoresed through an increasing linear denaturant gradient (for example by adding urea, formamide or having a temperature gradient). At a certain denaturant level, a region of the double-stranded DNA will become single-stranded thereby producing a branched structure. This becomes entangled in the gel matrix pores

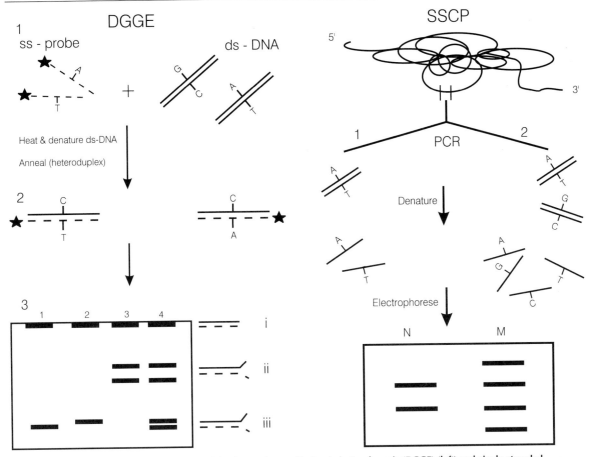

Fig. 2.17 Screening for point mutations in DNA by denaturing gradient gel electrophoresis (DGGE) (left) and single-stranded conformation polymorphism (SSCP) (right).

(Left) In DGGE single base changes are detected as points of mismatch in double-stranded (ds) DNA. Such mismatches will cause separation of the strands under denaturing conditions which is seen as altered gel mobility. **(1)** Radiolabelled and single-stranded (ss) DNA probe for a normal (wild-type) sequence (★) is mixed with ds target DNA which will contain both normal sequences (—A—, —-T—-) and mutant sequences (—-G—, —C—-). The latter results from a point mutation (A to G and so T to C in the antisense strand). The ss ^{32}P-probe contains the normal DNA base sequence (the critical region being —-A—- with the complementary strand —-T—-). The mixture is heated to make ss target DNA. **(2)** As the temperature falls, probe and target DNA will anneal and this forms *heteroduplexes* which contain the mismatched base pairs (A + G; T + C). *Homoduplexes* (A + T, G + C) which contain the normal base pair matches will also form but these are not shown here. **(3)** DNA is electrophoresed through a denaturing gradient increasing from i to iii. In position i, DNA will be in the double-stranded form. As the denaturing gradient increases, melting domains where mismatched bases occur will be affected and branched structures form. These reduce the DNA's mobility. Lanes 1 and 2 illustrate the mobility of wild-type (A + T) and mutant homoduplex (G + C) controls, respectively. In lane 3 the heteroduplexes formed will migrate less since they are unstable because of the mismatched bases. In lane 4 there is a mixture of homoduplexes and heteroduplexes. If only the wild-type sequence is labelled with ^{32}P and autoradiography is used to detect DNA, the mutant homoduplex would not be 'seen' in lane 4. It would be visible if DNA were stained with ethidium bromide.

(Right) DNA from a specific region is amplified. In **(1)** the region is normal; in **(2)** there is also a mutant allele. DNA is made single-stranded and then electrophoresed in a non-denaturing gel. Mobility of DNA is affected by its length and sequence. Two normal ssDNA bands are seen but the presence of DNA fragments with a different sequence is suggested by the finding of additional (novel) bands.

which retards mobility. The normal mobility will be shown by the migration pattern of the wild-type homoduplex which is run in parallel. Thus, variations in nucleotide bases will be distinguished by different mobilities since the denaturing domain is dependent on the underlying nucleotide sequence (Fig. 2.17).

Another, and perhaps more popular, screening method is SSCP. In this, small DNA fragments, e.g. 200–300 bp in size, generated by digestion of DNA or PCR are denatured into their single strands and then electrophoresed in

polyacrylamide gels under non-denaturing conditions. The movement of the single-stranded DNA is dependent on the size *and* sequence. Therefore, point mutations in DNA lead to shifts in mobility. SSCP is taking over from DGGE because it is easier to use and more consistently detects single base changes. CCM has also undergone recent modifications to utilise enzymes rather than the more toxic chemicals initially involved. This will make available another useful approach to screening DNA for single base changes.

Confirming point mutations

The 'gold standard' for mutation detection is DNA sequencing. However, this is a lengthy procedure and is usually only required for the initial identification and characterisation of a specific mutation. Thereafter, more rapid means are devised to detect the same mutation in other specimens. The detection of single base changes in DNA has relevance to a number of areas in medicine. They include genetics, to identify genetic disorders, oncology, to look for mutations in cancer-producing genes, and microbiology, for epidemiological purposes.

Known point mutations can be identified if, by chance, they occur at a restriction endonuclease recognition site. Thus, the point mutation produces a novel restriction enzyme fragment (Fig. 2.18). An alternative way to detect point mutations is to hybridise with *allele-specific oligonucleotides* (ASOs), i.e. oligonucleotide probes which are synthesised so that one probe will hybridise to the wild-type (normal) sequence and the second probe will hybridise to the mutant DNA. Hybridisation patterns will define the underlying genotypes, e.g. only mutant probe hybridises – homozygous for that mutation; only wild-type probe hybridises – normal; both mutant and wild-type probes hybridise – heterozygote (Fig. 2.19).

Oligonucleotides as DNA probes have widespread application in the diagnosis of genetic disorders (see Ch. 3) and have proven useful in the screening of DNA libraries. Because of their small size, oligonucleotide probes have a limited function when total genomic DNA is used. In this

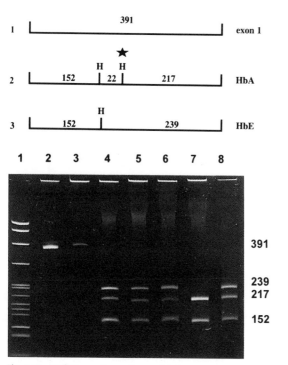

Fig. 2.18 Utilising a change in a restriction enzyme pattern produced by a point mutation.
Haemoglobin E (HbE) is a relatively common condition found in some communities. It involves a single nucleotide change in the first exon of the β globin gene. This alters the 26th amino acid from a glutamic acid (codon GAG) to lysine (codon AAG). In normal haemoglobin (known as HbA) DNA sequence at the site of the HbE defect contains a recognition site for the restriction enzyme *Hph*I. The HbE mutation destroys this recognition site. (1) By designing primers on either side of the HbE mutation site, it is possible to amplify a segment of DNA 391 bp in size. (2) In normal DNA there are two recognition sites of *Hph*I, one of which (★) is fortuitously located within the codon which is mutated in HbE. Therefore, digesting the 391 bp with *Hph*I will produce three smaller fragments 152, 22 and 217 bp in size. (3) In HbE, only two fragments result since the *Hph*I recognition site (★) is destroyed. HbE will be distinguished by finding a unique 239 (217 + 22) bp fragment. In the ethidium bromide-stained PCR gel, track 1 is the size marker, tracks 2 and 3 are undigested fragments of amplified DNA 391 bp in size, track 7 is a normal DNA control and tracks 4–6 and 8 are examples of heterozygous HbE. The ethidium bromide-stained gel does not show the 22 bp fragment which is too small. A homozygous-affected HbE would not have the normal 217 and 152 bp fragments but only the 239 bp pattern (see also Fig. 2.19).

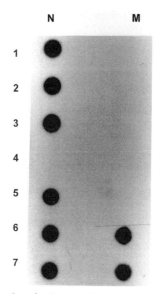

Fig. 2.19 Detection of point mutations – ASO probes.
Allele-specific oligonucleotide (ASO) probes are synthesised so that one probe hybridises to the normal (wild-type) sequence and the second to the mutant DNA sequence. ^{32}P or chemically labelled ASO probes hybridise to amplified, single-stranded DNA which is fixed onto nylon membranes in the form of dots (hence the name dot blots). In the example given, DNA has been amplified in the region of the β globin gene located at the HbE mutation (see Fig. 2.18). One ASO probe is specific for the normal DNA sequence and the second will differ by one nucleotide base (A instead of the normal G) where HbE occurs. The membrane containing the dot blots is simultaneously hybridised with both normal (N) and mutant (M) probes. Dot blots in tracks 1–3 and 5 hybridise only to normal, i.e. the DNA does not contain the HbE mutation; track 4 is a control (no DNA) used to check for contamination in the PCR; tracks 6 and 7 show that the dots hybridise to both normal and mutant probes, i.e. DNA is heterozygous for HbE. The ASO approach can be used when there is no change in the restriction enzyme pattern associated with a mutation. ASOs can also be more easily multiplexed, i.e. a number of mutations can be sought in one reaction.

circumstance, there is considerable background hybridisation. On the other hand, amplified DNA is an ideal target against which oligonucleotide probes can hybridise. The probes can be end-labelled with ^{32}P or linked to chemicals, enzymes or fluorochromes.

Protein truncation

Another method has recently been developed to identify a specific type of point mutation, i.e. one which involves the formation of a premature stop codon and so produces a shortened ('truncated') protein product. The mutation detection strategy utilises DNA but the end-point is the protein (Fig. 2.20). The protein truncation test has proven useful in diseases associated with a large gene and multiple nonsense or frameshift mutations, e.g. the *APC* gene in familial adenomatous polyposis and the *BRCA1* gene in breast cancer (see Ch. 6 for further discussion).

CHROMOSOME ANALYSIS

Chromosome analysis involves the morphological study of chromosomes to determine their number and any structural changes which may be present. Chromosomal analysis requires various combinations of: (1) cell culture, (2) spreading out of chromosomes according to their size, i.e. a karyotype, (3) DNA probe, radiolabelled or conjugated to fluorochromes, and (4) cell fusion procedures using human and non-human cells.

Karyotype

A karyotype describes an individual's chromosomal constitution. It was only in 1956 that the human diploid chromosome number was shown to be 46. During the 1970s, methods were developed to distinguish bands within individual chromosomes. Each of the 44 human autosome chromosomes and the X or Y sex chromosomes can now be counted and characterised by banding techniques. The most common is called G-banding and involves trypsin treatment of chromosomes followed by staining with Giemsa. G-banding produces a pattern of light and dark staining bands for each chromosome. The light bands (euchromatin) are more likely to be associated with genes while DNA comprising the darker bands (heterochromatin) is rich in repetitive sequences (Fig. 2.21).

The banding patterns, the size of the chromosome and the position of the centromere enable the accurate identification of each individual chromosome. The number of bands which can be detected per haploid chromosome set with G-banding varies with the technique used. An average number is 550. Each band is estimated to contain ~5–10 x 10^6 base pairs of DNA. One suggestion is that there are in the vicinity of 100 genes per band. However, this will vary considerably since some genes are small (e.g. globin genes are ~1 kb in size) and others are very large (e.g. Duchenne muscular dystrophy gene is over 2000 kb in size). Some chromosomes, e.g. 19 and 22, appear to be more gene-rich than others, e.g. chromosomes 13, 18 and Y.

Cytogenetic nomenclature is derived from the 1985 International System for Human Cytogenetics recommen-

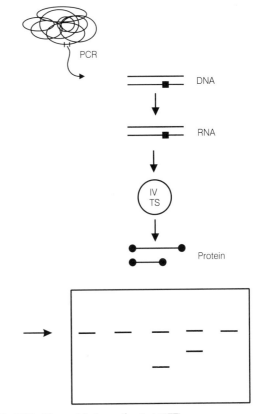

Fig. 2.20 The protein truncation test (PTT).
This is a new test which has developed since the first edition of *Molecular Medicine* was written. PTT is used if a point mutation leads to a premature stop codon so that one of the two alleles for a gene will produce a smaller peptide, i.e. it is truncated. A segment of the gene of interest is first amplified from mRNA by RT-PCR (see Fig. 2.2). Alternatively, individual exons from genomic DNA are amplified by PCR. The ■ indicates the position of a premature stop codon on one DNA strand. cDNA or genomic DNA is then made into mRNA with an in vitro transcription system. Two RNA species will result since one will also contain the premature stop codon. The final step involves an in vitro translation system (IVTS) which allows protein to be produced from the RNA. Two distinct proteins will result with the mutant one being smaller. The proteins are separated by electrophoresis with the smaller (truncated) one having a faster mobility (→ indicates the normal protein species). See Figure 6.17 for a PTT example. All the steps described in PTT can be undertaken with commercial kits.

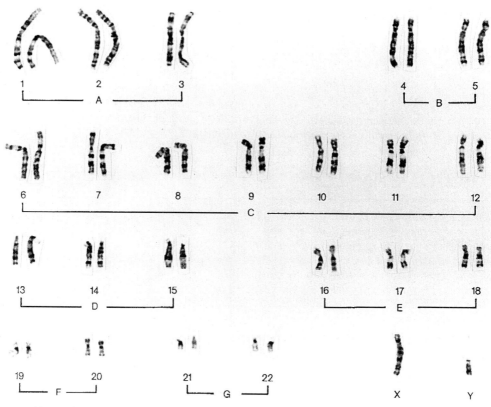

Fig. 2.21 A normal human karyotype (46,XY) which illustrates G-banding.
A normal individual has 44 autosomal (non-sex) and two sex chromosomes. These can be counted and characterised by staining techniques which produce bands. Note the light and dark bands on the chromosomes. (Karyotype provided by Dr A Smith, Department of Medical Genetics, New Children's Hospital, Westmead.)

dations. The centromere, a constricted portion of the chromosome where the chromatids are joined, divides the chromosome into a short arm designated 'p' (for petit) and the long arm or 'q'. Each arm is divided into regions which are marked by specific landmarks. Regions comprise one or more bands. Regions and bands are numbered from the centromere to the telomere along each arm (Fig. 2.22). Therefore, each band will have four descriptive components. For example, the cystic fibrosis locus on chromosome 7q31 defines a band involving chromosome 7, on the long arm at region 3 and band 1. Additional information is available by higher resolution banding techniques which enable sub-bands to be identified. In the case of the cystic fibrosis locus this becomes 7q31.3 where the .3 defines the sub-band (Fig. 2.22).

Fluorescence in situ hybridisation

Even greater resolution than was possible by chromosomal banding became available through the development in the late 1970s and early 1980s of in situ hybridisation. Radiolabelled DNA probes were now able to define the chromosomal localisation of single-copy DNA sequences in metaphase chromosomes. In situ hybridisation can be used to assign a chromosomal location for DNA probes. In the past few years the emergence of non-isotopic in situ hybridisation, particularly FISH, has greatly enhanced the utility of this technique. The potential to use a number of DNA probes each labelled with a different fluorochrome in the same procedure means that separate loci can be identified, comparisons can be made and relationships to the centromere and telomeres established. In the long term, FISH will have an important role, perhaps more so than PFGE, as the method for mapping and ordering DNA probes along a chromosomal segment. Since chromosomes are more extended in interphase compared with metaphase, the application of FISH during interphase will resolve even further the location of DNA probes and the characterisation of chromosomal rearrangements, such as microdeletions in genetic or malignant disorders.

Somatic cell hybrids

Human cells (e.g. fibroblasts) can be fused with tumour cells from other animals (e.g. rodents) using an agent such

CHROMOSOME

7

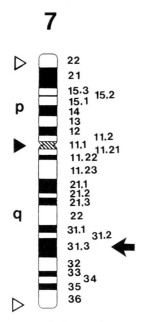

Fig. 2.22 Banding patterns for human chromosome 7.
The individual bands are designated by numbers. The short and long arms are shown by p and q, respectively, the centromere by a filled triangle and the telomeres by open triangles. An arrow marks position q31.3.

as the Sendai virus. After fusion there are both human and non-human chromosomes in the hybrid cells. Subsequently, chromosomes, particularly the human ones, will be gradually lost from the fused cells until after a few generations the cells become more stable. Hybrids can be selected for and their stability improved by using selectable markers so that hybrids with human chromosomes are more likely to survive following culture in a special medium. Each of the stable hybrids can then be propagated as an individual cell line. The cell lines will differ from each other in the numbers of human chromosomes which they have retained. Karyotypes of the cell lines will identify which human chromosomes are present. These cell lines can then be used for mapping purposes, e.g. a DNA probe is available but its location in the genome is unknown. This probe can be hybridised to a panel of somatic cell hybrids. Some hybrids will be positive and others negative following hybridisation. From this it will be possible to determine the chromosomal origin for the DNA probe: e.g. somatic cell line No. 1 has human chromosomes 1, 3, 7; line No. 2 has human chromosomes 1, 5, 8, 19; line No. 3 has human chromosomes 1, 12, 15, 18, 21; and line No. 4 has human chromosomes 3, 4, 6, 19. If the DNA probe hybridises to lines 1, 2 and 3 but not 4 it is likely that the probe is located on chromosome 1.

EXPRESSION OF RECOMBINANT DNA

Enabling a gene to function and so produce its associated protein product in an artificial environment is known as expression. Expression of recombinant DNA requires: (1) cloned DNA, (2) expression vector, and (3) microinjection.

In vitro expression

The most basic expression vector is a plasmid which has an origin of replication, a selectable DNA marker to allow it to be detected and also provide a growth advantage, and a multiple cloning site into which the gene to be expressed is inserted. The vector with the inserted gene is then transfected into a host cell such as *E. coli* and allowed to replicate. Selection, usually via an antibiotic resistance gene, ensures that only *E. coli* with the plasmid insert will remain viable (Fig. 2.23). In the prokaryotic system, bacterial proteins are expressed but eukaryotic-derived products will be degraded unless they are fused to a bacterial protein. Expressed protein is isolated, purified and able to be used as a therapeutic agent. More sophisticated in vitro expression systems utilise yeast, insect or mammalian-derived vectors. In the non-bacterial expression

systems, protein products can undergo post-translational changes. These can be significant for biological activity (see Ch. 7 for a description on how expression vectors are utilised to produce drugs and vaccines).

In vitro expression systems are also used to test gene function, particularly that involving promoter regions. For example, the enzyme chloramphenicol acetyl transferase can be included in an expression plasmid. It is then possible to insert eukaryotic promoter sequences 5' to the chloramphenicol acetyl transferase gene. The activity of these promoters can be tested by noting their effect on the production of the above substance. RNA transcripts can be produced from plasmid vectors which have special promoters located 5' to their cloning site. The promoters are recognised by DNA-dependent RNA polymerases and so produce RNA rather than DNA. RNA transcripts formed in this way can be used as RNA probes or to identify and quantify mRNA production in vitro.

In vivo expression

Fertilised oocytes contain two pronuclei. Into one of these a gene, in the form of cloned DNA, can be microinjected

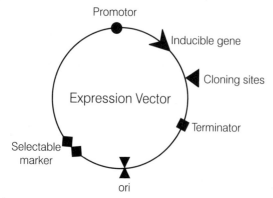

Fig. 2.23 An idealised bacterial expression vector.
The vector has an origin of replication (ori) which allows it to replicate autonomously. There is a selectable marker (usually antibiotic resistance) which gives a growth advantage to the host (e.g. *E. coli*) containing the vector. The bacterial promotor ensures that transcription will occur. An inducible bacterial-specific gene is useful since it allows expression to be controlled. The DNA sequence to be expressed is inserted into the multiple cloning site. Transcription is stopped by placing a termination signal downstream of the multiple cloning site. A fusion product (protein from the inducible gene plus protein from the cloned gene) is expressed which protects the non-bacterial protein from degradation. These two components need to be separated and the expressed protein purified from bacterial products.

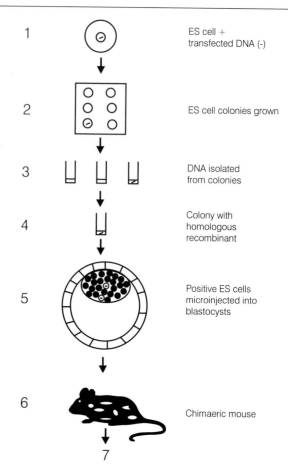

Fig. 2.24 Embryonic stem cells (ES cells) used for in vivo expression of recombinant DNA.
This method produces transgenic mice which can be used to test the function of genes in vivo. **(1)** ES cells are transfected with foreign DNA. Many ES cells will take up the DNA but this will occur into different sites in the mouse genome by a process of random integration. In a very rare case, the integration will have occurred into the correct part of the genome by a process of homologous recombination. **(2)** Colonies of ES cells are grown. **(3)** DNA is isolated from pools of colonies. **(4)** The colony which has DNA integrated into the *correct* position in the genome by homologous recombination can be identified by PCR. **(5)** ES cells which have the homologously recombined DNA are injected into mouse blastocysts. **(6)** Using different coloured mice as sources of ES cells (e.g. white mouse) and blastocysts (e.g. black mouse) will enable chimaeric mouse to be distinguished. **(7)** If the transgene has also integrated into the germline it will be possible to obtain a homozygote animal by appropriate matings. See also Chapter 10 for further discussion on ES cells.

using a finely drawn pipette. The injected (foreign) DNA becomes randomly integrated as multiple copies in a head to tail tandem arrangement in oocyte DNA. Expression of foreign DNA will occur if there are sufficient copy numbers and the environment into which DNA has become integrated is suitable. Transgenic animals (animals, usually mice, with the foreign DNA) can be detected by screening their DNA with Southern blotting or PCR. Expression of the transgene can be altered if an inducible promoter is included in the gene construct, e.g. the metallothionein promoter is inducible in the presence of heavy metals which can be added to the animal's drinking water when the transgene needs to be 'switched on'.

Transgenic mice provide very useful in vivo expression systems to test the function or significance of genes, gene sequences or promoters. Disadvantages of the technology include the relatively inefficient production of transgenics (e.g. only ~20% of microinjections will give a viable transgenic animal), the inability to control where the gene will integrate, and, frequently, low expression of the transgene. To overcome the integration problem, *embryonic stem cells* are becoming increasingly popular since it is possible to utilise *homologous recombination* to target the gene into its correct chromosomal position. Embryonic stem cells which have the appropriately integrated gene are identified. These are microinjected into blastocysts to produce chimaeric animals. These represent a model closer to the in vivo situation since inserted DNA is now in its 'natural' location in the genome (Fig. 2.24). Further discussion of embryonic stem cells and homologous recombination is given in Chapters 7 and 10.

FURTHER READING

General

Davies K E 1993 (ed) Human genetic disease analysis: a practical approach, 2nd edn. Information Press, Oxford

Rosenthal N 1994 Molecular medicine – Tools of the trade – recombinant DNA. New England Journal of Medicine 331: 315–317

Watson J D, Gilman M, Witkowski J, Zoller M 1992 Recombinant DNA, 2nd edn. W H Freeman, New York

DNA

Chelly J, Concordet J P, Kaplan J C, Kahn A 1989 Illegitimate transcription: transcription of any gene in any cell type. Proceedings of the National Academy of Science USA 86: 2617–2621

Lewin B 1994 Genes V. Oxford University Press, Oxford

Rosenthal N 1994 Molecular medicine – DNA and the genetic code. New England Journal of Medicine 331: 39–41

DNA amplification

Harper D R 1994 Molecular virology. Bios Scientific, Oxford

Lo A C, Feldman S R 1994 Polymerase chain reaction: basic concepts and clinical applications in dermatology. Journal of the American Academy of Dermatology 30: 250–260

Cloning DNA

Monaco A P, Larin Z 1994 YACs, BACs, PACs and MACs: artificial chromosomes as research tools. Trends in Biotechnology 12: 280–286

Old R W, Primrose S B 1994 Principles of Gene Manipulation – an introduction to genetic engineering, 5th edn. Blackwell Science, Oxford

Sequencing DNA

Rosenthal N 1995 Molecular medicine – fine structure of a gene – DNA sequencing. New England Journal of Medicine 332: 589–591

Positional cloning

Boguski M S 1995 Hunting for genes in computer data bases. New England Journal of Medicine 333: 645–647

Harper R 1995 World wide web resources for the biologist. Trends in Genetics 11: 223–228

Hearne C M, Ghosh S, Todd J A 1992 Microsatellites for linkage analysis of genetic traits. Trends in Genetics 8: 288–294

Schlessinger D 1990 Yeast artificial chromosomes: tools for mapping and analysis of complex genomes. Trends in Genetics 6: 248–258

Wicking C, Williamson B 1991 From linked marker to gene. Trends in Genetics 7: 288–293

Mutation analysis

Cotton R G H 1993 Current methods of mutation detection. Mutation Research 285: 125–144

Housman D 1995 Molecular medicine – human DNA polymorphism. New England Journal of Medicine 332: 318–320

Korf B 1995 Molecular medicine – molecular diagnosis. New England Journal of Medicine 332: 1218–1220, 1499–1502

Roberts R G, Gardner R J, Bobrow M 1994 Searching for the 1 in 2,400,000: a review of the dystrophin gene point mutations. Human Mutation 4: 1–11

Roest P A M, Roberts R G, Sugino S, van Ommen G-J B, den Dunnen J T 1993 Protein truncation test (PTT) for rapid detection of translation-terminating mutations. Human Molecular Genetics 2: 1719–1721

Chromosome analysis

Cohen M M, Rosenblum-Vos L S, Prabhakar G 1993 Human cytogenetics – a current overview. American Journal of Diseases of Children 147: 1159–1166

Wolman S R 1994 Fluorescence in situ hybridization: a new tool for the pathologist. Human Pathology 25: 586–590

Expression of recombinant DNA

Old R W, Primrose S B 1994 Principles of Gene Manipulation – an introduction to genetic engineering, 5th edn. Blackwell Science, Oxford

3

MEDICAL GENETICS

INTRODUCTION

Genetic diseases can be classified in a number of ways. For example:

- Single-gene
- Polygenic
- Multifactorial
- Chromosomal
- Somatic cell.

The first four will be discussed in this chapter and somatic cell disorders will be discussed in Chapter 6.

Mendelian Inheritance in Man (MIM) is a compendium of human genes and genetic disorders which has evolved into an encyclopaedia of gene loci. The first edition was published in 1966 with a total of 1487 entries and the 11th edition (1994) has 6678 entries (Fig. 3.1). In 1987, MIM became available online (OMIM – *Online Mendelian Inheritance in Man*) and in 1993 as a CD-ROM. In the preface to the 1994 edition, McKusick suggests that future editions of MIM might be organised into separate volumes for each chromosome. The latest entries relate to autosomal dominant disorders in ~67% of cases, autosomal recessive in ~26% and X-linked disorders in ~6%. The majority of loci involve single genes. A second group, known as the multifactorial disorders, represents traits which occur as a result of interactions between the environment and genes.

Some key words used frequently in this chapter will be defined. Different forms of a gene at a locus are called **alleles**. The **haplotype** refers to a set of closely linked DNA markers at one locus which are inherited as a unit. The **genotype** is the genetic (DNA) make-up of an organism. In the present context, genotype would also refer to the

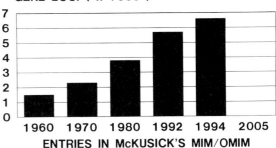

GENE LOCI (x 1000)

ENTRIES IN McKUSICK'S MIM/OMIM

**Fig. 3.1   Entries in McKusick's *Mendelian Inheritance in Man*
and its Online version.**
The increasing number of entries reflects the intensive work being undertaken in the area of genome mapping. Access to this compendium may be obtained through hard-copy, CD-ROM or the Internet. Contacts in different countries for the latter are provided in the preface to the 1994 MIM or in the reference by McKusick and Amberger (1994). The reference to the year 2005 relates to the completion date of the Human Genome Project when the estimated 80 000 human genes will have been characterised (see Ch. 10 for a discussion of the Human Genome Project).

genetic constitution of alleles at a specific locus, i.e. the two haplotypes. The **phenotype** reflects the recognisable characteristics determined by the genotype and its interaction with the environment. An individual is **homozygous** if both alleles at a locus are identical and **heterozygous** if the alleles are different. **Autosomal** inheritance involves traits which are encoded for by the 22 pairs of human autosomes. **X-linked** inheritance refers to genes located on the X chromosome. The products of both the normal (**wild-type**) alleles at a particular locus need to be non-functional in a **recessive** disorder, e.g. cystic fibrosis. On the other hand, a **dominant** disorder results if only one of the two wild-type alleles is mutated, e.g. familial hypertrophic cardiomyopathy or Huntington disease. Eponyms are frequently used in medical genetics. An inconsistent finding in the literature is the use of the possessive or non-possessive forms, e.g. Huntington's disease versus Huntington disease. Following general recommendations, the non-possessive form is used since an eponym provides a link to the disorder and does not necessarily mean it was first described by that individual (or individuals).

The contribution of genetic disease to ill-health has been assessed in a Canadian survey of one million consecutive live-births. This study found that before reaching the age of 25, ~53/1000 live-borns developed some disorder which had an identifiable genetic component (Table 3.1). In this chapter, genetic diseases will be described in terms of their inheritance patterns as well as underlying molecular defects. Emphasis will be placed on how a greater understanding of aetiology and pathogenesis has come from developments in molecular medicine. At the same

Table 3.1   Genetic contributions to ill-health before the age of 25.

Results of a Canadian study of one million consecutive live-births. The total number affected was approximately 53/1000 live-borns (Baird et al 1988).

Category	Group	Number/1000 with genetic components
Single-gene disorders	Autosomal recessive	1.4
	Autosomal dominant	1.7
	X-linked	0.5
Multifactorial disorders	–	46.4
Chromosomal anomalies	–	1.8
Undefined	–	1.2

time, DNA diagnosis of genetic disorders has rapidly expanded following the application of molecular biology techniques.

THALASSAEMIAS – MODELS FOR MOLECULAR GENETICS

Haemoglobinopathies are inherited disorders of globin. They are classified into the *thalassaemia syndromes*, e.g. α thalassaemia, β thalassaemia, and the *variant haemoglobins*, e.g. sickle-cell haemoglobin (HbS). The thalassaemias were one of the first human disorders to be characterised at the molecular level. As such they provided, and in many cases continue to provide, a useful starting point in understanding the molecular basis for human genetic disorders.

The haemoglobinopathies represent one of the commonest single-gene disorders. Estimates from the WHO predict a carrier rate of 7% by the year 2000. The high frequency of thalassaemia and sickle haemoglobin carriers reflects the protection from malaria provided by these disorders. Clinical and molecular classifications for the thalassaemias are given in Table 3.2.

Linkage analysis and mutation detection

Before the availability of PCR, it was difficult to detect single base changes in genes if mutations were multiple or heterogeneous in type. For example, there are in excess of 200 different DNA mutations which produce β thalassaemia. Thus, detection by an *indirect* strategy, which identified the segregation of DNA *polymorphisms* in members of a family, was developed. The thalassaemias became one of the earliest models to demonstrate the utility of what is called *linkage analysis*.

DNA polymorphisms associated with the β globin gene cluster were first described in 1978. Today, there are many known polymorphisms which lead to variations in

Table 3.2   Clinical and molecular classification of the thalassaemias.

Thalassaemias are inherited disorders of globin resulting from an imbalance in α and β globin chain production.

Classification	Genotypes[*]	Clinical consequences	Molecular defect(s)
β Thalassaemia	β^T/β^A	Mild anaemia	Point mutations
	β^T/β^T	Severe anaemia	
α Thalassaemia	$-\alpha/\alpha\alpha$	Nil, not usually detectable	Deletions
	$-\alpha/-\alpha$	Mild anaemia	
	$--/\alpha\alpha$	" "	
	$-\alpha/--$	Mild to severe anaemia (HbH disease)	
	$--/--$	Fatal (Hb Bart's hydrops fetalis)	
HPFH	Variable	Nil, not usually detectable	Point mutations or deletions

[*]Genotypes: [T]Thalassaemia genes, [A]normal genes. – = deletion. Normal genotype for the β globin cluster is β^A/β^A and for the α globin cluster $\alpha\alpha/\alpha\alpha$. HPFH, hereditary persistence of fetal haemoglobin.

the size of DNA fragments produced by digestion with restriction enzymes or following amplification by PCR (Fig. 3.2). The great advantage of a DNA polymorphism is that it provides an *indirect* marker for a particular locus or gene. A minimum requirement for a DNA linkage study is a family unit in which there is a key individual who is completely normal or is homozygous-affected. The key

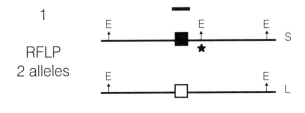

Fig. 3.2 The β globin gene cluster and its associated restriction fragment length polymorphisms (RFLPs).
The ↑ indicate the various RFLP sites and the names of the restriction endonuclease enzymes which digest DNA at these loci to give fragments of varying lengths. Although this diagram is now 'out of date' since RFLPs are rarely used for diagnosis of the β thalassaemias, the number and dispersal (both extragenic and intragenic) of the polymorphisms illustrate the extensive range of markers which can be accumulated for gene loci (see Fig. 1.4 for a description of the genes in the β globin cluster).

person is essential to allow assignment of the polymorphic markers to the wild-type or mutant alleles. DNA from each of the relevant family members is digested with the restriction endonuclease which is known to be polymorphic for a particular locus. After Southern blotting, the restriction enzyme digest is hybridised to the relevant DNA probe for further analysis (Fig. 3.3, see also Ch. 2). Many of the DNA polymorphisms are now characterised more easily by PCR.

In some cases, DNA polymorphisms are not informative, i.e. family members are homozygous for the polymorphic alleles. Therefore, the normal and mutant alleles cannot be distinguished. In this situation, additional DNA polymorphisms are sought until informative ones are found. Informativeness in linkage studies has been improved with the use of more variable polymorphisms, for example, VNTRs (variable *number* of *tandem repeats*), minisatellites or microsatellites. (The different types of polymorphisms will be discussed in greater detail in Chapter 8.) The number of alleles in the group of multiallelic polymorphisms is greater than the biallelic RFLPs (restriction *fragment length polymorphisms*) as illustrated in Figure 2.15 and the examples given in Chapter 8.

Studies based on a polymorphic linkage analysis strategy have a number of potential problems. These include: (1) the requirement for a family study, (2) non-paternity can lead to erroneous patterns, and (3) recombination is possible. The last is a function of the distance between a polymorphic marker and the gene of interest. Although there are many exceptions, a *physical* distance of 1 Mb is roughly equivalent to a genetic distance of 1 cM (cM = centiMorgan). One cM indicates a 1% recombination potential (i.e. in 100 meioses there will be one recombination event between the DNA polymorphism and the target DNA of interest). Thus, genetic distances are measured by family studies which allow the closeness of an association between a DNA polymorphism and the DNA of interest to be assessed.

The three problems described above make DNA polymorphic linkage analysis an unsatisfactory way to detect β thalassaemia mutations since there are now alternative and more direct strategies available, e.g. PCR (see Fig. 2.19). However, DNA polymorphisms and linkage analysis remain a basic component in most positional cloning

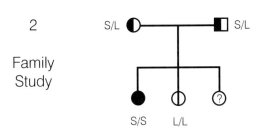

Fig. 3.3 The application of DNA polymorphisms in linkage analysis.
A single base change in the DNA sequence can affect the recognition site for a restriction enzyme, resulting in different DNA fragment lengths, i.e. a polymorphism. Using a key individual who is completely normal or is homozygous affected, i.e. both alleles are abnormal, the polymorphism can be linked to the disorder and so used as an *indirect marker*. **(1)** Recognition sites for a restriction enzyme 'E' are shown as ↑. One of the restriction sites is polymorphic (★), i.e. it may or may not be present. The fragments can be detected using a DNA probe (upper horizontal bar). The two allelic genes are shown by ■ and □. Digestion with a restriction enzyme (E) gives either a small (S) or large (L) band depending on whether the polymorphic site is present (S) or absent (L). **(2)** A pedigree is drawn in which both parents are informative for the polymorphic marker since they are heterozygotes (S/L). An *affected* offspring is homozygous S/S, thereby showing that the mutant allele for an autosomal recessive disorder in this family co-segregates with the 'S' marker. This is confirmed by a normal child who is homozygous 'L'. Subsequent offspring can have their genetic status identified on the basis of this polymorphic marker (L/L = normal; S/S = affected; S/L = carrier).

strategies, as will be seen from the examples which follow. DNA polymorphisms are also essential to diagnose disorders for which the gene has not been isolated or the multiple mutations present make it difficult to identify one which is unique to an individual, e.g. haemophilia.

Gene interactions – γ and β globin genes

Homozygotes for β globin gene mutations giving rise to β thalassaemia or sickle-cell anaemia usually develop a severe clinical disorder. On the other hand, homozygotes for a rare thalassaemia called *deletional hereditary persistence of fetal haemoglobin* (HPFH) have no β globin gene activity (just like β thalassaemia) because both β globin genes are deleted. However, their clinical phenotypes are mild compared to homozygous β thalassaemia. This occurs because there is greater γ globin gene activity and so increased amounts of fetal haemoglobin (HbF) in the blood (in normal adults, HbF comprises <1% of the total haemoglobin). It is now known that individuals who have the ability to produce an excess of HbF for whatever reason (see Ch. 1) will have milder forms of β thalassaemia and sickle-cell disease.

At the molecular level, there is considerable interest in characterising the 'mutants of nature' such as HPFH. An understanding of why HbF is elevated in these rare disorders will be invaluable in planning therapeutic strate-

gies which are based on HbF (the therapeutic implications of HbF are discussed further in Ch. 7).

Gene regulation

As well as showing how genes might interact to change the severity of genetic diseases, the thalassaemia syndromes have proven to be good models for the study of gene regulation in the eukaryote. The approach followed in the first instance has again centred around the experiments of nature, i.e. human genetic disorders whose changed clinical phenotypes are examined to determine, by molecular means, which elements in their genes are responsible for the altered phenotype. From this, the *normal* regulatory components can be deduced.

The increased output of HbF seen in the deletional HPFH disorders can occur by a number of mechanisms. An earlier hypothesis described the loss of putative inhibitors which were present in the deleted segment. A more recent theory implies that deletions enable the juxtaposition of the γ globin genes to distant 3' regulatory elements such as enhancer sequences, which are then able to increase the γ globin gene's activity. Which of the two is more significant remains to be determined. Other, as yet undefined, mechanisms may also be operational (Fig. 3.4).

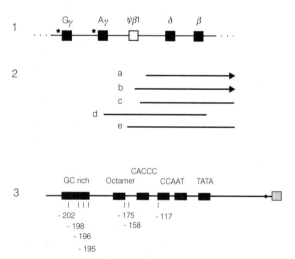

Fig. 3.4 Deletional and non-deletional molecular changes which can alter fetal haemoglobin (HbF) regulation.
(1) The normal β globin gene complex; ■ = functional genes; □ is a pseudogene; ★ define the γ globin gene promotor regions. (2) Five examples of deletional hereditary persistence of fetal haemoglobin (HPFH) are given: a and b – Black African, c – Italian, d – Haemoglobin Kenya, e – Indian HPFH. The arrow indicates that the 3' end of the deletion continues further downstream. (3) The γ globin gene promotor region (★ for 1). Six loci (including a duplicated CCAAT box) important for transcription are identified. The positions of the various non-deletional HPFH point mutations are shown. The (●) defines the gene's Cap site (initiation point for translation) (the −202 to −117 mutations are numbered from the Cap which is +1). Horizontal lines refer to the position of the mutations. The beginning of the protein (the ATG start codon) is indicated by a hatched box.

Box 3.1 Genetic control of fetal haemoglobin (HbF) synthesis

Different mechanisms have evolved for dealing with the low oxygen environment of embryonic and fetal life. For example, in primates and ruminates there is a separate high-oxygen affinity fetal haemoglobin (HbF) which is active in middle to late uterine life. The globin gene α-like and β-like clusters are arranged in order of their developmental expression with two switches occurring during embryonic and fetal development (Fig. 1.4, Table 1.3). The important switch in the present context involves HbF which is produced in utero and gradually changes to adult (HbA) by about 6 months of life. This explains why patients with homozygous β thalassaemia are normal when born (since the genes for HbF are normal and still contribute to total globin output) but become ill 6 months later, i.e. the time that HbF is replaced by the defective HbA. Despite a lot of research, the molecular basis for HbF-switching remained elusive. Important progress occurred when transgenic mice containing human globin genes were able to be made. From these animals, evidence has emerged to suggest that switching may involve a number of components including DNA-binding proteins which interact with DNA regulatory elements such as the *locus control region* situated on either side of the β globin gene cluster (Box 1.4). In the near future the complete regulatory mechanism involved in HbF-switching will be defined.

Another HbF-producing disorder is known as *non-deletional* HPFH. In this condition, the levels of elevated HbF are lower and the molecular defects involve point mutations. It is noteworthy that the point mutations associated with non-deletional HPFH occur within the gene's immediate 5' flanking region which has an important part to play in the regulation of transcription (Fig. 3.4). The regulatory regions associated with the human Gγ and Aγ globin genes are a GC-rich region, an octamer sequence, and CACCC, CCAAT and TATA boxes. It has been shown that these DNA motifs are involved in the binding of proteins to DNA. Many DNA-binding proteins have now been isolated. These can be ubiquitous (found in all cells) or specific to the red blood cells.

Evidence derived from DNA linkage analysis suggests that other gene loci which are *not* on chromosome 11 (where the β globin gene complex is found) are also involved in the regulation of γ globin gene expression. Further work is required to define the multiple molecular mechanisms involved in the HPFH disorders. Identifying the complex interactions which occur between DNA/DNA, DNA/proteins and proteins/proteins will enable the molecular controls of eukaryote genes to be better understood. Data are also starting to emerge from studies of the physiological HbF switch, mentioned earlier in Chapter 1, to illustrate potential ways in which proteins can modulate the expression of genes (Box 3.1). As well as explaining the pathogenesis of some genetic diseases, knowledge from the HbF models will identify potential targets for future therapeutic strategies.

AUTOSOMAL RECESSIVE DISORDERS

Clinical features

The appearance of an autosomal recessive disorder in a pedigree gives rise to a *horizontal* rather than *vertical* pattern. This occurs because affected individuals tend to be limited to a single sibship and the disease is not usually found in multiple generations (Fig. 3.5). Males and females are affected with equal probability. In specific populations with rare autosomal recessive traits, *consanguinity* can be demonstrated in affected families. The usual mating pattern which leads to an autosomal recessive disorder involves two heterozygous individuals who are clinically normal. From this union, there is a one in four (25%) chance that each offspring will be homozygous-normal or homozygous-affected for that trait or mutation. There is a two in four (50%) chance that offspring will themselves be carriers (heterozygotes) for the trait or mutation. The *same* risks apply to *each* pregnancy.

The inheritance patterns described may not be apparent, particularly in communities where the number of offspring are few. In these instances, the genetic trait or mutation can appear to be *sporadic* in occurrence. Therefore, the finding of a negative family history in autosomal recessive disorders

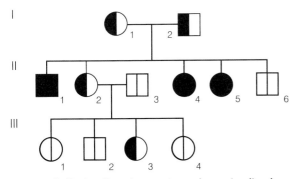

Fig. 3.5 Idealised pedigree for an autosomal recessive disorder. Females are represented by ○; males by □. Carriers of the genetic trait are indicated as half-filled circles or squares, affected as ● or ■. Note the horizontal distribution for the affected.

Box 3.2 Cystic fibrosis: a molecular model for autosomal recessive inheritance

Cystic fibrosis (CF) is a multisystem disorder of children and adults characterised by an abnormality in exocrine gland function manifesting as chronic respiratory tract infections and malabsorption. The first comprehensive description of this disorder was given in 1938. Approximately 1 in 2000 newborns of Northern European descent are affected, although considerable differences exist within ethnic groups, e.g. the incidence in oriental Hawaiians is 1 in 90 000. Many complications are associated with CF. (1) *Respiratory* – chronic bacterial infections which eventually lead to respiratory and cardiac failure. (2) *Gastrointestinal tract* – 10% of newborn infants with CF present with obstruction of the ileum (meconium ileus). More than 85% of children show evidence of malabsorption due to exocrine pancreatic insufficiency which requires dietary regulation and supplementation with vitamins and pancreatic enzymes. Rare gastrointestinal tract complications include biliary cirrhosis. (3) *Infertility* – affects 85% of males and to a lesser extent females. Long-term outlook for CF has improved dramatically over the years. Nevertheless, the median survival remains about 25 years. The availability of specialised CF clinics has enabled a multidisciplinary approach to follow-up and treatment of this disorder (see also Fig. 10.3). Support groups, such as the various CF Associations, have ensured that families are aware of recent developments which have followed from the cloning of the CF gene. These include expanded diagnostic options and gene therapy which is presently being evaluated as a form of treatment.

should not be ignored since the genetic defect can still be transmitted to the next generation, particularly if the mutant gene occurs at a high frequency in that population, e.g. cystic fibrosis.

Cystic fibrosis

Cystic fibrosis is the most common autosomal recessive disorder in Caucasians (Box 3.2). It affects ~ 1 in 2000–2500 live births, with a carrier rate in northern Europeans of 1 in 20 to 1 in 25. The high incidence of cystic fibrosis remains unexplained. Several hypotheses have been proposed, but as yet none are proven. They include: an increased mutation rate, genetic drift, multiple loci, reproductive compensation and selective advantage of the heterozygote carrier (see also Box 10.1 for further discussion).

After asthma, cystic fibrosis is the commonest cause of chronic respiratory distress in childhood and is responsible for the majority of deaths from respiratory disease in this age group. Clinical features relate to the thick tenacious secretions which can manifest with intestinal obstruction in the newborn (called meconium ileus), pancreatic insufficiency and chronic respiratory infections in childhood.

Although first described as a clinical entity in the 1930s, the pathogenesis of cystic fibrosis remained elusive despite clinical, electrophysiological and other conventional approaches used to study this disorder. The only clue to the underlying defect related to the elevated chloride levels in sweat. In the mid-1980s, chloride ion conductance across the apical membranes of respiratory epithelial cells or sweat ducts was shown to be decreased. Whether this represented an abnormal chloride channel or aberrant control of a normal channel was unknown. This remained the state of knowledge until 1989 when the cystic fibrosis gene was isolated by the DNA strategy of *positional cloning*.

Positional cloning

Linkage analysis
The first step in positional cloning is *chromosomal localisation* of the mutant gene. The clue for this can come from cytogenetic studies which identify a chromosomal rearrangement such as a deletion or translocation associated with the clinical phenotype. For example, the isolation by positional cloning of the Duchenne muscular dystrophy gene and the neurofibromatosis 1 gene was made easier by finding a deletion of Xp21.2 (Duchenne muscular dystrophy) and a number of balanced translocations involving chromosome 17q11.2 in the case of neurofibromatosis 1.

Initial attempts at chromosome localisation in cystic fibrosis were unsuccessful. This delayed isolation of the gene since a blind 'trial and error' approach was required to determine which DNA polymorphic markers would co-segregate with the cystic fibrosis defect within pedigrees. In contrast to the relatively simple linkage analysis steps described for the β thalassaemias, the procedures

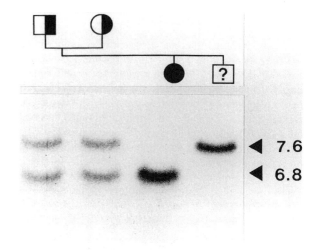

Fig. 3.6 DNA polymorphic (RFLP) patterns associated with the cystic fibrosis locus.
The probe is KM-19 and the restriction enzyme is *Pst*I which gives RFLP polymorphic fragments of 7.6 kb or 6.8 kb. The parents who carry the cystic fibrosis defect are heterozygous for the polymorphism since they have both fragments and so they are informative for this RFLP. Their affected child is homozygous for the 6.8 kb marker which indicates that *within this family* the cystic fibrosis defect co-segregates with the 6.8 kb restriction enzyme fragment. The second offspring has an unknown carrier status (?). However, DNA polymorphisms would predict that this individual is homozygous normal (with approximately a 1% error rate to allow for recombination).

involved in demonstrating co-segregation between an *unknown* locus and DNA polymorphisms are more complex. Both larger pedigrees and an increased number of DNA markers now become essential. Analyses of the linkage data also require sophisticated computer programs.

In 1985, linkage of cystic fibrosis to DNA markers on chromosome 7q31 was shown. Multiple markers were tested until the genetic map enabled the region containing the gene to be narrowed to a distance of ~1.5 Mb. The two closest markers were designated KM-19 and XV-2c. These were used in first trimester prenatal diagnoses for cystic fibrosis although there was an inherent error rate (about 1%) related to recombination between the markers and the actual cystic fibrosis gene. A second disadvantage of the linkage approach was the requirement for a family member (preferably a sibling) who had cystic fibrosis. This person was essential to allow assignment of polymorphisms to the normal or mutant alleles. Without this individual, linkage was not usually possible. Thus, at-risk couples having their first child were unable to have prenatal diagnosis in this way. Even if an affected person was available, not all families would turn out to be informative for the DNA polymorphisms (Fig. 3.6).

Chromosome walking
To identify the cystic fibrosis gene required *chromosome walking*. Cloned segments of DNA were ordered by using

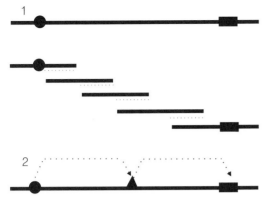

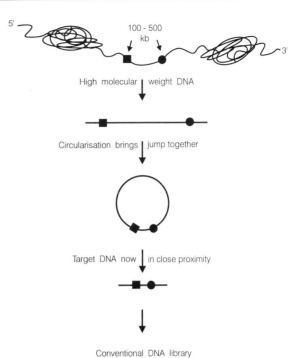

Fig. 3.7 Chromosome walking and chromosome jumping.
Identifying a gene within a segment of DNA can be achieved by mapping overlapping clones from the region so that DNA markers can be ordered both in their orientation and distance from the gene locus. The closer a marker is to the gene of interest the fewer recombinations which will occur. YACs have facilitated the process of chromosome walking by allowing large segments of DNA to be ordered in this way. **(1)** Illustrates a number of contigs or overlapping DNA clones, which cover a region between the starting DNA (●) and the target sequence (■) in a chromosome walk. .... indicates the segment which can be used as a DNA probe to isolate overlapping clones. **(2)** In comparison to walking, there are only two chromosome jumps required to reach the target DNA. A DNA segment (▲) between the start and the target allows a second jump to occur. Intermediate DNA can be bypassed by jumping from one segment to another.

the overlapping sections as probes to identify the adjacent segment. In other words, overlapping clones for the 1.5 Mb cystic fibrosis region needed to be arranged consecutively so that DNA markers and their orientation as well as their distance from the putative cystic fibrosis locus could be determined (Fig. 3.7). As indicated earlier, distance in genetic terms is measured by the number of meiotic recombinations (breakage and rejoining) which occur between a DNA marker and the clinical phenotype. The fewer recombination events, the closer the marker is to the gene of interest. Once the correct orientation is determined (by showing that the genetic distance for each marker is progressively reduced) chromosomal walking allows unidirectional progress until the gene of interest is reached. Two disadvantages of chromosome walking are: (1) it is a slow, tedious process since only short distances can be transversed at any one time; and (2) there are regions in the genome which cannot be cloned and so they interrupt the chromosome walk. *Chromosome jumping*, an alternative and more efficient strategy to chromosome walking, was developed and proved to be very successful in cystic fibrosis. In this technique a DNA segment between the start and the target is used to link the two. Intermediate DNA does not have to be characterised (Fig. 3.7).

Chromosome jumping
The approach to chromosome jumping is illustrated in Figures 3.7 and 3.8. The advantage of this strategy is that

Fig. 3.8 Construction of a DNA jumping library.
Two DNA sequences in total genomic DNA are depicted (■ and ●) which are about 100–500 kb apart. To 'walk' from one to the other would be difficult. The alternative is to 'jump' from one to the other by first isolating large segments from total genomic DNA. One of these will contain the two sequences within the one piece. All the DNA segments are circularised. This enables the sequences of interest to come into close proximity. A much smaller segment of DNA now contains the two sequences. A conventional library is prepared (see Fig. 2.9). This library will contain many jumping fragments. The one of interest will be identified by screening the library. If the ■ and ● sequences are used as probes it will be possible to identify within the library a segment which contains both. In this way, a jump has been obtained along a large chromosomal region.

linked probes, which can be located at considerable distances apart, e.g. 100–500 kb, are generated. Probes produced in this way allowed the extensive 1.5 Mb region of the cystic fibrosis locus to be analysed more rapidly than would otherwise have been possible by chromosome walking. Unclonable regions were bypassed. The probes generated from jumping libraries were used to screen conventional phage or cosmid libraries for *candidate* genes, features of which are summarised in Table 3.3. Today, chromosome jumping might not be the first alternative to chromosome walking since there are other 'shortcuts' to gene identification. One method which shows promise is known as *exon trapping*. This approach involves the screening of YAC contigs to identify individual exons. These are then used to look for the actual gene in cDNA libraries.

Finding the right gene
The search for the cystic fibrosis gene eventually narrowed to a region of DNA ~ 0.5 Mb in size. This area contained

Table 3.3 **Features which would indicate that a segment of DNA contains a candidate gene.**
Probes generated from chromosome walks, jumping libraries or exon trapping can be used to screen libraries for candidate genes.

Molecular findings	Significance
Deletions or gene rearrangements	A consistent finding in affected individuals but not in normal controls would suggest a possible functional role
Cross-hybridisation to DNA from primates, rodents and other species	Evolutionary conservation of DNA sequences would suggest functional significance (procedure is called a zoo blot)
Identification of CpG islands	Regions of DNA which are rich in hypomethylated cytosine (C) followed by the base guanine (G) are frequently found 5′ to vertebrate genes
Identification of open reading frames	Computer programs enable DNA sequences to be scanned for stop codons. If these are not present, the DNA sequence has the potential to encode for a gene
Identification of mRNA transcripts	Putative candidate genes identified through chromosome walking etc. are more likely to be significant if a corresponding mRNA can be isolated by northern blotting and/ or from a cDNA library. The DNA probe to do this would be derived from the candidate gene

a handful of genes. Clues which suggested that one of these was the cystic fibrosis gene included: (1) there was conservation of DNA sequence across a number of species, i.e. the gene has an important function; and (2) northern blotting showed that mRNA from this gene was present in tissues connected with cystic fibrosis, i.e. lung, pancreas, intestine, liver and sweat glands. Having decided that a piece of genomic DNA contained a likely candidate gene, it became necessary to take that DNA and use it as a probe to identify the corresponding segment from a cDNA library. In this way the long and complex exon/intron structure of genomic DNA could be simplified by looking at cDNA (exons alone).

Once cloned, the putative cystic fibrosis gene was demonstrated to code for a protein of 1480 amino acids. The gene was large, comprising 27 exons located over 250 kb of DNA. The mRNA transcript was 6.5 kb in size. Final proof of the gene's identity came with the demonstration of mutation(s) which correlated with the cystic fibrosis phenotype. The first mutation to be found involved a 3 bp deletion in exon 10. This resulted in loss of the amino acid phenylalanine at residue 508 (the mutation is called ΔF508 mutation – F is the biochemical abbreviation for phenylalanine). This was a causative mutation rather than a

---

Box 3.3 Determining the function of a gene

The function of a gene can be decoded in a number of ways. (1) An *expression vector* containing the gene of interest is constructed and from this a recombinant protein produced (see Fig. 2.23). Antibodies are raised against this protein. By in situ hybridisation it is possible to determine which tissues produce the protein. At the cellular level, the location and distribution of the protein can provide a clue to its function. An example of how this helped decoding is seen with Duchenne muscular dystrophy (DMD). Dystrophin, the 427 kDa protein encoded by a gene thought to be that responsible for DMD was shown to be localised to the inner surface of the sarcolemma in normal skeletal muscle. Thus, it was considered to be involved in the contractile apparatus of striated muscles. This provided preliminary evidence that the correct gene had been isolated. (2) *DNA sequence* from the gene of interest is obtained. A computer search is next undertaken of DNA databases to look for homology, i.e. similarity. There are three outcomes from this search. The gene's function is quickly determined since the database lists another substance which has been described previously and to which the DNA in question has 100% homology (unfortunately, this option also means that the investigator is too late since the gene has already been found!). At the other extreme, database searches produce a complete blank. Considerable work will now be required to identify the gene's function. The final and best option results in the finding of *partial homology* with one or more substances in the databases, i.e. there are now some clues relating to potential function. This occurred with the cystic fibrosis (*CFTR*) gene which showed homology to a group of proteins known as the ATP-binding cassette family of transport proteins. These proteins bind ATP and transport substances across membranes, i.e. the type of activity which would be expected for a gene which caused cystic fibrosis (see also Box 3.6 for another relevant example).

---

neutral polymorphism since it was consistently found in cystic fibrosis patients but not in the normal population.

When the cystic fibrosis gene was being sought by positional cloning, it was predicted that the number of underlying defects would be few. This has now been proven incorrect, with more than 600 different mutations described by 1996! The commonest is the ΔF508 mutation which is present in ~70% of cystic fibrosis chromosomes from northern Europeans. An additional 20 or so mutations occur at a frequency of only 1–5% in most Caucasian populations. The remainder are sporadic.

'Decoding'
An increasingly common consequence of molecular strategies such as positional cloning will be the identification of genes or gene-like sequences for which a function needs to

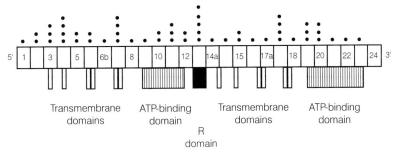

Fig. 3.9   The cystic fibrosis gene.
The diagram, which is not drawn to scale, shows the 27 exons of the *CFTR* gene (there are 24 exons numbered but the total is 27) since after the exons were first described and numbered, introns were detected dividing exon 6 (so that there are now exons 6a and 6b), exon 14 (exons 14a and 14b) and exon 17 (exons 17a and 17b). Above each exon is an indication of the number of mutations which affect that exon as well as its corresponding intron, e.g. exon 3 and intron 3. Each ● represents about 10 mutations. Below each exon is an indication of the major functional domains for *CFTR*, i.e. transmembrane domains, ATP-binding domains and an R (regulatory) domain (adapted from Tsui 1992). The ΔF508 defect is found on exon 10 and so interferes with ATP binding.

be found. This has been called 'decoding' by Collins. The function of a gene can be decoded in a number of ways (Box 3.3). The 168 kDa protein encoded for by the cystic fibrosis gene (the gene is also called *CFTR* or **c**ystic **f**ibrosis **t**ransmembrane conductance **r**egulator) was shown to have considerable similarity to a family of membrane-associated ATP-dependent transporter proteins which are found conserved throughout evolution since they are present in a wide range of species including bacteria, *Drosophila* and mammals. These proteins are involved in the active transport of substances such as ions and small proteins across membranes. Common structural findings in the above transporter proteins include one or two hydrophobic transmembrane domains and one or two nucleotide-binding folds for ATP attachment from which the energy for transport is obtained. *CFTR* has the above features and as well there is a highly charged central R domain (R for regulatory) which is considered to be involved in phosphorylation by protein kinase. The various components which make up the cystic fibrosis protein are depicted in Figure 3.9. There is now experimental evidence that *CFTR* codes for a *chloride ion channel*. Activation of this channel can occur by cyclic AMP or following phosphorylation by protein kinase. The latter involves the R domain and may work through a conformational change which then allows the passive flow of chloride ions. Whether *CFTR* has other functions remains to be determined.

Diagnosis of cystic fibrosis

Diagnosis of an individual affected by cystic fibrosis is possible by sweat testing, i.e. local sweating is induced and the chloride ion content measured. Those with cystic fibrosis have an elevated chloride level. Disadvantages of the sweat test are its inability to detect heterozygotes (carriers) and that it is not suitable for prenatal diagnosis. Once DNA polymorphisms became available it was possible to utilise a linkage study approach to undertake prenatal diagnosis and carrier testing within a family unit. Since 1989, direct identification of the ΔF508 mutation by PCR

has become the method of choice for laboratory detection. Diagnostic applications include prenatal diagnosis, carrier testing and helping to distinguish those disorders which resemble cystic fibrosis but have atypical features (Box 3.4).

As indicated previously, the ΔF508 mutation affects about 70% of the cystic fibrosis chromosomes in northern Europeans, with a lower frequency in other regions, e.g. 50% in southern Europe, 30% in Israel (Ashkenazic Jewish). The multiplicity of mutations makes detection of all cystic fibrosis defects an unrealistic proposal with present technology. Therefore, DNA-based diagnostic tests incorporate ΔF508 and a limited number of other mutations (e.g. six) which are selected on the basis of their prevalence in each population. This enables ~80% of the cystic fibrosis mutations to be detected. Tissues which are suitable for PCR include blood, hair follicles, chorionic villus, blood spots (such as those taken from neonatal heel pricks – 'Guthrie spots'), cells shed in amniotic fluid (amniocytes) and buccal cells which can be obtained from mouth washes.

The potential of PCR to be automated and so screen for the ΔF508 defect on a widespread basis has produced a controversy, i.e. whether there should be population-based (random) cystic fibrosis screening. The protagonists point out the importance that this knowledge would have on future reproductive decisions. Those against random population testing indicate that benefits and risks of population screening are uncertain unless 90–95% of carriers can be detected, which is presently not an economical or realistic goal with the *CFTR* gene (see Ch. 9 for a further discussion of screening).

Future directions

Considerable effort has gone into the correlation of *genotypes* (DNA defects) with *phenotypes* (clinical features) because of the genetic and clinical heterogeneity found in cystic fibrosis. For example, normal pancreatic function is often found in mild forms of the disorder. Molecular defects which are associated with pancreatic sufficiency can now be identified. These usually involve missense codon

Box 3.4 Diagnostic applications of DNA testing

Two cases involving cystic fibrosis (CF) related problems are described to illustrate the applications of DNA testing. In the first, the consultand (II_3) requested information about her CF carrier status since she was considering pregnancy. The consultand had a cousin with CF (II_1) and a brother (II_2) who was being treated for pancreatic malabsorption. Sweat tests in the latter were equivocal. After DNA testing, the cousin with CF was shown to be homozygous for the ΔF508 mutation. This mutation was not present in the consultand's mother, her spouse and brother (their genotypes are written as N/N for the ΔF508 defect). Two facts emerge from the DNA studies: (1) the risk for CF in future offspring of the consultand and her spouse is very low and (2) CF is an unlikely explanation for pancreatic malabsorption in the consultand's brother.

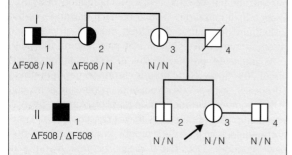

The second case involves a pregnant woman who is an obligatory carrier of the CF defect since she has an affected child (→). The story is more complicated in that it is not known which of two men is the father in the present pregnancy and the mother requests that neither is tested or approached. DNA studies show that the CF child is a double heterozygote for the ΔF508 and the G551D mutations and the mother has the less frequently found G551D defect. CF is excluded in the fetus since DNA from a chorionic villus sample does not have the G551D defect. In this circumstance the genetic status of the potential fathers is irrelevant. N = normal.

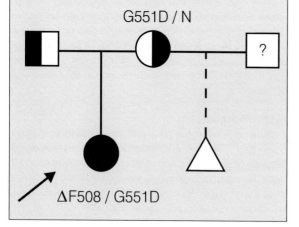

Box 3.5 The many 'faces' of cystic fibrosis

1. ***Congenital bilateral absence of the vas deferens***: A recent and interesting development in the field of in vitro fertilisation is the realisation that a number of men who present with infertility have a mild type of cystic fibrosis (CF). On retrospection this might not be considered an unexpected finding since most male patients with CF have congenital bilateral absence of the vas deferens and about 6% of normal males with obstructive azoospermia as the cause of their infertility demonstrate the same abnormality in their vas. Molecular analysis in the normal infertile males has now identified mutations in the *CFTR* gene. These mutations are 'mild' and so do not produce the complete CF phenotype but affect the vas deferens, an organ that is very sensitive to dysfunction of the *CFTR* gene (from Chillon et al 1995).

2. ***Mild lung disease:*** A similar correlation between clinical severity and *CFTR* mutation has been observed in the Dutch population which has the A455E mutation as its second most common defect after ΔF508. In this population a comparative study has shown that those who are compound heterozygotes for A455E/ΔF508 have milder lung disease than those homozygous for the ΔF508 mutation. Since lung disease is an important cause of morbidity and mortality in CF, the identification of a genetic marker which carries a better prognosis becomes a useful parameter in clinical management (from Gan et al 1995).

3. ***Sweat test negative lung disease:*** Another dilemma in clinical practice is the finding of respiratory and chest X-ray findings which are suggestive of CF but sweat tests are equivocal or negative. A study of these types of patients has characterised *CFTR*-specific mRNA from their nasal epithelium. Results showed that 13 of 23 patients from 8 unrelated families had an intron 19 defect which interfered with normal splicing and so interfered with the function of *CFTR* (from Highsmith et al 1994).

changes. In these circumstances, prognosis for cystic fibrosis is considerably improved. The molecular defects in those with pancreatic insufficiency (i.e. cystic fibrosis is of the severe type) are more likely to involve the ΔF508 deletion which is located within the first ATP-binding site (Fig. 3.9). Similarly, meconium ileus is frequently found in the newborn with the ΔF508 deletion. Other mutations associated with severe phenotypes produce premature stop codons, frameshifts or splicing defects. However, the story is not that simple since exceptions are found. Interesting genotype/phenotype comparisons in cystic fibrosis are now emerging with disorders such as congenital bilateral absence of the vas deferens and various types of mild respiratory disease (Box 3.5).

Research into the molecular defects, other genetic components (such as immune responsiveness) and the elucidation of interacting environmental factors will enable a more accurate assessment of prognosis in children with cystic fibrosis. Both in vitro and in vivo studies are under way to characterise further the role played by *CFTR* and the consequences of mutations in this gene. A recent important development has been the production of a mouse model of cystic fibrosis using the strategy of homologous recombination where an inserted gene is targeted to its correct position (see Chs 2, 10). Knowledge gained from the above studies will help to rationalise the therapeutic regimens and reduce the fear associated with cystic fibrosis. A number of gene therapy trials to treat cystic fibrosis are presently underway (discussed further in Ch. 7).

AUTOSOMAL DOMINANT DISORDERS

Clinical features

The characteristic feature in a pedigree with autosomal dominant inheritance is a *vertical* mode of transmission. This appearance comes from the fact that the disorder can appear in every generation of the pedigree. Both males and females are affected and their offspring are at 50% risk (Fig. 3.10). There are a number of additional features which need to be considered when dealing with autosomal dominant disorders. The following become important in counselling.

Sporadic cases occur and these become increasingly more common as the mutation in question interferes with fertility. This may represent a secondary effect of the disorder or because death occurs before reproductive age is reached. (1) Mutations in unrelated families with X-linked Duchenne muscular dystrophy are usually of independent origins because affected individuals are unlikely to survive to a reproductive age. (2) It is estimated that in achondroplasia, an autosomal dominant form of dwarfism, 50% of cases are spontaneous mutations since the disorder indirectly reduces the individual's reproductive capacity in terms of finding a partner. (3) At the other end of the spectrum, Huntington disease does not have a direct effect on reproduction. Thus, sporadic cases of Huntington

disease are rare and even some of these can now be shown to have inherited an unstable 'premutation' allele from a parent (discussed further on p. 58).

Penetrance is an all-or-nothing phenomenon that describes the clinical expression of a mutant gene in terms of its presence or absence. Thus, an individual carrying a mutant gene may not express the clinical phenotype, i.e. the condition is non-penetrant. From family studies it is possible to determine the number of obligatory heterozygotes for a mutant allele. If 7 out of 10 heterozygotes show the clinical phenotype, the disorder is described as being 70% penetrant. That is, there is a 70% probability that an individual carrying a mutant gene at a certain age will display the clinical phenotype. This aspect of autosomal dominant disorders is discussed further under familial hypertrophic cardiomyopathy and Huntington disease. Apart from spontaneous mutations and death before onset of symptoms, penetrance is an additional mechanism which would account for affected offspring having an apparently normal parent.

Expressivity refers to the severity of the phenotype. There are genes which can produce apparently unrelated effects on the phenotype or act through involvement of multiple organ systems. This is called *pleiotropy*. Such genes often show variable expressivity. An example of this is Marfan syndrome which has autosomal dominant inheritance and involves connective tissues in the skeletal system, the eye or the heart. Individuals with Marfan syndrome have any combination of manifestations which can also be present in different degrees of severity. Such variability can occur even within families in which it is presumed the same mutant allele is present. To date, the underlying basis for expressivity has not been defined but it is thought to represent either gene/environment or gene/gene interactions. Somatic instability, a recently described mechanism, may provide another explanation (see Fragile X syndrome, Box 3.11).

Severity of an autosomal disorder can also be influenced by the sex of the transmitting parent (see Huntington disease, p. 58) or the sex of the affected person. An example of the latter is otosclerosis (a cause of deafness in adults due to overgrowth of bone in the ear) in which the female to male ratio is $\sim$ 1.8:1. The reason for this is unknown. One hypothesis suggests that affected males may have a selective disadvantage compared to females and this selection is having its effect prenatally.

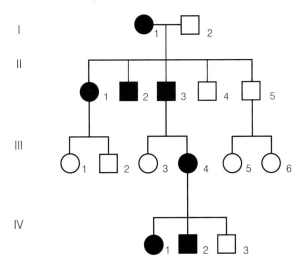

Fig. 3.10 Idealised pedigree for an autosomal dominant disorder.
Affected individuals are indicated by ● and ■. Note the vertical disease pattern (cf. Fig. 3.5) with the disease apparent in every generation and affecting males and females.

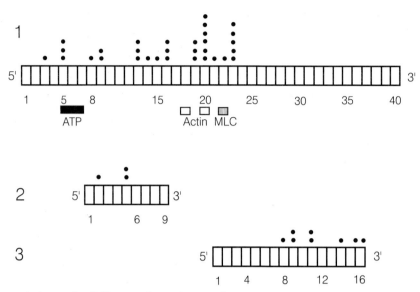

Fig. 3.11 Three genes which cause familial hypertrophic cardiomyopathy (FHC) and their associated mutations.
(1) The cardiac β myosin heavy chain gene with its 40 exons depicted as boxes (not drawn to scale). The numbers under the boxes identify the exons. Some of the missense mutations which cause FHC are indicated by ●. Functional domains include: ATP – ATP-binding site; Actin – actin-binding sites and MLC – myosin light chain-binding sites. Note that all missense mutations involve the first 23 exons of the gene, i.e. the head portion of this protein, which may reflect the functional domains present. The human structure of the troponin T gene (2) and α tropomyosin (3) are still in the process of being defined. Mutations are few in number and structure/function remains to be determined.

Familial hypertrophic cardiomyopathy

Unlike cystic fibrosis, which is well down the track in terms of our understanding of its molecular basis, the model chosen to illustrate autosomal dominant inheritance is still at the beginning of the molecular story, with more questions than answers at present!

Familial hypertrophic cardiomyopathy (FHC – also abbreviated to HCM) is an autosomal dominant disorder involving heart muscle. It was first described in 1958. The characteristic finding is ventricular hypertrophy which can lead to left ventricular outflow obstruction. Apart from cardiac failure, the major complication in this disorder is sudden death which can occur in adolescents or young adults. FHC is also a common autopsy finding in sportsmen and sportswomen who die suddenly. Its frequency in a general population has been estimated at 2 in 10 000. Diagnosis is made by a combination of clinical examination, ECG and echocardiography. However, this approach will not detect all affected individuals, particularly the young.

Positional cloning – candidate gene approach

A potential chromosomal location for FHC has never been obvious from cytogenetic studies. Thus, positional cloning to find the FHC gene would need to be undertaken blind in a similar way to that required for cystic fibrosis. However, an alternative approach was possible since *candidate genes* could be defined. Candidate genes in FHC would be those that encoded for muscle proteins, particularly genes which are expressed in heart, e.g. actin and myosin. Linkage studies were undertaken using DNA probes derived from

these genes' sequences. In 1990, linkage was established between the β isoform of the cardiac heavy chain myosin gene on chromosome 14 and FHC. The gene was further implicated in aetiology when it was shown that a number of individuals with FHC had point mutations in various exons (Fig. 3.11).

The connection between the β myosin heavy chain gene and FHC was only the start of the molecular story since there were families who did not show linkage to the chromosome 14 locus or mutations in the β myosin heavy chain gene. Therefore, FHC was heterogeneous at the DNA level and other candidate genes were sought. The remaining chromosomal loci to be implicated in FHC were 1q3 (gene for troponin T), chromosome 15q2 (gene for α tropomyosin) and chromosome 11p11.2 (gene for myosin-binding protein C) (Fig. 3.11). A fifth, as yet unidentified, locus is also present since there are families with FHC that do not map to the four known chromosomal loci (Table 3.4).

DNA diagnosis

There are two aspects of FHC which make diagnosis by conventional approaches difficult. (1) The disease can occur *sporadically* and so a family history will be missing. At first it was thought that this may occur in about half the cases but as molecular analysis has progressed it is now considered that sporadic forms are less common and perhaps rare. (2) *Variable penetrance* is a further confounding issue. Some mutations in FHC, e.g. arg403gln (the normal arginine at amino acid 403 is replaced by glutamine – see Table 2.1 for abbreviations used with amino acids),

Table 3.4 Molecular classification and mutations in familial hypertrophic cardiomyopathy (from Watkins et al 1995).
Heterogeneity is the key word to describe the range of mutations in this genetic disorder.

Type (% affected)	Chromosome	Gene	Mutations
CMH1 (~30%)	14q11	Cardiac β myosin heavy chain	Over 30 missense mutations reported – involve mainly the protein's head region
CMH2 (~15%)	1q3	Cardiac troponin T	Majority missense mutations, mainly in conserved regions
CMH3 (<3%)	15q2	α Tropomyosin	Missense mutations in conserved regions
CMH4 (~10%)	11p1.2	Myosin-binding protein C	Most recent gene found – duplication point mutations reported
CMH5 (?)	?	?	–
CMH + WPW* (?)	7q3	?	–

*Wolff–Parkinson–White syndrome – a relatively benign disorder associated with cardiac arrhythmias predominantly atrial in type.

are considered to have 100% penetrance in adults, i.e. all who have the mutation will eventually manifest the disease. Other mutations, e.g. leu908val, may only be 50% penetrant, i.e. an adult carrying that mutation may not develop features of the disease although he/she could pass on the mutant gene to the next generation. In both circumstances, the availability of additional diagnostic criteria such as a DNA mutation would be helpful in clinical evaluation.

DNA testing has progressed in FHC. Since most mutations involve *missense* changes in amino acids, it is not possible to use the more recently described DNA approaches, e.g. protein truncation. Conventional analysis by changes in restriction enzyme sites or ASO (allele-specific oligonucleotide) blots are required. Because there are many mutations which cause FHC, the detection rate depends a little on the laboratory's resources and interest in this disorder. If only peripherally involved, the laboratory may provide a limited DNA diagnostic service for those mutations which have been shown to recur in unrelated families. On the other hand, a research laboratory working in FHC may want to detect a more comprehensive range of mutations.

Box 3.6 Pathogenesis of the Long QT (LQT) syndrome

As well as providing an alternative approach to diagnosis, the molecular characterisation of genetic disorders has been crucial in understanding their pathogenesis. Another autosomal dominant cardiological disorder first described in 1963–1964 illustrates this point. The LQT syndrome (known also as the Romano–Ward–Long-QT syndrome) shares the common feature with familial hypertrophic cardiomyopathy that it is a cause of sudden death in otherwise healthy individuals. This occurs because of a predisposition to ventricular fibrillation. Two hypotheses were proposed to explain the repolarisation abnormality in the LQT syndrome: (1) a defect in ion channel regulation or (2) an alteration in the autonomic nervous system. Which mechanism was correct remained unresolved until 1994 when scientists were able to show linkage between the LQT syndrome and chromosomes 7 and 3. Genes were soon isolated from these loci. On chromosome 7, mutations in a gene known as *HERG* were detected in affected individuals. Similarity between the DNA sequence of this gene and one in *Drosophila* enabled *HERG* to be identified as a potassium channel. The chromosome 3-related gene was also characterised. It is known as *SCN5A* and codes for a cardiac sodium channel. The molecular basis for the LQT syndrome as a *cardiac ion channel defect* was thus clarified by finding and then characterising the underlying genes. The molecular pathology provided both a unifying mechanism for the disorder as well as an explanation for the variable clinical phenotypes. Treatment of affected individuals with β blockers has been effective in reducing the mortality although up to 30% of patients continue to die suddenly despite treatment. Knowledge of the molecular defects may assist in finding more effective drug approaches for this potentially fatal disorder. Not surprisingly, the LQT syndrome displays molecular heterogeneity, i.e. a number of genetic loci and multiple mutations in each gene. The LQT syndrome has now been classed into types LQT1 – chromosome 11p15.5 (gene still to be identified), LQT2 – chromosome 7q35 (*HERG* gene), LQT3 – chromosome 3p21 (*SCN5A* gene) and LQT4 – chromosome 4q25-q27 (gene still to be identified). It would not be unexpected to find LQT5, 6, etc. in the future.

Once a mutation is detected in a family, it can be looked for in other at-risk individuals. The benefits of this approach are two-fold. Those who do not carry the mutation can be reassured that they will not develop the disease. The degree of assurance will depend on the evidence available which confirms the mutation as being causative. The second advantage is that those who carry the mutant gene can be followed more carefully and, in some cases, preventative measures such as drugs and implantantable defibrillators considered. From a research point of view, prospective and long-term follow-up of individuals with FHC-related mutations will be important in determining the effects of medical treatment which is started before complications occur or the disease, in the form of left ventricular hypertrophy, is fully developed.

Phenotype/genotype correlations

An important benefit of DNA testing is that a mutation can be found and then the type of mutation will enable a prediction to be made about the likely medical outcome for that disorder. This has been illustrated in the cystic fibrosis example (see Box 3.5). When mutations were first described in the FHC gene, some were labelled as 'benign', e.g. leu908val which has previously been noted to have low penetrance. Others were considered to be 'malignant', e.g. arg403gln, which was so named because of frequent FHC-related cardiac deaths in these families. A reason proposed for the benign or malignant nature of missense mutations was whether these changes in amino acids produced no net charge difference (benign defect) or an altered overall charge (malignant defect).

As experience has been gained in FHC it is apparent that the phenotype/genotype correlation is less than straightforward. For example, the mutation val606met does not alter the overall charge and so, not surprisingly, was first reported in the context of a family in which sudden deaths had not occurred. More recently, another family with the same mutation has had four sudden deaths out of eight affected individuals. Clearly, there are additional factors involved in prognosis. These could include other genes, interacting genes and/or environmental effects. This dilemma will only be resolved as molecular characterisation allows a comprehensive profile to be established in FHC so that the various molecular changes (the genotypes) can be carefully measured against the clinical pictures (the phenotypes).

Pathogenesis

Myofibrils, which are made of repeating assemblies known as *sarcomeres*, comprise the contractile elements in mus-cle. A sarcomere has seven major proteins and several minor ones organised into thin and thick filaments. Muscle contraction and force generation results from the relative sliding between these filaments. *Thin filaments* consist predominantly of actin with lesser amounts of the regulatory proteins tropomyosin and troponin. *Thick filaments* are composed principally of myosin which is a Y-shaped molecule with two globular heads and a tail. Each cardiac-specific myosin molecule forms a hexamer comprised of: (1) two identical heavy chain subunits, (2) two subunits of an alkali light chain, and (3) two subunits of a regulatory light chain. The C-terminal end of the two heavy chains coil to form a helix (the myosin tail). The N-terminal end of each heavy chain, together with two of the light chains, forms the globular head. Although the sarcomere is the key element in muscle contraction, the unit itself is susceptible to a number of influences such as ion channels, calcium and adrenergic receptors to name a few. Thus, muscle contraction is a complex process. Not surprisingly, there are numerous muscle genes of potential significance to FHC and these are located on many chromosomes. There are also different forms depending on whether skeletal, cardiac or smooth muscle is involved. This complexity makes an understanding of pathogenesis in FHC particularly difficult.

Since the first three muscle genes associated with FHC (β myosin heavy chain, α tropomyosin, troponin T) are critical to the function of the sarcomere, FHC has been described as a 'disease of the sarcomere'. The more recent identification of another sarcomere-related gene (myosin-binding protein C) is consistent with this hypothesis. Thus, in a relatively short time of 5 years, our understanding of pathogenesis in FHC has progressed rapidly and it is likely that the future will see other important data emerging which will influence how FHC might be treated in a more effective manner. Similar rapid and exciting developments have occurred in another autosomal dominant cardiological disorder known as the Long QT syndrome (Box 3.6).

X-LINKED DISORDERS

Clinical features

X-linked disorders result from abnormal gene function associated with the X chromosome. Males, who have only one X chromosome, i.e. they are *hemizygous*, will fully express a mutant gene on the X chromosome. On the other hand, females, who have two X chromosomes, will be carriers of the defect in the majority of cases. Although females have two X chromosomes to the male's one, products from this chromosome are quantitatively similar in both sexes because one of the two X chromosomes in females is inactivated.

Lyonisation (named after M Lyon) describes the random X-inactivation of an X chromosome which occurs during embryonic development. Because of the early onset and randomness of the process, female carriers of X-linked disorders can demonstrate variable amounts of the gene product, i.e. a protein, which will depend on the proportion of normal to mutant X chromosomes which remain functional. The majority of the X chromosome is inactivated although there are some segments which escape this process. The molecular basis for X-inactivation is unknown. Methylation may play some role (methylation is discussed further on p. 65).

The 'shape' of a pedigree illustrating X-linked inheritance is shown in Figure 3.12. The X-linked pedigree has an oblique character through involvement of uncles and nephews related to the female consultand. The usual mating pattern involves a heterozygous female carrier and a normal male. Each son has a 50% risk of being affected

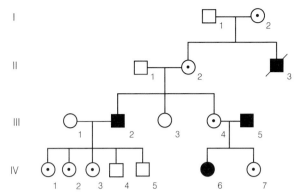

Fig. 3.12 Idealised pedigree for an X-linked disorder.
Females are represented by ○ and males by □. Female offspring of affected males are obligatory carriers. IV_6 is an affected female since both her parents carry the X-linked disorder. Affected males are shown by ■; carrier females are shown by ☉ ; individual II_3 is deceased. Because the disease is X-linked, a male cannot transmit the disorder to his sons (to whom he contributes a Y chromosome). Construction of pedigrees are key steps in the understanding of genetic diseases and how they are transmitted (see also mitochondrial DNA inheritance and imprinting, which reinforce the value of a pedigree).

Table 3.5   Clinical, laboratory and molecular features of the haemophilias.
These are inherited coagulation disorders associated with bleeding problems.

	Haemophilia A	Haemophilia B (Christmas disease)
Frequency	1–2 in 10 000 males; all ethnic groups	1 in 50 000 males; all ethnic groups
Defect	Clotting factor VIII – complex protein which circulates bound to von Willebrand factor. Produced in the liver	Clotting factor IX – serine protease. Produced in the liver
Clinical	Prolonged bleeding spontaneously or after minor trauma involving joints, muscles, subcutaneous tissues and organs. Approximately 50% have a severe disorder (factor VIII <1%), 10% are moderately severe (factor VIII 2–5%), 30-40% are mild (factor VIII 5–30%)	Bleeding manifestations as for haemophilia A
Genetics	X-linked, 10–30% spontaneous mutations	As for haemophilia A
Gene size	26 exons over 186 kb	8 exons over 34 kb
Chromosomal location	Distal to Xq28	Xq27

by inheriting the mutant maternal allele. Similarly, each daughter has a 50% chance of inheriting the mutant gene from her mother but will remain unaffected since she has her father's normal X chromosome. Male-to-male transmission is not seen but may appear to occur if the trait is sufficiently common that by chance the mother also carries the mutant gene. An example of this would be glucose-6-phosphate dehydrogenase deficiency in those of black African origin. Approximately 10–20% of blacks in the United States are carriers or hemizygous for this defect.

Females can be symptomatic carriers or develop X chromosome-related disorders in a number of ways. (1) A disproportionate number of normal X chromosomes are inactivated. This can be a chance event or following a translocation between an X chromosome and an autosome. In the latter, X inactivation appears to be non-random since the normal X chromosome is preferentially inactivated. This may represent a selective process as cells with the normal X inactivated are least imbalanced and so will have a survival advantage. (2) Hemizygosity in the female, e.g. Turner syndrome or 45,X. (3) Inheritance from both parents of a frequently occurring X-related gene, e.g, glucose-6-phosphate dehydrogenase deficiency. (4) The recently defined heritable unstable DNA repeats which are described on p. 58.

Just as for autosomal dominant conditions, the frequency of *spontaneous mutations* in the X-linked disorders needs to be considered, particularly when counselling females who are potential carriers. Haemophilia does not interfere with the reproductive capacity of the affected individual. In contrast, Duchenne muscular dystrophy is usually fatal in the 2nd decade of life. Therefore, spontaneous mutations occurring in the latter disorder would be greater in

number and correspondingly the proportion of females who are carriers will be less.

Haemophilia

The clinical, laboratory and molecular features of the haemophilias are summarised in Table 3.5. In the context of molecular medicine, the haemophilias will be used to illustrate the difficulties which can arise in identifying carriers particularly when the X chromosome is involved. Much has been written about *positional cloning* strategies to detect genes. The haemophilias also illustrate the value of *functional cloning*.

Carrier detection in X-linked disorders

Protein assays
Protein levels for coagulation factor VIII (deficiency of which produces haemophilia A) and coagulation factor IX (deficiency gives haemophilia B) demonstrate a wide normal range in blood. Because of random X-inactivation, the levels of factors VIII and IX can vary considerably in females who are carriers of haemophilia. This scatter makes an accurate assessment of carrier status difficult if the subject being tested demonstrates a normal or borderline result for the coagulant protein (Fig. 3.13). The level may

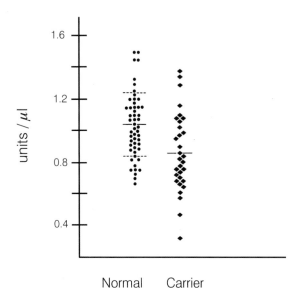

Fig. 3.13 The levels of factor VIII as measured by its coagulant activity in a normal population and in obligatory female carriers for haemophilia A.

Levels of the factor VIII protein have a broad normal range in blood. The levels in carriers can also vary considerably because of random inactivation of an X chromosome. This makes accurate assessment of carrier status difficult. The mean values are indicated, as well as the standard deviation for the normal range. There is considerable overlap between factor VIII coagulant levels in normals and carriers. Better discrimination can be obtained by measuring ratios, e.g. factor VIII coagulant/factor VIII antigen, although there is also overlap with ratios. Thus, a woman who wants to know her carrier status will only get an unequivocal result if her factor VIII level is very low. Any other result cannot be definitive.

reduce the individual's a priori risk but does not provide definitive proof of her carrier status. In addition to X-inactivation there are physiological fluctuations seen with the coagulation factors, e.g. pregnancy (or taking the oral contraceptive), at which times the baseline levels for coagulation factors can increase. Finally, there is the not infrequent problem of assessing whether an affected relative is an example of a spontaneous mutation rather than the transmission of a haemophilia defect within a family. This occurs when there is only one affected male in the family (see Factor VIII inversion, p. 54, which shows how a molecular marker will help in this dilemma).

DNA linkage analysis

Testing for DNA mutations has advantages over protein assays: (1) access to DNA is unlimited whereas an abnormal protein may not be easy to obtain, and (2) unlike protein, DNA is not affected by physiological fluctuations. The former is not a problem in haemophilia in which a blood sample suffices. The latter is an important consideration. An indirect DNA linkage approach for diagnosis has been used in haemophilia since: (1) the majority of defects are point mutations, (2) the genes are large, and (3) there are many mutations (Table 3.6). A number of DNA

Table 3.6 Molecular defects in the factor VIII gene which produce haemophilia A (from Antonarakis et al 1995).
A mixture of major DNA rearrangements and point mutations is found in this gene.

Mutations	Comments
Inversions	Detected by Southern blotting; found in ~50% of severe haemophilia A (see also Fig. 3.16)
Deletions	Very heterogenous; usually produce severe haemophilia A; detectable with Southern blotting
Insertions, duplications	Less frequently described involving insertion of retrotransposons or duplication of exon 13
Chromosomal rearrangements	Unusual cases; may interfere with random X chromosome inactivation
Point mutations, small deletions/ insertions	The most frequently found defects. Point mutations produce missense, nonsense or splicing defects. Best detected with PCR after appropriate screening procedures, e.g. SSCP

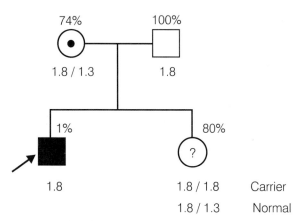

Fig. 3.14 Utility of the DNA linkage analysis approach in detecting female carriers for haemophilia.

The *consultand* is the mother of the child with severe haemophilia B (→) who has a factor IX level of 1%. The mother is an obligatory carrier because she has an uncle who is affected (not shown in the pedigree). Factor IX coagulant levels are given as percentages in the pedigree (normal is >50%). Factor IX levels for the mother and her daughter are within the normal range (74% and 80% respectively) but as indicated previously this does not exclude the carrier state because of random X-inactivation in females. From the DNA polymorphism patterns, it is evident that the haemophilia B defect co-segregates with the 1.8 kb DNA polymorphism since this is the marker present in the haemophiliac son. Therefore, the daughter's carrier status can be determined on the basis of which DNA polymorphism she inherits from her mother, i.e. if the daughter is homozygous for the 1.8 kb marker (she will always inherit one 1.8 kb marker from her father) she is a carrier. If the daughter has both the 1.8 and the 1.3 kb markers then the latter must have come from her mother. The daughter cannot be a carrier since the 1.3 kb polymorphism is a marker for the normal maternal X chromosome. The *proband* in this family is the haemophiliac child (→) who has only one polymorphic marker compared to his female relatives, i.e. he is hemizygous since he does not inherit an X chromosome from his father.

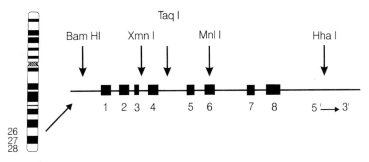

Fig 3.15 The structure of the factor IX gene.
The eight exons of the factor IX gene are shown. Five polymorphic restriction enzyme sites giving restriction fragment length polymorphisms (RFLPs) are indicated by ↓. Three occur within the gene (intragenic) and two (*Bam*HI and *Hha*I) are extragenic. Some of these polymorphisms are inherited in a preferential association known as linkage disequilibrium, e.g. *Xmn*I and *Mnl*I, and are therefore less useful in DNA testing.

polymorphisms have been described which are located within (intragenic) and in close proximity to (extragenic) the factor VIII and factor IX genes. These polymorphisms allow DNA diagnosis (prenatal or carrier) to be made in ~70–80% of families (Fig. 3.14). Intragenic polymorphisms have the advantage that recombination is unlikely to occur since the markers are located within the gene.

The disadvantages of DNA testing must also be considered. The DNA polymorphic approach requires key family members (who may be deceased or unavailable). It is very difficult to determine if mutations are spontaneous events. Germline mosaicism, where an individual has two or more cell lines of different chromosomal content derived from the same fertilised ovum, cannot be excluded. (This is discussed further on p. 66.) An additional problem with the DNA linkage approach in haemophilia B is the effect that *linkage disequilibrium* (preferential association of linked markers) can have on the informativeness of polymorphisms. For example, there are five biallelic DNA polymorphisms associated with the factor IX gene (Fig. 3.15). Some of these polymorphisms are inherited in a preferential association, i.e. the *Xmn*I and *Mnl*I polymorphisms are in linkage disequilibrium, which means that results obtained with either are similar since one allele of the polymorphism is nearly always inherited with the same allele of the other. Therefore, not all five polymorphisms will necessarily be informative. This is a particular problem with the factor IX gene locus in Chinese and Asian Indian populations. Non-paternity and its effect on DNA polymorphisms is not an issue if male offspring are studied because the father does not contribute his X chromosome to males. However, the source of the paternal X chromosome is important if a female is being assessed for carrier status. Direct detection of mutations would overcome many of the above problems.

Direct detection of mutations

Two novel approaches have been described which enable haemophilia mutations to be detected. In the first, factor

---

Box 3.7 Direct detection of mutations using 'ectopic' mRNA analysis

When it comes to diagnosis of genetic disease, a difficulty of using mRNA compared to DNA is the former's tissue specificity, i.e. mRNA is only found in tissues where a particular gene is functional. However, it has been shown recently that mRNA for some genes can be isolated from peripheral blood lymphocytes despite the fact that these cells are not *specifically* expressing those genes. This type of mRNA production has been called *'illegitimate transcription'* or *'leaky RNA'* or *'ectopic RNA'*. The amount of mRNA produced in this way is very small, but detectable with PCR. An additional step needs to be incorporated to convert mRNA to cDNA and then the cDNA can be amplified. This process is known as 'RT-PCR' since reverse transcriptase is first used to make cDNA and then it becomes possible to use PCR to amplify the DNA. Once the DNA is amplified it can be screened for mutations by using techniques such as SSCP or DGGE (see Fig. 2.17) and suspicious areas of DNA are then characterised more carefully. 'Illegitimate transcription' in peripheral blood lymphocytes has been found for a number of medically important conditions, e.g. haemophilia, Duchenne muscular dystrophy and familial hypertrophic cardiomyopathy. The advantages of the ectopic mRNA approach are: (1) cells from blood are suitable for assay and (2) mRNA is considerably smaller than genomic DNA (there are no introns). This simplifies mutational analysis of large genes, e.g. mRNA for the factor VIII gene is 9 kb cf. 186 kb found with the genomic structure. A disadvantage of this technique is that mRNA only codes for exons. Therefore, mutations in introns or potential regulatory regions of the gene are less likely to be found except for splicing defects which produce different sized mRNA products.

VIII-specific mRNA from peripheral blood lymphocytes is characterised using PCR. This overcomes the size problem since mRNA is considerably smaller than genomic DNA

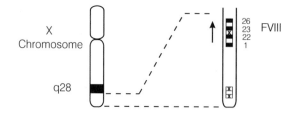

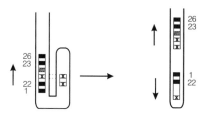

Fig. 3.16 The factor VIII inversion mutation.
The region of the X chromosome below band q28 which contains
the factor VIII gene is magnified. Only relevant exons (1, 22, 23, 26)
in the gene are shown as ■. The x indicates the location within
intron 22 of an inverted DNA repeat known as F8A. DNA
homologous to F8A and located more telomeric is also displayed
(xx). The vertical ↑ indicates the direction that the factor VIII gene is
transcribed. The lower part of the diagram shows an
intrachromosomal crossing-over event between the two F8A
homologous regions (...). The additional (banded) section in intron
22 above F8A is a second gene which has been found in intron 22.
The final result from the cross-over is a factor VIII gene which has
been flipped around (inverted) and is now in two sections: exons 1
to 22 and one F8A segment transcribe in a telomeric direction; two
F8A segments and exons 23–26 transcribe towards the centromere.
This gross structural rearrangement has a major effect on factor VIII
production resulting in a severe form of haemophilia A. The
cross-over depicted has occurred between the intron 22 F8A and
the more distal of the F8A homologous regions. Proximal cross-overs
can also occur. This mutation is detected by Southern blotting (see
Fig. 4.4 which gives an example). A PCR-based method is not yet
available.

and it now becomes technically possible to look for a
mutation (Box 3.7).

A second development occurred in 1993 when it was
reported that a *factor VIII inversion* was frequently found in
patients with severe haemophilia A. The inversion involved
exons 1–22 and 'flipped' this part of the gene. A lot more
is now known about the inversion. It occurs because there
is another 'gene' (called F8A) located within intron 22 of
the factor VIII gene. There are two additional regions on
the X chromosome situated more distally which have
homology to F8A and these areas predispose to homologous recombination (Fig. 3.16). The inversion occurs in
about 50% of severe haemophilia A patients and is
detectable by Southern blotting. The inversions originate
almost exclusively during male meiosis. Therefore, it is not
surprising that mothers of boys (even if there is only one

**Box 3.8 Identification of the Duchenne
muscular dystrophy gene**

Duchenne muscular dystrophy (DMD) and Becker muscular dystrophy (BMD) affect 1 in 3000 newborn males.
DMD leads to progressive muscle wasting with death
occurring in the 2nd decade from the complications of
muscle degeneration. BMD is a milder disease of late
onset and slower progression but is otherwise identical
to DMD. The two are allelic, i.e. they involve the same
gene. An elevation in muscle enzymes such as creatine
phosphokinase (CPK) is found in all affected patients.
However, because of random X-inactivation, female
carriers are difficult to identify since only 70–80% show
increases in CPK and these are modest. Therefore,
DNA testing has been developed. The gene responsible
for these dystrophies was one of the first to be isolated
by positional cloning. The location of the DMD gene to
chromosome Xp21 was made in 1977 on the basis of a
translocation which produced the muscle disorder in a
female. In 1982, DNA polymorphisms linking DMD to the
short arm of the X chromosome were described. Carrier
detection and prenatal diagnosis using a number of DNA
polymorphisms became possible in 1986. By 1987, the
gene was cloned and characterised. It is the largest
described in humans (~1% of the X chromosome itself)
extending over 2300 kb and comprising 79 exons. The
encoded protein is called 'dystrophin'. The large size of
the gene may explain in part its high spontaneous
mutation rate since one-third of cases are considered to
be new defects. Mutations in unrelated families are
usually different abnormalities. In ~65% of patients with
DMD or BMD, the molecular abnormality involves exon
deletions. Present methods for DNA detection utilise
PCR to look for mutations in DNA or ectopic mRNA and,
more recently, the protein truncation test (Box 3.7,
Fig. 2.20). At the molecular level, an explanation can be
provided why the same gene produces different clinical
severities. In DMD, the mutations usually involve frameshifts so that protein production is considerably
impaired. In BMD there is less disruption to the reading
frame and more dystrophin protein is present, hence the
disorder is milder in severity.

---

affected male in the family) with this inversion are themselves *usually* carriers.

Functional cloning

In contrast to positional cloning where a gene is found on
the basis of its location in the genome, *functional cloning*
allows a gene to be cloned because some information is
already known about the gene's structure or function (see
Fig. 2.14). This approach to cloning is well illustrated by
the haemophilias. The factor VIII and factor IX proteins had
been isolated and characterised in human and other
species by the late 1970s and early 1980s. These factors
form part of the middle phase of the intrinsic clotting
cascade and are serially activated and involved, in association with calcium and phospholipid, in the activation of

Table 3.7   The major functional domains of factor IX.
Structure/function of the factor IX coagulant protein as determined
from its DNA sequence and gene organisation.

Site	Activity
Exon 1 (or exon a)	Hydrophobic signal peptide: allows secretion from the hepatocyte into the bloodstream
Exons 2, 3 (b, c)	Contains a propeptide and undergoes post-translational modification (required for correct folding and calcium binding)
Exon 4 (d)	Epidermal growth factor-like domain; also binds additional calcium
Exon 5 (e)	Second epidermal growth factor-like domain
Exon 6 (f)	Activation domain for factor IX
Exons 7, 8 (g, h)	Serine protease (catalytic) domain; important for proteolysis of factor X to its active form

clotting factor X. Since the protein structure was known it was possible to utilise functional cloning to look for the relevant genes.

Oligonucleotide probes were synthesised from protein sequences of human and porcine factor VIII and bovine factor IX. Thus, each individual amino acid in the protein sequence was able to be reproduced in the form of a triplet codon in the oligonucleotide probe. The problem of a degenerate DNA code (i.e. there can be more than one codon for most of the amino acids – Table 2.1) was overcome by synthesising the codon which is more commonly used or alternatively making a mixture of 'degenerate' oligonucleotides. These comprised a cocktail of the possible codon combinations. DNA libraries were screened with these oligonucleotides and the genes isolated. Not surprisingly, from its complex protein structure, the factor VIII gene is large (discussed further in Ch. 7). DNA probes from within and around the two genes were isolated and are now used for diagnostic purposes.

Knowledge of the structure and function of factor VIII and factor IX has been enhanced following cloning of the relevant genes and characterisation of their DNA sequences. For example, the eight exons of the factor IX gene encode for six major functional domains in the 415 amino acids of the glycoprotein. The domains are summarised in Table 3.7. Information gained from structure/function comparisons has been invaluable in our further understanding of the protein's biology. It has proven useful in the production of a recombinant DNA-derived product (see Ch. 7).

At the molecular level, similar developments described for haemophilia have also occurred in Duchenne muscular dystrophy, another important X-linked disorder (Box 3.8).

MULTIFACTORIAL DISORDERS

Phenotypes

The term *polygenic* can have a number of meanings. It is most often used to describe genetic diseases which arise from the interaction of multiple genes at different loci. The propensity to develop a polygenic disorder follows a normal distribution in a population and results from the additive effects of these genes. Traits in the population such as blood pressure, height and intelligence are frequently used to illustrate polygenic inheritance. However, there is no evidence that multiple genes are involved and the role played by the environment can be considerable. Therefore, the traits described may be better classified as *multifactorial inheritance*. At the molecular level polygenic or multifactorial traits have been extensively studied but one limiting factor to success has been the difficulty in defining accurately the phenotype to be tested, e.g. where is the division drawn between normal and elevated blood pressures?

Multifactorial disorders comprise a very significant proportion of the genetic diseases and are considered to result from an interaction between genes and the environment (Table 3.1). Study of the single-gene disorders, as illustrated above, has provided significant insight into their pathogenesis. The multifactorial disorders are now starting to be investigated. They are considerably more complex but will have far wider implications because of their association with *common* diseases. Some examples of likely multifactorial disorders include:

- Coronary artery disease
- Hypertension
- Psychiatric illness
- Dementia
- Diabetes (insulin-dependent)
- Cancer
- Mental retardation
- Congenital malformations, e.g. cleft lip/palate, congenital dislocation of the hip, pyloric stenosis.

Schizophrenia

Clinical

Schizophrenia is a relatively common and debilitating psychiatric disorder. The lifetime expectancy rate (chance of manifesting symptoms sometime during life) is ~1%. Schizophrenia can develop during adolescence or early adult life and can be difficult to diagnose because there are no universally accepted criteria. Overlap or schizophrenic-like illnesses occur and the relationship between these disorders provides a source of further confusion. Clinical features include hallucinations, delusions, cognitive impairment, emotional lability, disordered thought processes and social deterioration.

Genetics

Despite controversy over the years, few would now doubt that there is a genetic component in schizophrenia. This is based on family, twin and adoption studies (Box 3.9). Risks in first- and second-degree relatives of individuals affected with schizophrenia are higher than the general population.

Box 3.9 The utility of twin studies in assessing multifactorial traits

Monozygotic (MZ) or identical twins develop following division of a single fertilised ovum and so share 100% of their genes. In contrast, dizygotic (DZ) twins result from the fertilisation of two ova by different sperms. Thus, on average, DZ twins share half of their nuclear genes. Generally, the environment shared by DZ and MZ twins is similar. These associations in twins have made them an increasingly popular target in which molecular techniques can be used to assess the relative contributions of genes and environment to a disease. This approach in schizophrenia research has been productive. Concordance (both of the twins are affected or unaffected) has shown that about 50% of MZ twins will both develop schizophrenia. In contrast, the same risk for DZ twins is ~15%. These results suggest that there is a genetic component in the pathogenesis of schizophrenia. To counter the criticism that MZ and DZ twins may not necessarily be exposed to a comparable environment, e.g. MZ twins share a single placenta in utero and are more likely to be 'closer' than their DZ counterparts, a study in schizophrenia has looked at a small number of MZ twins who were raised apart. Concordance for the development of schizophrenia was not reduced, again strengthening the genetic association (from McGuffin et al 1995).

The utility of twin studies in cancer, another group of multifactorial disorders, is illustrated by Hodgkin disease, a haematological malignancy. Familial clustering and an increased risk ($\times 3$–$\times 7$) for siblings of affected individuals has suggested a genetic, environmental or multifactorial aetiology. A twin study has shown that none of 187 DZ twins became concordant for Hodgkin disease compared to 10 of 179 MZ pairs. Affected individuals developed the malignancy in their late 20s and had the same histological type. The conclusion was that a form of Hodgkin disease affecting young adults has a genetic component in its aetiology (from Diehl and Tesch 1995).

Concordance rate in monozygotic twins is ~50% compared to ~15% in dizygotic twin pairs. There is a higher frequency of schizophrenia in the separated blood relatives of affected individuals compared to control adoptees whose parents have no known psychiatric history.

The mode of inheritance is unknown. Autosomal dominant with reduced penetrance, autosomal recessive, multifactorial, unstable DNA triplet repeats (see p. 58) and a number of other combinations have all been proposed. In this discussion schizophrenia is considered a multifactorial disorder as the inheritance pattern is complex and could involve a number of genes and/or interactions between genes and the environment. Since clinical and biochemical studies had failed to explain the genetics or pathogenesis of schizophrenia, the next obvious step was to try positional cloning.

Positional cloning

A clue where to start looking in the genome was provided when it was observed in one family that two schizophrenic males were partially trisomic for chromosome 5q11.2-q13.3. Therefore, DNA polymorphic markers for this region were used in linkage analysis. In 1988, *lod* scores between 3–6 were obtained in two British and five Icelandic families. The lod score is a statistical measure of an association or linkage. It represents the $\log_{10}$ of the odds favouring linkage, i.e. a lod score of $+3$ means the odds in favour of linkage are 1000:1. A lod score of -2 indicates a 100:1 or greater odds against linkage. Thus, DNA studies were consistent with the cytogenetic observation and pointed to a schizophrenia gene in association with chromosome 5q11–q13. Subsequently, multiple pedigrees have been investigated and these have failed to confirm the chromosome 5 findings. One initial explanation was that schizophrenia is caused by a number of mutations which involve other loci. However, reassessment of the chromosome 5 linkage studies with more families and additional DNA probes suggests that the positive lod score may have been a chance finding and the association between schizophrenia and chromosome 5 is probably not significant.

Other loci have been implicated in schizophrenia. They are chromosome 15, through an association with Marfan disease, the dopamine D2 receptor on chromosome 11q, chromosomes 6p, 8p, 9, 20 and 22 as well as the pseudoautosomal telomeric portion of the X chromosome. To date no single locus has been consistently highlighted, although recent data would suggest that chromosome 6p merits more intensive investigation. *Exclusion maps* of the human genome are also being constructed, i.e. linkage analysis will enable certain loci to be excluded in relation to schizophrenia. This is a laborious strategy and less likely to be successful, particularly if heterogeneity is a significant factor in schizophrenia.

An alternative approach to identifying the location of genetic factors in complex disorders, including schizophrenia, involves *sib-pair studies*. This requires the identification of shared alleles among pairs of affected siblings and then comparing these results with a representative population. A departure from the expected distribution of alleles might be indicative of an association between the disease and the locus being tested. This approach does not require a mode of inheritance to be assumed in calculating the association. The disadvantage is that a large number of sib-pairs need to be studied but this is less of a problem if multicentre trials are conducted.

Molecular genetics

Schizophrenia illustrates the difficulties which are inherent in recombinant DNA techniques when these are used to study complex disorders without an apparent genetic mode of inheritance. It is possible that there are multiple genetic loci involved in schizophrenia and chromosomal localisations described above are simply one of many options. Alternatively, the equivocal data may reflect the use of linkage analysis programs which are more appropriate for single-gene disorders that follow traditional Mendelian-type inheritance. More sophisticated computer programs are presently being developed to deal specifically with the multifactorial disorders.

Some reasons for discrepant results when applying linkage analysis studies to multifactorial disorders include:

- Genetic heterogeneity
- Variable penetrance
- Variable expressivity
- Late age of onset
- Existence of non-genetic cases
- Assortive mating (more than one disease gene in a family).

The example of schizophrenia also illustrates the importance of accurately defining phenotypes in linkage studies. Clinical criteria have been proposed to assist diagnosis of schizophrenia. However, the criteria are broad and there is overlap with other psychiatric disorders. The confounding effects of drug or alcohol abuse and the potential for some neurological disorders to produce schizophrenic-like features must also be considered when criteria for inclusion and exclusion are being determined in linkage studies.

Despite the shortcomings described above it is still appropriate to utilise molecular strategies since there is a reasonable chance that success will follow. In this respect, the experience from sporadic cancers is encouraging since the same DNA changes initially observed in the less

---

Box 3.10 Hypertension as an example of a multifactorial disease

Approximately 20% of the population has essential hypertension. Long-term complications involve the brain (stroke), heart (ischaemic heart disease, cardiac failure) and the kidney (renal failure). Family, twin and adoption studies have provided evidence that hypertension is a multifactorial trait. Like schizophrenia, isolating the many genes that are likely to be involved in the regulation of blood pressure will not be easy. An advantage with hypertension is the availability of animal models. Constructing genetic maps in laboratory animals is relatively easy compared to humans because of the breeding options available. One example is the *spontaneously hypertensive rat* which was obtained by in-breeding animals shown to have elevated blood pressures. Genome analysis then identified a region on rat chromosome 10 near the angiotensin-converting enzyme as a potential locus in hypertension. If candidate genes are found in the rat it would be possible to look for the corresponding ones in the homologous (syntenic) region in humans. *Congenic* animal strains provide a greater level of sophistication since they are genetically identical except for a single chromosomal segment. In this way, the 'noise' from the total genome (and so the other putative genes which are involved in blood pressure) can be avoided and a more focused study of a chromosomal segment becomes possible. The most direct assessment of blood pressure will come from *gene targeting*. These animal models require the identification of potential candidate genes. Function can then be tested more specifically by making a series of transgenic animals with various perturbations in the relevant genes (see Ch. 10 for further discussion of gene targeting and transgenic models).

---

common familial cancer syndromes are now being seen in the genetically more complex sporadic cases (discussed further in Ch. 6).

The complexities to be expected in the multifactorial disorders are well illustrated by the example of type 1 diabetes (insulin-dependent diabetes mellitus). Since monozygotic twins of affected individuals have only a 36% risk of developing diabetes, the environmental component in this disorder is important. Nevertheless, a genetic predisposition is also present. An early locus identified to be a risk factor in this type of diabetes is the MHC on chromosome 6p21. Extensive genome mapping has now identified a total of 12 distinct chromosomal loci which are thought to be involved in type 1 diabetes. Another of the multifactorial disorders which is actively being pursued by molecular techniques is hypertension (Box 3.10).

NON-TRADITIONAL INHERITANCE

Unstable DNA repeats

Huntington disease

Huntington disease is a neurodegenerative disorder with autosomal dominant inheritance. It can present in various ways including a progressive movement disorder (typically chorea), psychological disturbance and dementia. Disease onset is usually in the mid-30s and there is complete penetrance by the age of 80. Because of the relatively late appearance of Huntington disease, reproductive and other life decisions have often been made before knowledge of genetic status is known. Offspring of affected individuals have a 50% risk of inheriting the disease. In the early 1980s, the only hope for an advance in our knowledge of Huntington disease rested with recombinant DNA techniques.

Positional cloning

An important consideration in deciding to utilise positional cloning was the availability of a number of very large Huntington disease pedigrees. Since cytogenetic data had not indicated the likely chromosomal location for this disorder, a trial and error approach was required to look for DNA polymorphisms linked to the clinical phenotype. Success with this approach would not have been possible without the large pedigrees to test by linkage analysis. In 1983, a DNA marker name G8 (locus D4S10) located on chromosome 4p16.3 was found to co-segregate with Huntington disease. From 1983, chromosomal walking strategies were used in many laboratories to find the Huntington disease gene. This was successful in 1993 when a gene called *IT15* (IT – interesting transcript) was isolated. The related protein was called 'huntingtin'. Although it took 10 years to find the right gene, the molecular approaches enabled DNA testing to be started soon after 1983.

Predictive testing – linkage analysis

By using DNA polymorphisms linked to the Huntington disease locus it became possible to undertake *predictive testing* within the confines of a family unit. Individuals with a family history of Huntington disease now had an opportunity to alter their a priori risks by DNA studies. Two important issues emerged from the Huntington disease predictive testing programmes started in many centres. First, key individuals could be lost through death (including suicide) and this prevented a number of families from having access to predictive testing. The concept of a *DNA bank*, which will be discussed below, assumed increasing importance in this circumstance.

A second consideration related to the comprehensive clinical, counselling and support facilities which were necessary in a predictive testing programme. These had major resource implications. It should also be noted that

in some instances DNA testing placed further stress on individuals and/or their families. The potential ethical/social issues resulting from DNA testing will be discussed further in Chapter 9. Prenatal detection also became possible and will be described in Chapter 4.

Predictive testing – direct mutation analysis

Once the Huntington disease gene was found, DNA testing took on a new direction since mutation detection, rather than a family linkage study, became possible. Predictive testing programmes were able to expand although still requiring the intensive counselling and support described earlier. DNA testing for the gene mutation became an option to assist physicians in the differential diagnosis of a neurological disorder, e.g. gait disturbances or dementia.

The underlying defect in Huntington disease proved to be another example of the novel mechanism which had been shown 2 years earlier to cause the fragile X syndrome (Box 3.11). This involved expansion of DNA triplet repeats. In Huntington disease the triplet was a $(CAG)_n$ with the normal value for n being 6–37 (Fig. 3.17). Expansions over 37 repeats were associated with the development of disease. Statistically it was also shown that the greater the number of repeats the earlier was the onset of the disorder. Instability with expansion of the repeats was more likely to occur when transmitted through sperm which explained why cases of juvenile Huntington disease invariably inherited the mutant gene from their fathers.

Today, there are an increasing number of both rare and relatively common neurological disorders which have an associated triplet repeat as the underlying aetiology (Table 3.8). It is expected that more will be described. Triplet repeat expansion as a mechanism for genetic disease was novel and explained a number of unusual observations such as the failure to identify a traditional inheritance pattern in some disorders and *anticipation*, i.e. the earlier onset and more severe phenotype as the mutant gene passed through succeeding generations (Fig. 3.18).

Although sizing of the triplet repeat made DNA testing more accurate and available to a larger group of at-risk individuals, some results remained equivocal. For example, an expanded repeat in the 40s had *nearly* a 100% probability of leading to Huntington disease and similarly a repeat value in the 20s or less indicated a normal allele (Fig. 3.19). However, repeat expansions in the range 30–37 are described as 'intermediate' and require more careful consideration. The significance of these remains topical although there is some evidence that they may represent premutations which could expand, particularly if passed through the male germline, in future generations. The concept of a premutation helped to explain some cases of 'sporadic' Huntington disease. Where asymptomatic parents could be tested, it was sometimes found that one of

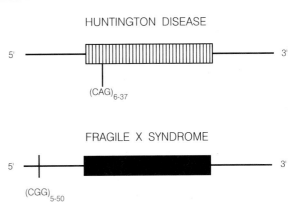

Fig. 3.17 DNA triplet repeats and disease pathogenesis.
For Huntington disease, the (CAG) triplet repeat is located within the gene's early coding region. In contrast, the fragile X syndrome (CGG) repeat is in the 5′ flanking region of the gene.

Box 3.11 Fragile X mental retardation – a disease of unstable DNA

The fragile X syndrome is the most common inherited form of mental retardation, affecting approximately 1 in 2000 children. The syndrome derives its name from the finding that there is a fragile site in the X chromosome at band Xq27.3 which can be observed if cells for cytogenetic analysis are cultured under special conditions. By positional cloning, the gene for this disorder, known as *FMR1*, was isolated in 1991. DNA at the 5′ end of the gene has two interesting features. (1) There is a CpG island which is methylated in patients with fragile X but hypomethylated in normals. This observation may have functional significance since it is frequently found that actively transcribing genes are hypomethylated whilst inactive genes are methylated (Box 3.12). (2) In this region is also found a $(CCG)_n$ trinucleotide repeat where the number (n) can vary. In normals, $n < 50$. The $(CCG)_n$ repeat plays a role in pathogenesis since amplification of the repeat ($n > 200$) is associated with the fragile X mental retardation phenotype. The $(CCG)_n$ sequences are usually stable and transmitted in the same way as is seen for other DNA polymorphisms. In families with fragile X, the number of repeats can increase, particularly if transmitted by females. Once the amplified segment reaches what appears to be a critical size, around 200 copies, the CpG island becomes methylated, the *FMR1* gene is silenced and no protein is produced. At high copy number, the $(CCG)_n$ also demonstrates *somatic instability* as evidenced by tissue mosaicism (cell lines of different genetic content in one individual). Therefore, affected members of the same family will inherit fragile X but can manifest different phenotypes because of the superimposed differences in copy number in their somatic cells. Although the caption suggests a straightforward causative effect between phenotype and triplet repeat expansion, the mechanism is likely to be more complex. The key change is inactivation of *FMR1*. In some rare situations, what would seem to be adequate expansion of the $(CCG)_n$ triplet repeat does not produce fragile X. Therefore, it is proposed that the expansion represents a continuum with some overlap between the critical '*n*' for normal and abnormal. Another mechanism by which *FMR1* can be silenced is via mutations or deletions in the gene itself.

Table 3.8 Unstable trinucleotide repeats and genetic disorders (from MIM, 1994).
Amplification of these repeats can produce a range of neurological disorders.

Disorder	MIM*#	Repeat	Normal, n	Mutation, n
Fragile X syndrome	309550	$(CGG)n$	5–50	50–90*, >90
Myotonic dystrophy	160900	$(CTG)n$	5–10	19–30*, >30
Huntington disease	143100	$(CAG)n$	6–37	37–121
Spinocerebellar ataxia 1	164400	$(CAG)n$	6–39	41–81
Dentatorubral-pallidoluysian atrophy	125370	$(CAG)n$	7–34	54–75
Kennedy spinal & bulbar muscular atrophy	313200	$(CAG)n$	12–33	40–62

*# MIM, *Mendelian Inheritance in Man* (or Online MIM) reference number.
* Premutation repeat sizes. The fragile X syndrome triple repeat is located 5′ to the gene and the myotonic dystrophy triplet at the 3′ end of the gene. The remainder are intragenic.

---

them, usually the father, had a triplet repeat size in the intermediate range, i.e. a premutation.

Future directions
At present, it is not known how the expansion of the triplet repeat causes gene dysfunction or how *IT15* works in relation to the pathogenesis of Huntington disease. All that can be said is that the expanded $(CAG)_n$ represents a long sequence of polyglutamines. This is considered to result in a 'gain of function' that eventually leads to damage and so

disease. Current therapy for Huntington disease is empirical, e.g. drugs for depression. Therefore, a research priority is the identification of gene function. Following from this will come more rational therapeutic measures ranging from drugs to gene therapy to treat or even prevent the development of this devastating genetic disorder. Animal models which have been genetically engineered to contain expanded repeats are being produced and will provide critical information (see Ch. 10).

Mitochondrial DNA

The nucleus is not the only organelle in eukaryote cells that contains DNA. Mitochondria have their own genetic material in the form of a 16.6 kb double-stranded circular

DNA molecule. Mitochondrial DNA is characterised by a high mutation rate (5–10 times that of nuclear DNA), few non-coding (intron) sequences, a slightly different genetic code and maternal inheritance. The last occurs since spermatozoa make a negligable contribution to the con-

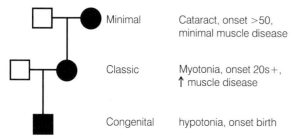

Fig. 3.18 **The phenomenon of anticipation.**
Myotonic dystrophy is an autosomal dominant, multisystem disorder which is the most common form of adult muscular dystrophy. A feature of this disease is its variable expressivity, including a very severe congenital form. Molecular characterisation has now explained the phenomenon of anticipation which is seen in myotonic dystrophy. The diagram illustrates the increasing severity and earlier onset of symptoms expected in anticipation. A corresponding expansion in the myotonic dystrophy $(CTG)_n$ triplet as it is passed through the female germline would parallel the clinical changes.

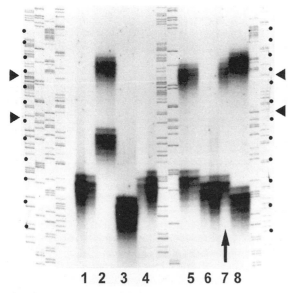

Fig. 3.19 **Measurement of the Huntington disease triplet repeat.**
DNA on either side of the Huntington disease-specific $(CAG)_n$ is amplified. ^{32}P is added to the PCR mix and so radioactivity becomes incorporated into the amplification products. The two alleles are separated on a gel by electrophoresis. The top triangle indicates where $n = 38$ and the lower triangle $n = 30$. The normal value for n would be <30, the intermediate range 30–37 and the abnormal (Huntington disease) range >37. These values can vary by 1–2 in different laboratories. All samples except for 7 are tested in duplicate. Size ladders are present on either side of the gel and in the centre. Samples 2, 5, 7 and 8 have one allele which is >37 and so these individuals are likely to develop Huntington disease. The remaining samples have alleles within the normal range.

ceptus in terms of mitochondrial DNA. Each mitochondrion contains 2–10 DNA molecules and in each cell there can be hundreds of mitochondria. This allows the situation known as *heteroplasmy* to develop (see the next section for further discussion).

Although most mitochondrial proteins are encoded by nuclear DNA, a few are encoded only by mitochondrial DNA. These include 13 proteins, two rRNA and 22 tRNA species, all of which are involved in the respiratory chain required for oxidative phosphorylation. This pathway allows mitochondria to play a vital role in the cell's energy requirements through the generation of ATP.

Mitochondrial DNA and genetic disease

It is only since 1988 that some genetic disorders, particularly those affecting organs with high energy requirements such as the brain, skeletal and heart muscles, have been proven to result from mutations in mitochondrial DNA. Prior to that it was suspected that mitochondria were involved on the basis of maternal inheritance, biochemical abnormalities and abnormal morphology on microscopy. However, definitive proof required DNA characterisation.

Features which suggest a mitochondrial DNA origin for an underlying disease are: (1) maternal inheritance, i.e. both males and females can be affected but the disorder is only transmitted by females, (2) the pathophysiology involves defects in mitochondrial oxidative phosphorylation, i.e. energy production, (3) there can be heterogeneity in affected individuals (this reflects *heteroplasmy* – the finding of a mixture of mutant and wild-type mitochondrial DNA species in the same cell), and (4) tissues will be affected differentially on the basis of their energy requirements. Thus, the central nervous system, skeletal and cardiac muscle fibres are at highest risk.

The types of defects in mitochondrial DNA are considerable, ranging from deletions and duplications to point mutations. It is interesting that the more severe mutations demonstrate heteroplasmy since they would otherwise be lethal. Because of their effect on reproductive fitness, these mutations are very heterogeneous suggesting independent origins. On the other hand, the milder point mutations can be found in all cells, i.e. *homoplasmy*. Examples of some genetic disorders which arise from mitochondrial DNA defects are given in Table 3.9.

It is likely that the list of mitochondrial DNA-associated defects will grow considerably as PCR allows the mitochondrial genome to be studied with greater ease. Previously the heterogeneity of the clinical phenotypes and the difficulty in using conventional biochemical approaches for study of mitochondria has meant that there has been little understanding of aetiology or pathogenesis. Even DNA technology before the availability of PCR was demanding since large quantities of a tissue rich in mitochondria, e.g, the placenta, were required to enable sufficient DNA to be isolated. This is not a limitation to PCR and strategies can be developed which allow the 16.6 kb genome to be

Table 3.9   Some examples of mitochondrial DNA-associated genetic diseases and their underlying mutations (see Johns 1995 for a more extensive list).
Since mitochondrial proteins can also be encoded by nuclear DNA it is possible to have what appears to be autosomal inheritance for a mitochondrial disorder. The phenotype is similar to that described below under Kearns–Sayre syndrome. Nuclear DNA changes have yet to be detected in these circumstances.

Disease	Clinical phenotype	DNA mutation(s)
MELAS syndrome	Myopathy, encephalomyopathy, *lactic acidosis* and stroke-like episodes	Point mutations affecting tRNA for leucine; heteroplasmy
Leber hereditary optic neuropathy	Late onset optic neuropathy giving rise to blindness in young adults (usually males)	Missense mutations involving different genes; homoplasmy
Kearns–Sayre syndrome (a type of chronic progressive external opthalmoplegia)	Progressive neuromuscular disorder; visual impairment, opthalmoplegia, retinal degeneration, ataxia, muscle weakness and deafness	Deletions, duplications and point mutations; heteroplasmy

amplified in segments using peripheral blood as a source of mitochondrial DNA. These are then screened for mutations by methods such as SSCP. Segments which are shown to contain differences in nucleotide bases are confirmed to be abnormal by DNA sequencing.

Mitochondrial DNA and ageing

The high mutation rate in mitochondrial DNA reflects a combination of inadequate DNA repair mechanisms, a high mutagenic environment secondary to the free radicals produced during respiration, absent (protecting) histone proteins, a rapid turnover rate and a paucity of non-coding segments so that chance mutations are likely to involve coding regions of DNA. The observations of a decline in mitochondrial respiratory activity with senescence and a concomitant accumulation of DNA mutations has led to the hypothesis that mitochondria play a role in the normal ageing process, particularly in cells such as neurons which have a limited capacity for cell division. Unlike the genetic effects described earlier, these changes are *somatic* in origin and not passed on to the next generation. The relationship between mitochondrial DNA and ageing is presently the focus of much research. Neurological disorders which are more common in the aged, e.g. Parkinson disease and Alzheimer disease would be good models for further investigation. To date, deletions in mitochondrial DNA have been observed in the basal ganglia of patients with Parkinson disease and there is an excess deletion rate in the brain cortex of patients with Alzheimer disease compared to normal

controls. However, these changes are presently considered to be non-specific effects. The significance of mitochondrial DNA in ageing awaits further molecular characterisation. Unfortunately, there are no good animal models for the mitochondrial DNA disorders.

Mitochondrial DNA and population markers

The high mutation rate of mitochondrial DNA makes it more useful than nuclear DNA for evolutionary studies. In addition, the strictly maternal inheritance removes confounding effects such as recombination between the maternal or paternal alleles in these comparisons. Thus, extensive studies of the mitochondrial genome by population geneticists and molecular anthropologists have been reported. For example, a nine base pair deletion in one of the few non-coding mitochondrial DNA regions is a polymorphic marker for individuals of East Asian origin.

Apart from the obvious anthropological uses of mitochondrial DNA markers to trace the origin and dissemination of the human species, the ethnic background of populations may explain predisposition to certain diseases. For example, those of Asian origin have a high carrier rate for the hepatitis B virus. Whether this represents a genetic or environmental effect is not entirely clear. The genetic component may be important since Polynesians, a geographically distinct community, are derived from East Asians on the basis of their mitochondrial DNA markers and they also have a predisposition to becoming chronic carriers for hepatitis B.

Genomic imprinting

Contrary to Mendel's original theory that genes from either parent have equal effect, it is now clear that expression of some genes is dependent on their *parent of origin*. The mechanism involved is called *genomic* (or *genetic*) *imprinting*, which refers to the differential effects of maternally and paternally derived chromosomes or segments of chromosomes or genes. Genomic imprinting implies that during a critical time in development, some genetic information can be marked temporarily so that its two alleles undergo differential expression. The critical period is considered to be the time of germline formation. Imprinting can be erased or re-established in the germ cells of the next generation. The actual mechanism(s) involved in imprinting remain to be defined. It is not clear whether imprinting has a *positive* effect, thereby activating a gene which would normally be silent or if imprinting exerts a *negative* influence on a gene's function.

An imprinted locus will be inherited along Mendelian lines while the expression of that locus will be dependent on the parent of origin (Fig. 3.20). Imprinting occurs in certain parts of the genome and has been shown to play a fundamental role both in normal development and in some pathological states such as genetic disease and cancer. Evidence for imprinting in placental mammals comes from

1 Paternal allele inactive

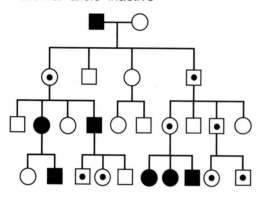

2 Maternal allele inactive

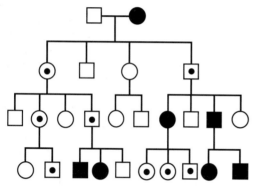

Fig. 3.20 Hypothetical pedigrees which illustrate imprinting (parent of origin effects).
An imprinted locus is inherited in a normal manner but the expression of the two alleles will depend upon the parent of origin. **(1)** The paternal allele is inactive. There will be no expression of the mutant allele when transmitted by the father. For the mutant gene to express its phenotype it must pass through the maternal line. **(2)** In this case it is the maternal allele which is inactive and the disease phenotype only becomes apparent after paternal transmission of the mutant allele. In both cases there are carriers (indicated with a dot in the circle or square) who have normal phenotypes but can transmit the trait depending on their sex. There are equal numbers of affected and unaffected males and females in each generation.

a number of observations which are summarised in Table 3.10.

Imprinting is more accurately detected and its implications have become better understood with the utilisation of molecular technology. This has enabled accurate assessment of the parental origin for chromosomal abnormalities such as deletions, aneuploidies or uniparental disomies (see the next section for further description of uniparental disomy). At the DNA level, the parental source for a deletion can be defined. Genetic disorders and cancers whose inheritance is unclear are now being reassessed to

Table 3.10 Evidence for genomic imprinting in placental mammals.
Chromosomes of maternal or paternal origin can demonstrate differential expression of certain genes as a result of imprinting, a process which probably occurs during germline formation.

Observation	Parent of origin effect
Pronuclear transplantation gives zygotes with both sets of haploid chromosomes either maternal or paternal in origin	*Androgenetic* (paternally derived nuclear material): embryos have relatively normal development of membranes and placentas but very poor development of embryonic structures *Gynogenetic* (maternally derived nuclear material): the opposite occurs in development
Human chromosomal triploids	*Android* (two paternal and one maternal haploid components): large cystic placentas *Gynoid* (two maternal and one paternal haploid components): small underdeveloped placentas
Uniparental chromosomal disomies	*Mice*: some loci on chromosomes 2, 6, 7, 11 and 17 display different phenotypes depending on whether there is maternal or paternal uniparental disomy *Human*: uniparental disomy chromosome 7 (cystic fibrosis), chromosome 15 (Prader–Willi syndrome or Angelman syndrome)
Chromosome deletions	*Genetic disease*: deletion chromosome 15q11-q13: Prader–Willi syndrome (paternal deletion) and Angelman syndrome (maternal deletion) *Cancer*: maternal deletions of chromosome 13q (sporadic osteosarcoma) or chromosome 11p (Wilms tumour)
Transgenic expression	There can be differences in the expression of a foreign gene by transgenic mice over a number of generations. Function or non-function of the transgene can be dependent on the sex of the transmitting parent. Methylation/hypomethylation of the transgene in association with the above has also been observed
Expression of specific genes	Parent of origin effects on phenotype, age of onset or severity, e.g. uncommon severe, rigid, juvenile form of Huntington disease (paternal transmission); severe, congenital form of myotonic dystrophy (maternal transmission)

look for *parent of origin effects*, which may help to explain their irregular inheritance patterns.

Table 3.11 Clinical, cytogenetic and DNA features of the Prader–Willi and Angelman syndromes.

Although the genes involved have been localised to the same chromosomal region, the phenotypes are completely different. Both disorders show imprinting – in the Prader–Willi syndrome a functional paternal allele is critical, in the Angelman syndrome it is the maternal allele which is important for normal development.

Parameter	Prader–Willi syndrome	Angelman syndrome
Clinical features	Obesity, waddling gait	Thin, ataxic gait with jerky involuntary movements, epilepsy with characteristic EEG
	Mental retardation (mild to moderate)	Mental retardation (severe), microcephaly
	Behavioural problems	Happy, sociable mood, paroxysms of laughter
	Characteristic facies (narrow bifrontal diameter, almond shaped eyes, triangular mouth)	Characteristic facies (prominent lower jaw with tongue protrusion)
	Small hands, feet and stature, hypogonadism	
	Floppy with feeding problems in the newborn period	Can be floppy at birth
	Hypopigmentation*	Hypopigmentation*
Cytogenetic findings	Deletions ~60%, normal ~33%, other anomalies ~7%	As for Prader–Willi syndrome
DNA findings	Paternal deletion 73%, uniparental disomy 25%, imprinting mutation 2%, other rare	Maternal deletion 73%, uniparental disomy 2%, imprinting mutation 5%, other 20%

*This feature is common to both syndromes but is not always present. It is now known to involve a gene for pigmentation located adjacent to but separate from the putative Prader–Willi and Angelman syndrome loci. Together the clinical features and genomic structure would be consistent with the concept of a contiguous gene syndrome (see p. 67).

Parent of origin effects

Genetic disease

Two rare syndromes with overlapping but different phenotypes have been localised by cytogenetic analysis to chromosome 15q11-q13. The two are the Prader–Willi syndrome and the Angelman syndrome. Characteristic features of these disorders are summarised in Table 3.11. The aetiology of both remained unknown until cytogenetic and then molecular analysis identified atypical modes of genetic inheritance. It is now considered that distinct but adjacent segments of chromosome 15q11-q13 are critical for normal development. Loss of the *paternal* segment of this chromosome leads to the Prader–Willi syndrome and loss of the *maternal* segment produces the Angelman syndrome, i.e. the two exhibit oppositely imprinted chromosomal segments.

What genes are involved and how they are imprinted remain to be fully determined. Molecular analysis has

clarified three ways in which an imprinted locus can be disrupted. The first is a *deletion* of DNA, e.g. the paternal segment in the case of the Prader–Willi syndrome. The second requires an additional copy of the non-critical parental allele, e.g. maternal *uniparental disomy* in the case of the Prader–Willi syndrome. A third involves an alteration in the methylation patterns for the maternal and paternal alleles which occur with a few genes in the Prader–Willi/Angelman syndrome loci. The underlying mechanism in the last example is considered to be a defect of the *imprinting process* itself.

Uniparental disomy

Uniparental disomy occurs when two copies of a chromosome are inherited from the *one* parent. There are two types of uniparental disomy: *isodisomy* – both chromosomes from the one parent are identical copies – and *heterodisomy* – the two chromosomes represent different copies of the same chromosome. Uniparental disomy has been reported in both the Prader–Willi and the Angelman syndromes (Fig. 3.21). Cytogenetic analysis will not detect uniparental disomy because the chromosomal numbers are preserved. It requires molecular analysis to show that the two chromosomes originated from the one parent.

There are two explanations for uniparental disomy. Either there is fertilisation between disomic (diploid content) and nullisomic (no chromosomal content) gametes or a trisomic conceptus (formed from a normal gamete and a disomic gamete) loses one of its chromosomes (Fig. 3.22). The latter would seem more likely since chromosome 15 is one of the more common trisomies associated with spontaneous miscarriages. If

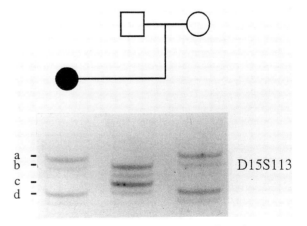

Fig. 3.21 Uniparental disomy in the Prader–Willi syndrome. Microsatellite (CA)$_n$ repeat analysis at locus D15S113. DNA polymorphic markers for the two paternal (b,c) and maternal (a,d) chromosomes 15 regions are distinguishable as two bands (with fainter stutter bands below). The affected child has inherited both her DNA markers (a,d) from the mother, i.e. she has maternal uniparental heterodisomy which is consistent with a diagnosis of Prader–Willi syndrome.

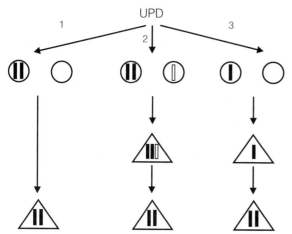

Fig. 3.22 Three ways in which uniparental disomy can occur.
Gametes are depicted as circles, zygotes as triangles. A chromosome is shown as a bar: it is present as one copy (monosomy – the normal situation), two copies (disomy) and no copies (nullisomy). **(1)** One gamete has two copies of a chromosome and the other no copies. This situation can arise following non-disjunction. Fertilisation between these two gametes would produce the normal diploid number but both chromosomes have come from the one parent, i.e. either iso- or heterodisomy. **(2)** Fertilisation in this case is between a disomic gamete and a normal monosomic one. The zygote is trisomic and is unlikely to survive unless one of the three chromosomes is lost. By chance (33% of the time) the one lost will be the normal gamete, i.e. the zygote is again diploid for that chromosome but both come from the same parent. **(3)** A third scenario involves fertilisation between a normal gamete and a nullisomic one. One way for the zygote to survive involves duplication of the single chromosome. In this case, uniparental isodisomy results. The mechanism depicted in **(2)** is considered the most likely since trisomy has been reported in chorionic villus samples but the newborn has the Prader–Willi syndrome secondary to uniparental disomy. The initial trisomic situation has been 'corrected' which allows the fetus to survive but disomy results.

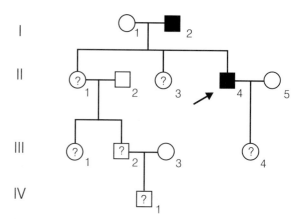

Fig. 3.23 A pedigree illustrating imprinting in cancer.
Four generations of a family are shown. Two males (I_2, II_4) presented with bilateral carotid body tumours when aged in their 30s. The consultand (→) has asked advice concerning risks in his daughter and other family members. Although the tumour is transmitted as an autosomal dominant trait, risk assessment is more complex since the tumour demonstrates imprinting with full expression only occurring when there is transmission through the *male* germline. The risk for the consultand's daughter is 50% *and* she will develop the tumour if she inherits the mutant allele from her *father*. The same situation holds for the consultand's two sisters. The a priori risk for III_1 and III_2 is 25%. However, neither individual will develop a tumour if he/she inherits a mutant gene from the *mother* but the offspring with the mutant gene will be a carrier. The a priori risk for IV_1 is 12.5% and tumour development would be expected since there has been transmission from a male.

required for normal development. It has been estimated that perhaps 1 in 500 cases of cystic fibrosis is due to uniparental disomy.

Cancer

There is evidence that imprinting is likely to play a role in the development or progression of human tumours. In these circumstances, there can be preferential loss of one parental allele, e.g. it is usually the paternal *RB1* (retinoblastoma) allele which becomes mutated in sporadic osterosarcoma.

Another example is glomus body tumour (also called hereditary paraganglioma) which is usually benign and involves the head and neck. Although the underlying gene has not been found, linkage analysis has identified a locus for this tumour on chromosome 11q23–qter. The parent of origin effect in glomus tumour relates to the clinical phenotype which only occurs if there has been paternal transmission, i.e. to function the gene must pass through the paternal line. Therefore, a mutation in the paternally derived gene produces a functional deficit and so tumour. On the other hand, transmission of the mutant gene through the maternal line does not lead to tumour formation because the gene is not normally active when it is transmitted by the mother. However, offspring will still be carriers and at 50% risk of having affected children if

trisomy were to follow from non-disjunction (uneven division of chromosomes during meiosis), a viable disomic conceptus is only possible if one of the three chromosomes is lost. If this were to occur, one-third of the concepti will have the two remaining chromosomes originating from the same parent. Thus, one mechanism for the Prader–Willi syndrome is maternal non-disjunction giving a trisomic conceptus which is 'rescued' when the paternal chromosome is lost.

Uniparental disomy is not unique to chromosome 15. Cystic fibrosis occurring in the case of a carrier mother but normal father has been explained by uniparental disomy. In this situation, non-paternity was excluded and it was shown that affected children had inherited two copies of the mutant chromosome from their carrier mothers, i.e. isodisomy had occurred. It is noteworthy that the cystic fibrosis phenotype was also associated with developmental abnormalities, e.g. moderate to severe intrauterine and postnatal growth retardation. Thus, it is possible that paternally derived gene(s) located on chromosome 7 are

they are males (Fig. 3.23). A further discussion of imprinting, with its potential effects in carcinogenesis, is given in Chapter 6.

Diagnostic implications

The ability to distinguish individual parental contributions by DNA testing is vital in studying imprinting. In a practical sense, diagnosis of the Prader–Willi syndrome or the Angelman syndrome is difficult until late childhood when the full clinical phenotypes become apparent. However, demonstration that there is abnormal paternal or maternal expression through a deletion, uniparental disomy or disruption of the normal methylation pattern enables an early diagnosis to be made.

Compared to cystic fibrosis and Huntington disease, positional cloning will be more difficult in these two syndromes because *genetic maps* cannot be constructed since the disorders are usually sporadic. Nevertheless, *physical maps* are being prepared for this region of chromosome 15. One development has been the isolation of the gene (called 'P') responsible for hypopigmentation in the two syndromes. It is interesting that the hypopigmentation gene is not imprinted and so this component of the phenotype will only become apparent if there is an underlying deletion.

Biology of imprinting

A number of fundamental questions about imprinting remain to be answered. For example, how and when does imprinting occur? The mouse model, including transgenic mice, has provided some information, e.g. a number of imprinted genes and chromosomal loci have been identified. Methylation has also been implicated as an important factor modifying the phenotype in the imprinting process (Box 3.12). Whether this is a cause or consequence of imprinting remains to be determined.

Answers on imprinting are being sought through molecular characterisation of loci shown to be imprinted in mice. Breeding experiments, which are easily undertaken in small animals, are impractical in humans. In this circumstance, the most useful strategy to follow involves an analysis of natural models, i.e. human genetic disorders in which there is a parent of origin effect on the individual's development or phenotype.

Mosaicism

Mosaicism refers to the presence in an individual or tissue of two or more cell lines which differ in genotype or chromosomal constitution but have been derived from a single zygote. Mosaicism is the result of a mutation which occurs during embryonic, fetal or extrauterine development. The time at which the defect arises will determine the number and types of cells (somatic and/or germ cells) which are affected. It is likely that mosaicism will be found

---

Box 3.12 Activity of genes and methylation

Over 60% of vertebrate DNA is methylated at the 5' position in cytosine where it is found in association with guanine in the genome (this is abbreviated to CpG). A small amount of DNA is hypomethylated and clustered into ~30 000 regions of DNA 1–2 kb in size and called CpG islands. CpG islands are frequently found at the 5' end of genes and so are useful to identify in a positional cloning strategy since they indicate the potential location of genes. The methylation status of DNA correlates with its functional activity, i.e. inactive genes are methylated and actively transcribing genes are hypomethylated. While there is an association between the methylation status and a gene's activity, the dilemma arises as to which came first, e.g. it is possible that methylation is an epiphenomenon which occurs once a gene has been inactivated by some other mechanism. One way in which methylation can modulate a gene's activity is through interference with protein–DNA binding which is essential for transcription to occur. Methylation plays a role in the maintenance of X-inactivation. Methylation may also be involved in imprinting. The methylation status of DNA is investigated by digesting it with a restriction enzyme (e.g. *Hpa*II) that recognises the 'CG' sequence (5'-CCGG-3') *but* only digests DNA which is not methylated. Therefore, two patterns are possible with *Hpa*II: (1) DNA is digested, i.e. it is not methylated, (2) DNA is not digested, i.e. it is methylated.

---

in all large multicellular organisms to some degree. Mosaicism is now able to be studied in a greater number of circumstances and in more depth because DNA techniques allow an accurate genotypic assessment of multiple tissues. In this way the identity of individual cells can be established.

Chromosomal mosaicism

Females are examples of chromosomal mosaicism since there will be random inactivation of one of the two X chromosomes in all tissues. As mentioned previously, this can confuse carrier testing for X-linked genetic disorders. It should also be noted that X-inactivation is not complete since there are regions on the X chromosome which are spared, e.g. the pseudoautosomal region on the end of the short arm. Thus, normal females do not have the phenotype of Turner syndrome in which one of the two X chromosomes is missing. Both Turner syndrome (45,X) and Down syndrome (trisomy for chromosome 21) have had chromosomal mosaicism demonstrated by cytogenetic analysis of cultured lymphocytes. The presence of a normal cell line in the mosaic Turner syndrome makes the clinical phenotype less severe.

There are a number of explanations for chromosomal mosaicism observed at prenatal diagnosis: (1) maternal contamination of sampled tissue, (2) a laboratory artefact,

(3) confined placental mosaicism, and (4) true fetal mosaicism. Chromosomal mosaicism usually results from non-disjunction occurring in an early embryonic mitotic division leading to the persistence of more than one cell line. With early fetal sampling made possible by chorionic villus biopsy, it has become apparent that chromosomal mosaicism affecting the placenta occurs more frequently than previously considered (~1–2% of samples). Chromosomal mosaicism confined to the placenta can produce false diagnostic results particularly in karyotypes obtained from chorionic villus sampling (see Ch. 4). Retarded intrauterine growth in fetuses with normal karyotypes may result from aneuploidy (the addition or subtraction of single chromosomes) confined to the placenta. Chromosomal mosaicism also explains why some aneuploid fetuses can survive to term since mosaicism allows a normal cell line to be present in the placenta.

Somatic cell mosaicism

Somatic cell mosaicism is proposed as a mechanism which can lead to phenotypic variation in single-gene disorders. Clues to the presence of mosaicism may come from the finding in sporadic genetic disorders of marked tissue dysplasia which is patchy in distribution. Alternatively, mild phenotypic manifestations in a person with an apparent spontaneous single-gene mutation or a mild phenotype in an individual whose offspring or parent is severely affected may represent examples of somatic cell mosaicism.

Somatic cell mosaicism may also be involved in neoplastic change. A number of tumour suppressor genes (for example, the retinoblastoma gene) are implicated in the development of malignancy once they are deleted or inactivated by a somatic event (see Ch. 6). It is hypothesised that focal areas of mutation (i.e. mosaicism) may lead to homozygosity for a mutant allele and so produce neoplastic change locally.

Gonadal (germ cell) mosaicism

From animal studies it has been estimated that the proportion of mosaicism in germ cells can vary from a few percent to 50%. Animal studies have also shown that germline cells separate from somatic cells at an early stage in development. Therefore, post-zygotic mutations which give rise to mosaicism can affect either somatic cells, germ cells or occasionally both.

Germ cell mosaicism is one explanation why parents, who are apparently normal on genetic testing, can have more than one affected offspring with an X-linked (e.g. Duchenne muscular dystrophy, haemophilia A or B) or dominant (osteogenesis imperfecta, tuberous sclerosis, achondroplasia, neurofibromatosis type 1) genetic

---

> **Box 3.13 Gonadal mosaicism**
>
> Neurofibromatosis type 1 (also called von Recklinghausen disease) is a common autosomal dominant disorder characterised by cutaneous neurofibromas and pigmented skin lesions known as café au lait spots. The gene for this disorder (NF1) was found in 1990. Some pertinent features include: (1) there is almost 100% penetrance, (2) expressivity, within the same family, can vary; and (3) ~50% of mutations represent new cases, most of which are in paternally inherited alleles. These observations can now be explained by gonadal and somatic cell mosaicism. In one case report, two affected children with phenotypically normal parents were shown to have a 12 kb deletion of the NF1 gene. DNA from the father's white blood cells, i.e. somatic cells, was normal but 10% of his sperm had the same deletion. Thus, it was proposed that a post-zygotic mitotic mutation had affected germ cells in the father. The father had no features of the disease which is consistent with normal DNA in his white blood cells. Therefore, the mutation had not occurred early enough to affect a more primitive cell which would give rise to both somatic and germ cells. The paternal sex bias in sporadically derived neurofibromatosis type 1 mutations may reflect the greater mitotic activity in sperm compared to ova (from Lazaro et al 1994).
>
> Another interesting case report has described a family with haemophilia B. The underlying DNA mutation was identified in the mother and grandmother of the affected male. Sequence analysis failed to show the mutation in the peripheral blood from the mother's four sisters and so it was presumed that they had not inherited the mutant gene. However, an intragenic DNA polymorphism clearly identified the mutant haplotype which was shared by the grandmother, the mother, her son *and* the four sisters. Therefore, it was proposed that germinal mosaicism had occurred in the grandmother so that all five of her daughters had inherited the mutant haplotype but only one the actual mutant gene. In this case DNA diagnosis would have been misleading since sequence analysis and intragenic DNA polymorphisms gave different results (from Sommer et al 1995).

---

disorder. Therefore, the suspicion of germ cell mosaicism means that recurrence of a genetic disorder needs to be considered when individuals are being counselled.

Gonadal mosaicism affecting sperm has been sought using PCR. Normal DNA patterns obtained from somatic cells, such as peripheral blood, are compared with sperm DNA patterns. The latter would show both normal and mutant DNA forms if there is germline mosaicism. From the frequency of the mutant form, a theoretical recurrence risk can be estimated (Box 3.13).

CONTIGUOUS GENE SYNDROMES

Clinical features

Another cluster of disorders which could be considered as 'polygenic' are the *contiguous gene syndromes*. At the molecular level, it is apparent that the contiguous gene syndromes represent disorders which arise from microdeletions (the deletions are large by DNA standards but small by cytogenetic analysis, hence the 'micro'). For this to occur the genes must be physically in close proximity (Fig. 3.24). Contiguous gene syndromes have some common features which are summarised in Table 3.12. The clinical phenotype produced in these syndromes will depend on the size and position of the underlying microdeletion.

Tuberous sclerosis

Tuberous sclerosis is a genetic disorder associated with unusual tumours in a variety of organs including the skin, eye, brain, heart and kidney. Additional features include mental retardation, intracranial calcification and seizures. The mode of inheritance is autosomal dominant with variable expressivity. Sporadic cases are also described.

There are two forms of tuberous sclerosis: *TSC1*, which is linked to chromosome 9, and *TSC2*, which is linked to chromosome 16.

Renal disease is an important cause of morbidity in tuberous sclerosis. In a small number of patients the renal disease takes on a form which is very similar to autosomal dominant polycystic kidney disease. Apart from the gene loci involved, there were no further advances at the DNA level in tuberous sclerosis until work to identify the gene for adult polycystic kidney disease progressed on chromosome 16.

Adult polycystic kidney disease

Adult polycystic kidney disease is characterised by bilateral enlargement of the kidneys as a result of multiple cysts. Cysts may be found in other locations (e.g. pancreas, liver, spleen). Hypertension is an important complicating factor and there is progressive renal failure. Approximately 10% of patients who are awaiting renal transplantation for end-stage renal failure have adult polycystic kidney disease. Affected individuals usually present for medical

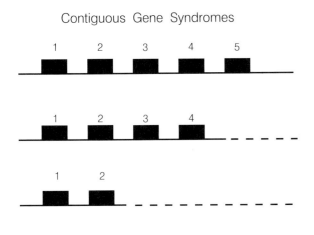

Fig. 3.24 A model for the contiguous gene syndromes.
The boxes represent five genes which are adjacent to each other but could have unrelated activities. Loss of one of the five genes through a deletion (-----) will produce a certain clinical phenotype. This phenotype will progressively become more florid as the number of deleted genes increases. An extensive deletion which removes all five genes will produce the complete syndrome. Other phenotypes are possible depending on which genes are deleted, e.g. 2, 3. The mental retardation found in the contiguous gene syndromes may not represent a direct effect of a specific gene deletion but can accompany any large DNA deletion.

Table 3.12 Characteristic features of the contiguous gene syndromes.
These involve different sets of genes producing a phenotype which may overlap but also show some distinctive features.

Parameter	Observations
Clinical features	Dysmorphic appearance and defective organ involvement Mental and growth retardation Heterogenous phenotypes since several functioning and unrelated genes are involved to variable degrees The severity of the disorder reflects the extent of the underlying mutation(s)
Genetic findings	Chromosomal deletions Normal chromosomal patterns shown on DNA testing to be microdeletions/microduplications or uniparental disomy if the involved region is imprinted Inheritance is usually sporadic although recurrences are possible if there is a mutation which can be transmitted from the parents and epigenetic factors, such as imprinting, can be satisfied. Affected individuals are unlikely to reproduce
Possible examples	Prader–Willi syndrome, Angelman syndrome, Beckwith–Wiedemann syndrome, retinoblastoma, Wilms tumour, Miller–Dieker syndrome, Langer–Giedion syndrome, DiGeorge syndrome

advice in their 40s or later decades of life. Presymptomatic detection of those at risk is possible by showing renal cysts with ultrasonography or nuclear imaging techniques. The accuracy of these studies is of limited value in childhood when renal cysts may not have developed fully. Moreover, simple cysts are found in ~10% of the normal population and can complicate the interpretation of imaging studies.

DNA detection of adult polycystic kidney disease became possible in 1985 when linkage between the disease phenotype and a DNA polymorphism located on the end of the short arm of chromosome 16 near the α globin gene locus was shown. This polymorphic marker was called 3'αHVR and it enabled presymptomatic detection of adult polycystic kidney disease (PKD1) with a recombination risk of 1–5% (Box 3.14). Subsequently, closer markers became available including polymorphisms located on either side of the putative gene locus. Flanking markers were useful in

these circumstances because they made it more likely to detect a recombination event (Fig. 3.25).

Nearly 3 years after the PKD1 locus was identified, it became apparent that a second locus for the disease must be present (PKD2). This conclusion was reached on the

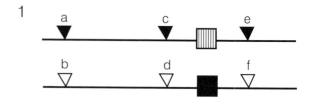

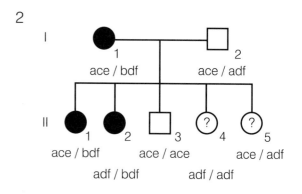

Box 3.14 Detection of adult polycystic kidney disease

The consultand (→, who is also the proband) has requested confirmation that his three children (aged 12, 14, 16 years) have adult polycystic kidney disease. The latter diagnoses were made a few years earlier on the basis of abdominal ultrasound examination. The consultand has the disease, one of his two sisters has undergone renal transplantation and the second sister has chronic renal failure and is awaiting transplantation. The pedigree illustrated below is an abbreviated version of a much larger family study which showed linkage of the adult polycystic kidney disease defect to the α globin gene complex on chromosome 16p13.3. DNA polymorphism results are listed in the pedigree and indicate that the disease locus co-segregates with DNA marker 2.2 in the consultand and his siblings. Therefore, one of the consultand's offspring (II1) has DNA evidence for the disease but the other two lack this marker and have a low probability for the disease. The major source of error would reflect the recombination risk for the DNA probe used. Review of the original ultrasound pictures confirmed that the reports for the last two children were probably incorrect.

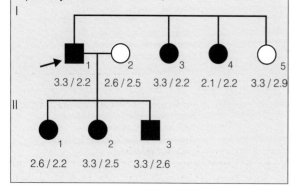

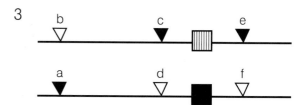

Fig. 3.25 Detecting recombination using flanking DNA markers in the adult polycystic kidney disease locus (*PKD1*).
It is assumed that in this family the *PKD1* defect involves the chromosome 16p locus. **(1)** The three polymorphic markers and their alleles for the *PKD1* locus are a or b, c or d, e or f. The hatched box is the normal gene, the solid box is the mutant. **(2)** The pedigree illustrates the segregation patterns for the above three polymorphisms. I₁ (female) has *PKD1*. Two of her children (II₁, II₂) are affected and so they allow the mutant-specific haplotype to be identified as bdf/ since this is what the three have in common. The one male offspring (II₃) has not inherited the maternal bdf/ haplotype which is consistent with his normal phenotype at age 50 years. The remaining two siblings are a problem. The adf/adf genotype in II₄ does not fit. Non-paternity can be excluded since it is the maternal haplotype which is a problem. This is an example of recombination which has occurred somewhere between the a/b and the c/d loci (shown in panel 3). The mutant-specific haplotype has now become adf/. Therefore, going back to the pedigree it can be seen that II₄ has inherited the *PKD1* mutation. Individual II₅ is even more difficult to assess since the ace/ haplotype may have come from either parent and so it is not possible to determine whether haplotype adf/ is paternal in origin or a maternal recombinant. The latter cannot be excluded although two recombination events would be statistically less likely.

basis of linkage studies which failed to show co-segregation with the chromosome 16-specific probes in a few families. The second locus was subsequently found to reside on chromosome 4. From 1985 a number of laboratories were actively involved in positional cloning strategies to find the *PKD1* gene with success occurring in 1994.

A contiguous gene syndrome

Nine years after the chromosome 16 location for *PKD1* had been described an important observation in a Portugese family enabled the relevant genes to be isolated. This family had both tuberose sclerosis and autosomal dominant polycystic kidney disease. The underlying defect was a balanced translocation between chromosomes 16 and 22. The translocation breakpoint on chromosomes 16 provided the clue for the likely location of the *PKD1* gene and so this region was cloned. A 14 kb transcript was found. Studies of a number of *PKD1* patients subsequently identified mutations affecting this gene and so proved that it was the cause of adult polycystic kidney disease. How dysfunction in the *PKD1* gene leads to cysts in the kidney and other organs remains to be determined.

Adjacent to the *PKD1* gene was another smaller gene (5.5 kb transcript) which was expressed in many tissues and had sequence homology to the GTPase-activating protein GAP3. This gene has now been confirmed as the one involved in tuberose sclerosis. It has features to suggest it might function as a tumour suppressor gene, e.g. close association with tumour development and loss of heterozygosity has been reported in tumour tissue. Microdeletions identified by pulsed field gel electrophoresis or fluoresence in situ hybridisation (FISH) explain how it is possible to disrupt both the *PKD1* and *TSC2* genes and so produce a complex phenotype. On the other hand, discrete deletions or rearrangements in either of the genes will give a more selective spectrum of clinical disorders.

As indicated in Table 3.12 there are a number of disorders which are proposed as examples of contiguous gene syndromes. In most cases the underlying genes have not been identified and so definitive proof is awaited. One such case is the DiGeorge syndrome which is the focus of much interest as candidate genes have recently been cloned (Box 3.15).

Box 3.15 The CATCH 22 syndrome

The acronym CATCH 22 stands for *c*ardiac, *a*bnormal facies, *th*ymic hypoplasia, *c*left palate, *h*ypocalcaemia and *22*nd chromosome. For some time the DiGeorge syndrome was considered an '*immunological disease*' since a predominant feature was the T cell deficiency and thymic hypoplasia. Abnormalities of the heart, hypocalcaemia and facial deformities were also noted as part of the syndrome. From the point of view of a *cardiologist*, heart diseases are important findings in congenital anomalies. A form of congenital heart disease involves conotruncal defects, e.g. tetralogy of Fallot, truncus arteriosus and aortic arch anomalies. These were found to occur more frequently than would be expected in the DiGeorge syndrome patients. The *clinical geneticist* interested in dysmorphology would have been faced with another overlapping disorder known as the Velocardiofacial syndrome (palatal abnormalities, learning disabilities, unusual facies and congenital heart disease). The three, apparently distinct groups, were brought together when chromosome 22 microdeletions were noted in each. It is presently proposed that the CATCH-22 constellation represents a contiguous gene syndrome involving chromosome 22. The minimal size of the DiGeorge syndrome region has been 'narrowed' to 1.5 Mb (a very large segment at the molecular level) and candidate genes are presently being isolated. It is interesting that a similar observation which was crucial to the identification of the *PKD1* gene may also be relevant to the DiGeorge syndrome, i.e. the breakpoint involving a balanced translocation between chromosomes 2 and 22 has been used to narrow the region where potential candidate genes may be found. Two conclusions will be possible once the region is fully characterised. Either the contiguous gene concept will be proven and a number of adjacent genes found or a single gene will be implicated in aetiology. In the latter, the gene will have to perform some basic function in terms of developmental biology because of the broad clinical spectrum found.

CHROMOSOMAL DISORDERS

Cytogenetics

The study of chromosomes is known as *cytogenetics*. This discipline has relied heavily on a number of standard techniques such as cell culture, chromosomal isolation, staining and recognition patterns to identify normal and abnormal karyotypes (karyotype – the number, size and morphology of chromosomes in a cell, individual or species). More recent developments have enabled automation of karyotype analysis.

Fluorescence in situ hybridisation

A very important advance in cytogenetics has been the development of FISH during the late 1980s and early

1990s. With FISH, cytogenetics has entered the molecular era since the recognition and characterisation of chromosomes can now utilise DNA probes (some technical details for FISH are given in Ch. 2). In future, the disciplines of cytogenetics and molecular genetics are likely to merge into a single entity.

Chromosomal abnormalities may be classified as:

- *Numerical*, e.g. aneuploidies (monosomy, trisomy), polyploidies (triploidy, tetraploidy)
- *Structural*, e.g. translocations, deletions, inversions, isochromosomes
- *Cell line mixtures*, e.g. mosaicism, chimaerism.

Although the great majority of these abnormalities are detectable by conventional cytogenetic approaches, some remain unresolved and these provide an important use of FISH. The advantage of FISH is that DNA probes can be used and these enable specific chromosomes or segments of chromosomes to be accurately identified.

It has already been mentioned that diagnosis of the contiguous gene syndromes has relied heavily on FISH because the deletions are too large for conventional DNA mapping and too small for traditional cytogenetics. The diagnostic options in this situation are either FISH or PFGE. The place of FISH in prenatal diagnosis is the subject of much debate although it is likely that as the technology improves and becomes less expensive, FISH will play an important role in screening for common chromosomal disorders, e.g. trisomy 21. In the area of research, FISH has enjoyed a wide range of applications, not the least of which has been its utility in allowing accurate localisation of DNA probes or segments which have been isolated during positional cloning.

Chromosome 21

An important genetic disorder on chromosome 21 is Down syndrome which occurs in approximately 1 in 600 to 1 in 700 births. The chromosomal abnormalities in this condition include: (1) free trisomy ~95%, (2) translocations ~3%, and (3) mosaicism ~2%. Most trisomy cases involve an additional maternal chromosome 21 which has arisen by non-disjunction (see also uniparental disomy, p. 63, for other examples of non-disjunction). In the trisomy cases, causation and recurrence risks relate to maternal age. Considerable work is presently being undertaken to screen for Down syndrome during pregnancy by use of biochemical markers in maternal blood.

Genome mapping

Because of its relatively small size and the association with Down syndrome, chromosome 21 was an early target for inclusion in the Human Genome Project (see Ch. 10). The long arm of this chromosome is ~40 Mb in size and the short arm 10–15 Mb. It is estimated that the chromosome could have up to 1000 genes.

Two important genes identified on chromosome 21 are (1) *SOD1* – superoxide dismutase 1 – which is involved in a small proportion of familial autosomal dominant amylotrophic lateral sclerosis (motor neuron disease); and (2) *APP* – amyloid precursor protein – which is associated with some rare types of autosomal dominant Alzheimer disease cases. The gene(s) for Down syndrome has yet to be isolated although the critical area has been narrowed to <5 Mb. As well as explaining the mental retardation and the characteristic phenotype of Down syndrome, molecular characterisation is expected to provide additional information on the mechanism for non-disjunction, an important cause of autosomal trisomies.

DNA BANK

Three aspects of recombinant DNA technology have led to the concept of a *DNA bank* being developed. These are: (1) the rapid advances in gene identification which have meant that what is a genetic defect of unknown aetiology today is very likely to have an associated DNA marker tomorrow, (2) the necessity in linkage analysis to have key family members available for testing, and (3) the relative ease with which DNA can be stored long term.

Purpose of a DNA bank

Genes or DNA polymorphisms associated with human genetic disorders are being identified at an exponentially increasing rate. The information generated from the Human Genome Project will accelerate this process. The multifactorial disorders are starting to provide DNA data so that the genetic component for common diseases will be identifiable in the near future. Interesting or esoteric genetic diseases can be the focus of a research study but as the molecular defect becomes characterised the research emphasis can shift to a service (diagnostic) component. Therefore, individuals or families in which there is a genetic component to disease (be it single gene or multifactorial) should be informed that future technological developments may allow further definitive study of the disorder even though nothing tangible can be offered at present. In these circumstances, it is essential that health care professionals are aware of the potentials of recombinant DNA technology so that counselling given to individuals or families is accurate and permits them access to subsequent developments. The end-result is DNA (in the form of a tissue or DNA itself or immortalised cell lines) stored in a professional DNA banking facility which will

Table 3.13 Some disorders for which DNA banking could be useful.
Storage of DNA in these circumstances might prove useful for other family members.

Cystic fibrosis

Huntington disease

Myotonic dystrophy

Neurofibromatosis type 1

Tuberous sclerosis

Osteogenesis imperfecta

Adult polycystic kidney disease

Familial adenomatous polyposis

Fragile X syndrome

Duchenne/Becker muscular dystrophy

X-linked immunodeficiency

Haemophilia A, B

Familial hypertrophic cardiomyopathy

Familial breast cancer

make it available if required at some future date. A list of some disorders for which DNA banking might be considered appropriate is found in Table 3.13.

A number of professional societies have proposed guidelines for a DNA bank. These cover: actual physical facilities; the relationship between depositors, their families and health professionals; confidentiality of information; safety precautions; and quality assurance measures. The word *depositor* rather than donor is used because the individual giving the sample maintains ownership and is not acting as a donor in the broadest sense. Depositors need to have clear statements on the length of banking, the potential problems and their rights in respect of the banked DNA.

The strategy of linkage analysis for identification of genes requires a family study in which one or more key individuals allow the phase of polymorphisms to be determined. The types of key individuals required in the various genetic disorders have been described above. Similarly, parents are important to help identify inheritance patterns for the DNA polymorphisms. The unavailability of family members through death, separation or loss of contact may not stop a linkage study proceeding provided there are appropriate family members available which will allow DNA markers for that parent to be predicted. However, the amount of work is now considerably increased and the more meioses which have occurred the greater the risk for recombination and non-paternity. Thus, it is essential to store blood or DNA from key family members if it is likely to be useful in the future or if there is the potential for that individual to become unavailable for whatever reason.

Long-term storage of DNA
DNA can be stored in the form of whole blood or as DNA for years at $-20^{\circ}C$ or $-70^{\circ}C$. Transformation of lymphocytes with the Epstein–Barr virus produces an immortalised cell line which can be cryopreserved over many years in liquid nitrogen. Aliquots can be thawed and propagated as required. The advantage of immortalising lymphocytes is the availability of an unlimited amount of DNA which is also suitable for techniques such as PFGE and PCR of ectopic mRNA. One disadvantage of immortalised lymphocyte cell lines is the technical demands required to prepare and then maintain the lines. The utility of PCR which can identify DNA mutations in the smallest amount of tissue has also meant that it may not be necessary in future to transform DNA. Material which is suitable for DNA banking includes blood, hair follicles, liver/spleen and other tissues from a deceased individual, abortus specimens, buccal cells from mouth washes and dried blood spots ('Guthrie spots'). The last are collected as part of a neonatal metabolic screening service (see Ch. 4).

Potential problems
A DNA bank is not simply a diagnostic laboratory which keeps a number of DNA specimens in the refrigerator for 'future purposes'. It is a planned activity with very strict operating guidelines. There are legal requirements, defined above, that the depositors will need to understand. What can be done with the DNA, particularly in terms of research, brings out both legal and ethical issues which need definition. For example, can the banked DNA be used for research and where does research end and clinical service (diagnosis) begin? Can the courts direct that a banked DNA specimen be tested for legal purposes? Confidentiality becomes an important consideration particularly when other family members may need access to information derived from the banked DNA. These issues will be discussed further in Chapter 9. Guidelines for DNA banking have been published by a number of professional societies. The American College of Medical Genetics has recently issued a policy statement on the storage and use of DNA involving a range of laboratory-based activities.

FURTHER READING

General
Baird P A, Anderson T W, Newcombe H B, Lowry R B 1988 Genetic disorders in children and young adults: a population study. American Journal of Human Genetics 42: 677–693

Deitz H C, Pyeritz R E 1994 Molecular genetic approaches to the study of human cardiovascular disease. Annual Review of Physiology 56: 763–796

McKusick V A, Amberger J S 1994 The morbid anatomy of the human genome: chromosomal location of mutations causing disease. Journal of Medical Genetics 31: 265–279

Mueller R F, Young I D 1995 Emery's elements of medical genetics, 9th edn. Churchill Livingstone, Edinburgh

Weatherall D J 1991 The new genetics and clinical practice, 3rd edn. Blackwell, Oxford

Thalassaemias – models for molecular genetics

Flint J, Harding R M, Boyce A J, Clegg J B 1993 The population genetics of the haemoglobinopathies. In: Higgs D R, Weatherall D J (eds) Bailliere's clinical haematology – The haemoglobinopathies. Bailliere Tindall, London, p 215–262

Higgs D R 1993 α-Thalassaemia. In: Higgs D R, Weatherall D J (eds) Bailliere's clinical haematology – The haemoglobinopathies. Bailliere Tindall, London, p 117–150

Thein S L 1993 β-Thalassaemia. In: Higgs D R, Weatherall D J (eds) Bailliere's clinical haematology – The haemoglobinopathies. Bailliere Tindall, London, p 151–175

Autosomal recessive disorders

Chillon M, Casals T, Mercier B et al 1995 Mutations in the cystic fibrosis gene in patients with congenital absence of the vas deferens. New England Journal of Medicine 332: 1475–1480

The Cystic Fibrosis Genetic Analysis Consortium 1994 Population variation of common cystic fibrosis mutations. Human Mutation 4: 167–177

Gan K-H, Veeze H J, van den Ouweland A M W et al 1995 A cystic fibrosis mutation associated with mild lung disease. New England Journal of Medicine 333: 95–99

Highsmith W E, Burch L H, Zhou Z et al 1994 A novel mutation in the cystic fibrosis gene in patients with pulmonary disease but normal sweat chloride concentrations. New England Journal of Medicine 331: 974–980

Sferra T J, Collins F S 1993 The molecular biology of cystic fibrosis. Annual Review of Medicine 44: 133–144

Tsui L-C 1992 The spectrum of cystic fibrosis mutations. Trends in Genetics 8: 392–398

Wicking C, Williamson B 1991 From linked marker to gene. Trends in Genetics 7: 288–293

Autosomal dominant disorders

Fananapazir L, Epstein N D 1994 Genotype–phenotype correlations in hypertrophic cardiomyopathy: insights provided by comparisons of kindreds with distinct and identical β-myosin heavy chain gene mutations. Circulation 89: 22–32

Fananapazir L, Epstein N D 1995 Prevalence of hypertrophic cardiomyopathy and limitations of screening methods. Circulation 92: 700–704

Nadal-Ginard B, Mahdavi V 1989 Molecular basis of cardiac performance: plasticity of the myocardium generated through protein isoform switches. Journal of Clinical Investigation 84: 1693–1700

Towbin J A 1995 New revelations about the Long-QT syndrome. New England Journal of Medicine 333: 384–385

Watkins H, Seidman J G, Seidman C 1995 Familial hypertrophic cardiomyopathy: a genetic model of cardiac hypertrophy. Human Molecular Genetics 4: 1721–1727

X-linked disorders

Antonarakis S E, Kazazian H H, Tuddenham E G D 1995

Molecular etiology of factor VIII deficiency in hemophilia A. Human Mutation 5: 1–22

Gardner R J, Bobrow M, Roberts R G 1995 The identification of point mutations in Duchenne muscular dystrophy patients by using reverse-transcriptase PCR and the protein truncation test. American Journal of Human Genetics 57: 311–320

Giannelli F, Green P M, Sommers S S et al 1994 Haemophilia B: database of point mutations and short additions and deletions, fifth edition. Nucleic Acids Research 22: 3534–3546

Graham J B, Kunkel G R, Egilmez N K, Wallmark A, Fowlkes D M, Lord S T 1991 The varying frequencies of five DNA polymorphisms of X-linked coagulant factor IX in eight ethnic groups. American Journal of Human Genetics 49: 537–544

Hoyer L W 1994 Hemophilia A New England Journal of Medicine 330: 38–47

Rossiter J P, Young M, Kimberland M L et al 1994 Factor VIII gene inversions causing severe hemophilia A originate almost exclusively in male germ cells. Human Molecular Genetics 3: 1035–1039

Multifactorial disorders

Cloninger C R, Adolfsson R, Svrakic N M 1996 Mapping genes for human personality. Nature Genetics 12:3–4

Cordell H J, Todd J A 1995 Multifactorial inheritance in type 1 diabetes. Trends in Genetics 11: 499–504 (the December 1995 volume of Trends in Genetics is a special issue on multifactorial inheritance)

Diehl V, Tesch H 1995 Hodgkin's disease – environmental or genetic. New England Journal of Medicine 332: 461–462

Harrap S B 1994 Hypertension: genes versus environment. Lancet 344: 169–171

Kurtz T W 1994 Genetic models of hypertension. Lancet 344: 167–168

Lander E S, Kruglyak L 1995 Genetic dissection of complex traits: guidelines for interpreting and reporting linkage results. Nature Genetics 11: 241–247

McGuffin P, Owen M J, Farmer A E 1995 Genetic basis of schizophrenia. Lancet 346: 678–682

Moises H W, Yang L, Kristbjarnarson H et al 1995 An international two-stage genome-wide search for schizophrenia susceptibility genes. Nature Genetics 11: 321–324

Thomson G 1994 Identifying complex disease genes: progress and paradigms. Nature Genetics 8: 108–110

Non-traditional inheritance

Bernards A, Gusella J A 1994 The importance of genetic mosaicism in human disease. New England Journal of Medicine 331: 1447–1449

Driscoll D J 1994 Genomic imprinting in humans. Molecular Genetic Medicine 4: 37–77

Guidelines for the molecular genetics predictive test in Huntington's disease 1994 Neurology 44: 1533–1536

Johns D R 1995 Mitochondrial DNA and disease. New England Journal of Medicine 333: 638–644

Lazaro C, Ravella A, Gaona A, Volpini V, Estivill X 1994 Neurofibromatosis type 1 due to germ-line mosaicism in a clinically normal father. New England Journal of Medicine 331: 1403–1407

Martienssen R A, Richards E J 1995 DNA methylation in eukaryotes. Current Opinion in Genetics and Development 5: 234–242

Nance M A 1996 Huntington disease – another chapter rewritten. American Journal of Human Genetics 59:1–6

Orr H T 1994 Unstable trinucleotide repeats and the diagnosis of neurodegenerative disease. Human Pathology 25: 598–601

Rainier S, Feinberg A P 1994 Genomic imprinting, DNA methylation and cancer. Journal of the National Cancer Institute 86: 753–759

Schapira A H V 1995 Nuclear and mitochondrial genetics in Parkinson's disease. Journal of Medical Genetics 32: 411–414

Sommer S S, Knoll A, Greenberg C R, Ketterling R P 1995 Germline mosaicism in a female who seemed to be a carrier by sequence analysis. Human Molecular Genetics 4: 2181–2182

Sutherland G R, Richards R I 1993 Dynamic mutations on the move. Journal of Medical Genetics 30: 978–981

Wallace D C 1995 Mitochondrial DNA variation in human evolution, degenerative disease and aging. American Journal of Human Genetics 57: 201–223

Contiguous gene syndromes

Brook-Carter P T, Peral B, Ward C J et al 1994 Deletion of the *TSC2* and *PKD1* genes associated with severe infantile polycystic kidney disease – a contiguous gene syndrome. Nature Genetics 8: 328–332

Hall J G 1993 CATCH 22. Journal of Medical Genetics 30: 801–802 (an editorial to the October 1993 volume of the Journal of Medical Genetics which deals with chromosome 22 deletions)

Ludecke H-J, Wagner M J, Nardmann J et al 1995 Molecular dissection of a contiguous gene syndrome: localization of the genes involved in the Langer-Giedion syndrome. Human Molecular Genetics 4: 31–36

Chromosomal disorders

Antonarakis S E 1993 Human chromosome 21: genome mapping and exploration, circa 1993. Trends in Genetics 9: 142–148

Buckle V J, Rack K A 1993 Fluorescent in situ hybridization. In: Davies K E (ed) Human genetic disease analysis: a practical approach, 2nd edn. Oxford University Press, Oxford, p 59–82

Le Beau M M 1993 Detecting genetic changes in human tumor cells: have scientists "gone fishing". Blood 81: 1979–1983

DNA bank

ACMG Statement – Statement on storage and use of genetic materials 1995 American Journal of Human Genetics 57: 1499–1500

DNA banking and DNA analysis: points to consider 1988 American Journal of Human Genetics 42: 781–783

Yates J R W, Malcolm S, Read A P 1989 Guidelines for DNA banking. Journal of Medical Genetics 26: 245–250

4

FETAL AND NEONATAL MEDICINE

INTRODUCTION

The 1980s were the beginning of a new era in reproductive technology. Developments included the availability of in vitro fertilisation, fetal blood and tissue sampling and ultrasound examination. By the 1990s, the fetus could be visualised with impressive clarity using modern ultrasound equipment. Biopsy of fetal tissue became a more acceptable procedure with chorionic villus sampling (CVS). Recombinant DNA techniques enabled fetal-specific DNA to be characterised. The last two developments made it possible to undertake many types of *prenatal diagnoses*. The long-term diagnostic potentials have become endless with the availability of PCR and, more recently, FISH. DNA technology has entered the traditional 'biochemical world' of *newborn screening*. The scope for DNA testing in the newborn has correspondingly increased. *Fetal therapy* has started. Blood transfusions and surgical corrections of some defects in the fetus are possible. Attempts at in utero gene therapy are conceivable options for the future. The *neonate* as a recipient of bone marrow has been explored using cord blood as a source of stem cells. Following on from the many changes described above has come the development of *fetal medicine units* which are primarily responsible for the care of the fetus in utero.

PRENATAL DIAGNOSIS

Sources of fetal tissues

Detection of genetic disorders by DNA testing in adults has been simplified since a blood sample serves as a universal source of DNA. In contrast, access to the fetus because of size and location is more difficult. Fetal tissues which can be used for prenatal diagnosis include fetal blood, amniocytes, chorionic villus and specific biopsy material, e.g. liver.

Fetal blood sampling

Fetal blood sampling is undertaken in the 2nd trimester of pregnancy since at this time access to fetal blood vessels becomes possible. In the 1970s and early 1980s, fetoscopy (fetal blood sampling from the chorionic plate or umbilical cord controlled by direct vision via a fetoscope) or ultrasound-guided puncture of the chorionic plate (placenta) were the usual means to obtain fetal blood. The latter was not always pure but mixed to a variable degree with maternal blood. Today, an experienced operator can consistently obtain pure fetal blood samples by ultrasound-guided umbilical vein puncture (called cordocentesis). This has been possible because of improved ultrasound imaging. Fetal blood can be tested in a number of ways including the isolation of DNA for prenatal diagnosis of genetic disorders, the obtaining of a karyotype or the detection of infectious agents.

Disadvantages of fetal blood sampling for prenatal diagnosis include: (1) technical skills and expensive ultrasound equipment required to obtain the sample, and (2) delay inherent in this technique. Thus, termination of pregnancy for a fetal abnormality is only possible during the 2nd trimester. The first prenatal diagnosis for thalassaemia on the basis of fetal blood sampling was reported in 1975. Fetal blood sampling is now only used if alternative diagnostic approaches such as amniocentesis or CVS (discussed below) are unable to be performed.

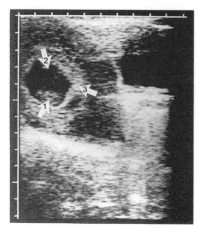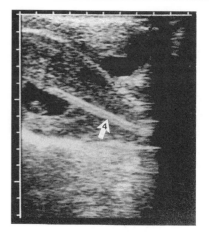

Fig. 4.1   Ultrasound pictures of a chorionic villus sampling.
The (→1) indicates the fetus in the amniotic sac (→2). An echo-dense region around the fetus (→3) defines the chorionic tissue. Under ultrasound guidance, a sampling probe (→4) has been placed into the chorion. The probe can be inserted through the mother's abdomen or per vaginam.

Amniocentesis

Amniocytes are shed fetal cells present in amniotic fluid. Amniocentesis is usually performed at the 15th week in pregnancy (i.e. the 2nd trimester) although there are presently trials under way to determine if earlier amniocentesis (~13 weeks) is possible. Amniocytes provide a source of fetal tissue for enzyme analysis (prenatal diagnosis of metabolic disorders), DNA characterisation (prenatal diagnosis of genetic disease) and a karyotype (chromosomal disorders in pregnancy). Amniocytes can be cultured if insufficient numbers are obtained from the amniocentesis sample. The use of PCR with amniocyte-derived DNA requires care since there will be the worry that contaminating maternal cells might be present in the amniotic fluid. Although safer and simpler to perform than either fetal blood sampling or CVS, amniocentesis has the major disadvantage of a late prenatal diagnosis during the 2nd trimester of pregnancy.

Chorionic villus sampling (CVS)

Chorion frondosum is tissue which surrounds the developing embryo. It is fetal in origin and will eventually become the placental site. It can be biopsied by the technique of CVS which was described in the late 1960s and became a routine procedure in China during the 1970s. CVS has now been used in many centres throughout the world and, despite some concern about possible side-effects including miscarriage and damage to the fetus, it has proven to be a reliable and safe procedure in the hands of the experienced operator.

A number of trials have looked at the safety issues and results suggest that women undergoing CVS have a slightly lower chance of a successful pregnancy outcome than those who have had 2nd trimester amniocentesis. However, actual figures for the miscarriage rate following CVS are difficult to obtain because of the spontaneous miscarriages which normally occur during the 1st trimester of pregnancy. Some cases of limb deformity following CVS have been reported. At this stage, this does not appear to be a significant problem although CVS at 8 weeks or earlier may be more risky and consequently is discouraged. The usual time for CVS is the 1st trimester of pregnancy at ~10 weeks (see Figs 4.1, 4.2).

A CVS is an excellent source of DNA. Provided the operator is experienced and the sample is carefully dissected under a microscope, there should be no contaminating maternal tissue. This is relevant since DNA amplification by PCR is frequently used (Fig. 4.3).

DNA testing in the fetus

Genetic disorders

The indications for prenatal diagnosis can vary in different communities. In general they exclude genetic defects which are not sufficiently life-threatening (e.g. adult polycystic kidney disease) or for which effective forms of treatment

Fig. 4.2   The appearance of chorionic villus tissue under a dissecting microscope.

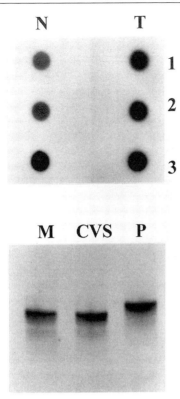

Fig. 4.3   Maternal contamination as a source of error in prenatal diagnosis.
Top gel: A 'dot blot' has been used to look for the −28 mutation which produces a form of Chinese-specific β thalassaemia (see Fig. 2.19 and Ch. 2 for further explanation of this technique). The parents (1, 2) hybridise to the normal (N) and thalassaemia (T) probes and so they are both carriers. The CVS (3) has the same pattern. In this circumstance, either the fetus is a carrier or the CVS taken was contaminated with normal DNA (which in most cases would have come from the mother). To distinguish the two possibilities, a polymorphic marker ('microsatellite') is used to re-test the three key samples **(lower gel).** In the example given, M – maternal DNA, P – paternal DNA. There is no paternal-specific band present in the CVS and so the heterozygous result is erroneous and is likely to represent maternal contamination.

are available (e.g. phenylketonuria). Nevertheless, circumstances arise when prenatal diagnosis (even if termination of pregnancy is not being considered as an option) is useful to allay anxiety in a couple. A list of some genetic disorders which can be detected prenatally is given in Table 4.1.

DNA strategies for genetic diagnosis have been described in Chapters 2 and 3. They fall into three groups:

• *DNA mapping* for deletional disorders (e.g. α thalassaemia) or DNA linkage using polymorphisms for disorders in which the mutant gene has not been found or the defect is non-deletional (Figs 4.4, 4.5). Another option for DNA testing is known as *prenatal exclusion.* In this situation, the parent at risk for a genetic disorder (the usual one being Huntington disease) does not want to know his/her carrier status but requests that the fetal risk is reduced (Box 4.1).

Box 4.1 Prenatal exclusion testing

A similar scenario as illustrated in Figure 4.5 occurs here but in this situation the consultand (→) has not developed features of the disorder and so remains at 50% risk. In this circumstance, her fetus is at 25% risk. Results of DNA polymorphisms remain the same although the interpretation changes. The paternal allele inherited by the consultand is 'a' but it is not possible to determine on the limited study undertaken whether the 'a' or 'b' paternal allele carries the mutant gene. Therefore, the genetic status of the consultand remains unknown. However, if her fetus has the 'cc' genotype, he/she is excluded from being at risk for the genetic disorder since the fetus has inherited the grandmaternal allele and nothing from the grandfather. On the other hand, if the fetus had the 'ac' pattern, he/she assumes the same risk as the mother, i.e. 50%. Prenatal exclusion testing is occasionally requested in the circumstance of Huntington disease when the at-risk parent does not wish to know his/her genetic status.

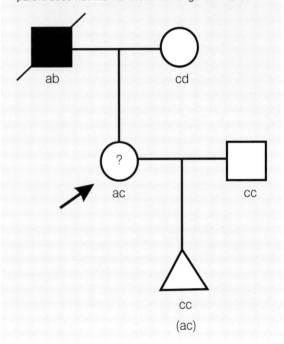

Table 4.1   Some genetic disorders for which prenatal diagnosis is available.

Thalassaemia α, β
Haemophilia A, B
Cystic fibrosis
Huntington disease
Fragile X mental retardation
Myotonic dystrophy
Duchenne muscular dystrophy
Tay–Sachs disease
α₁-antitrypsin deficiency
Dystrophic epidermolysis bullosa
Ornithine transcarbamylase deficiency
Retinoblastoma
Other less common disorders

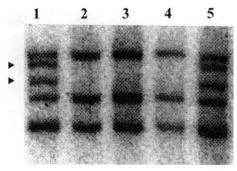

Fig. 4.4 Identifying the carrier state for a pregnant woman with a family history of haemophilia A.
An example of DNA mapping to detect a mutation is provided by a large inversion involving the factor VIII gene on the X chromosome (see Fig. 3.16 for a description of how the rearrangement occurs). In this particular case, a woman with a deceased haemophiliac brother requested assessment of her carrier state because she was pregnant. DNA linkage analysis was not an option because there were no males (haemophiliac or normal) available for testing. DNA testing for the factor VIII inversion is shown in the Southern blot. Lanes 1 and 5 contain DNA from the pregnant woman, the remaining lanes are normal controls. The ▶ identifies two abnormal bands which occur because of the inversion. Therefore, the woman is a carrier and would have the option of prenatal diagnosis with the same mutation sought in a male fetus. The band fragments look more 'grainy' than usual because the fragments were detected by a phosphorimager rather than the traditional, but slower, autoradiography.

- *DNA amplification* by PCR to detect small deletions (e.g. cystic fibrosis) or point mutations (e.g. β thalassaemia) (Figs 4.6, 4.7). PCR is also useful as a rapid means of detecting DNA polymorphisms and in many cases is replacing conventional DNA mapping for this purpose (Fig. 4.8).
- *FISH* to identify chromosomal abnormalities. FISH is a composite of DNA hybridisation and conventional cytogenetic techniques, i.e. cells are prepared along the lines required for cytogenetics using either interphase (non-dividing) nuclei or dividing cells in metaphase. DNA probes specific for various chromosomes are labelled with a fluorochrome and hybridised to the cells. The presence of an aneuploidy, e.g. trisomy 21, can be seen by an additional hybridisation signal (Fig. 4.9). FISH will increasingly become a useful addition to the options which are available for prenatal diagnosis. It may eventually replace conventional cytogenetic analysis which is time-consuming and labour-intensive. FISH can also be used to detect or confirm translocations and marker chromosomes.

Fetal sexing

Fetal sex can be determined from the DNA obtained by CVS at the 10th week in pregnancy. For this, it is possible to use Y-specific DNA probes for gene mapping or, more conveniently, Y-specific oligonucleotide primers for PCR. Sexing by DNA means has proven to be very useful in the X-linked genetic disorders such as haemophilia and

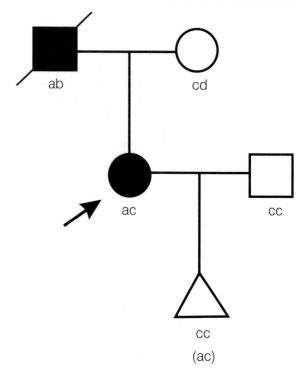

Fig. 4.5 A hypothetical prenatal diagnosis for a lethal genetic disorder using DNA polymorphisms.
DNA polymorphisms can be used to detect a disorder for which the mutant gene has not been found (or there are multiple non-deletional defects which cannot be detected individually). The consultand (→) (aged 20 and 8 weeks pregnant) has an autosomal dominant genetic disorder. With this disease she is unlikely to live beyond the third decade. She requests prenatal diagnosis since she will not proceed with her pregnancy unless it is possible to show that the fetus does not carry the mutant gene. DNA polymorphisms are given for the consultand (ac); her spouse (cc), affected father (ab) and normal mother (cd). Since the 'a' polymorphism in the consultand could have only been inherited from her affected father, that marker indirectly identifies the mutant gene. DNA from a chorionic villus biopsy has the polymorphisms 'cc'. Therefore, the fetus is normal. On the other hand, the finding of 'ac' in the fetus will identify the mutant gene. Potential errors in this type of indirect diagnosis include: (1) non-paternity, e.g. if the male indicated was not the biological father the polymorphic patterns could be incorrect and so lead to an error and (2) DNA recombination. This will be determined by the distance between the DNA polymorphism and the associated gene. An estimate of the error rate based on recombination can be made and discussed with the couple before prenatal diagnosis.

Duchenne muscular dystrophy since identification of the fetus as a female excludes a severely affected individual unless X-inactivation has not been random (see Fig. 4.10, Ch. 3).

Immune disorders

DNA testing can now be used to monitor a pregnancy at risk for immune haemolytic anaemia secondary to Rh (rhesus) incompatibility. Two genes for the Rhesus (Rh) antigens (D and C/E) have been isolated and characterised. PCR primers have been constructed from sequence infor-

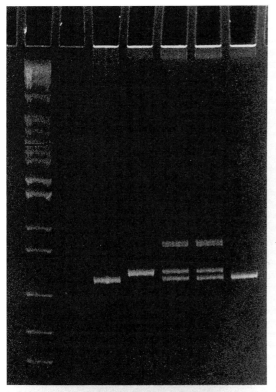

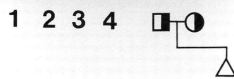

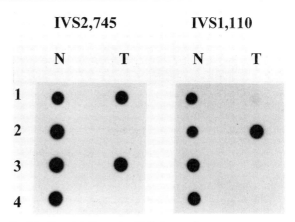

Fig. 4.7 Prenatal diagnosis of β thalassaemia.
How PCR can be used to identify point mutations in DNA. 'Dot blot' approach to mutation analysis shows DNA spotted onto a filter and hybridised to normal (N) and thalassaemic probes (T). One filter is testing for the IVS2,745 and the second the IVS1,110 β thalassaemia mutations. **IVS2,745 filter:** The mother (1) has the IVS2,745 mutation. Her spouse (2) does not have this defect. Twin #1 from the pregnancy (3) has the IVS2,745 mutation. Twin #2 (4) does not have the mutation. **IVS1,110 filter:** The IVS1,110 mutation is present in the father of the twins (2) and as expected his spouse (1) is normal. Both twin #1 and twin #2 (3, 4) do not have the IVS1,110 mutation. From this prenatal diagnosis it is evident that the twins are dizygous since twin #1 is heterozygous for the IVS2,745 mutation and twin #2 did not inherit either mutation from the parents. Because of the DNA patterns obtained, it would be necessary to exclude maternal contamination as a source of the DNA for the results obtained in twin #1 (see Fig. 4.3).

Fig. 4.6 Prenatal diagnosis of cystic fibrosis.
The use of PCR to detect small deletions. The figure shows an ethidium bromide-stained gel on which are amplified DNA products. The oligonucleotide primers were constructed so that amplification occurred at the locus affected by the ΔF508 mutation. Lane 1, DNA size markers; lane 2, no DNA control; lane 3, homozygote ΔF508 control; lane 4, normal control. The two parents heterozygous for the ΔF508 mutation have both the upper (normal) and the lower (ΔF508-specific) band. The difference between these two bands is three nucleotide bases. The CVS from the fetus (Δ) has only the lower band and so is homozygous for the ΔF508 mutation. DNA from the heterozygous parents shows a larger band (indicated by ◄). This is an artefact known as a heteroduplex which results from annealing of normal and mutant amplified products.

mation and it is now possible to study a CVS early in pregnancy to determine if the fetus from an at-risk pregnancy (i.e. the mother is RhD-negative and the father is RhD-positive) is RhD-positive or negative (Fig. 4.11). If the fetus is RhD-negative, the couple can be reassured that all is well and no further follow-up is necessary. An RhD-positive fetus makes it necessary to monitor the pregnancy carefully. Development of an anti-D antibody by the mother requires fetal assessment which is now best

undertaken with fetal blood sampling. If the fetal haemoglobin is low, intrauterine blood transfusions can be life-saving (see p. 86).

Congenital infections

Infections which can be transmitted from mother to fetus or neonate include bacteria, viruses and protozoans. Maternal infection is often mild or non-specific and can easily be missed. Effects on the fetus or neonate are variable, ranging from nothing to a fatal disorder. Thus, the screening of pregnant women for infections which can produce fetal abnormalities is a complex issue. A number of factors need to be considered: (1) the prevalence of the infection in each population, (2) the risk to the fetus or neonate, (3) the availability of treatment, (4) the sensivity and specificity of screening tests, (5) access to the screening tests, and (6) availability of counselling. Some infections which can be transmitted vertically from the mother to her fetus and cause damage to the latter are: rubella, syphilis, hepatitis B virus, cytomegalovirus, *Toxoplasma gondii*, human immunodeficiency virus (HIV), herpes simplex virus, group B streptococcus, *Chlamydia trachomatis* and *Neisseria gonorrhoeae*. In many communities the first three are routinely screened for during pregnancy. Whether it is appropriate to screen for others will depend on local requirements.

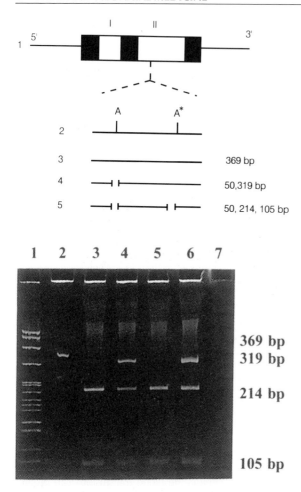

Fig. 4.8 β Globin (β thalassaemia) restriction fragment length polymorphism (RFLP) detectable by PCR.
Top: (1) A restriction fragment length polymorphism (RFLP) for the restriction enzyme *Ava*II is found in the second intron of the β globin gene (I, II identify the two introns of this gene). (2) 'A' marks two recognition sites for *Ava*II within this region of DNA. One site is constant and the second denoted by * is polymorphic. (3) A segment of DNA (369 bp in size) which includes the two *Ava*II sites is amplified by PCR. If this segment is digested with *Ava*II, a constant 50 bp band will always result. In addition, individuals can either have (4) the polymorphic recognition site missing (a fragment of 319 bp will result) or (5) it is present and so the 319 bp will be cut into two fragments of 214 bp and 105 bp. **Bottom:** DNA has been amplified in the region of the RFLP and digested with *Ava*II. Track 1 – size marker; track 2 – uncut DNA; tracks 3 and 5 – DNA from a homozygous-affected individual with β thalassaemia which shows only the 214 and 105 bp bands, i.e. the *Ava*II RFLP from both chromosomes is present; tracks 4 and 6 DNA from the individual's parents who are both carriers. They show three bands – 319 bp, 214 bp and 105 bp, i.e. in each parent the *Ava*II RFLP is present on one chromosome but not the other. Track 7 – no DNA control. The RFLP results in this family study show that the β thalassaemia mutation goes with the 214/105 bp markers. Therefore, for prenatal diagnosis, DNA from a CVS can be tested in this way to detect a fetus affected with homozygous β thalassaemia. This result can be obtained without the necessity to identify an individual mutation. However, recombination can be a problem but unlikely here because the RFLP is intragenic (i.e. within the gene). Non-paternity would need to be excluded as a source of error.

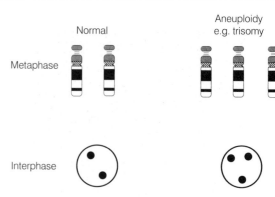

Fig 4.9 Fluorescence in situ hybridisation (FISH).
In conventional cytogenetic analysis, chromosomes must be examined during metaphase since it is at this time that they can be distinguished individually. However, chromosomes in most cells are in interphase and so they are condensed into focal regions. To get cells into metaphase usually requires culture. With FISH, either metaphase or interphase chromosomes can be studied since detection is based on DNA hybridisation rather than visual inspection of size, shape and staining characteristics of a chromosome. The example here shows chromosome 21 with the normal hybridisation pattern for metaphase (top) and interphase chromosomes (bottom), i.e. there are two chromosomes 21. In the case of trisomy 21 (Down syndrome) an additional hybridisation signal will be detectable in both metaphase and interphase spreads.

During pregnancy, acute infections in the mother which might involve any of the above pathogens require careful investigation. Conventional microbiological detection systems, described in Chapter 5, may or may not lead to a definitive diagnosis. The fetus introduces an additional complexity, i.e. infection in the mother, even if proven definitively, does not necessarily mean that the fetus will also be infected. Some examples which illustrate the potential use of recombinant DNA tests in dealing with infections acquired during pregnancy follow.

Rubella
Fetal abnormalities associated with rubella infection, particularly early in pregnancy, are numerous (Table 4.2). Although these complications have been reduced by

Table 4.2 Fetal complications associated with rubella infection in utero.

System	Complications
Cardiac	Patent ductus arteriosus, ventricular septal defect, pulmonary artery stenosis, aortic arch anomalies
Neurological	Mental retardation, seizures, deafness
Eye	Cataracts, chorioretinitis, micropthalmia, glaucoma
Immunological	Humoral and/or cellular immunodeficiency
Miscellaneous	Hepatosplenomegaly, intrauterine growth retardation, thrombocytopenic purpura, interstitial pneumonia, metaphyseal bone lesions

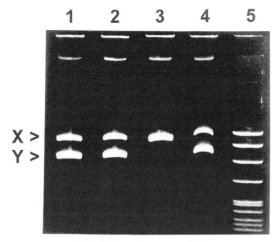

Fig. 4.10   Determining fetal sex by DNA analysis.
Amelogenin is a gene for tooth enamel. Amelogenin genes are
located on the short arms of both the X and Y chromosomes. PCR
enables a segment from each gene to be amplified separately. From
the X chromosome a 500 bp fragment is obtained; from the Y
chromosome a smaller 350 bp is seen. Using primers for the
amelogenin gene it is possible to determine the fetal sex. In this
example, lanes 1, 2 and 4 are from male fetuses and 3 is DNA from
a female. (Amelogenin primers were kindly provided by Dr David
Bailey, Cambridge University.)

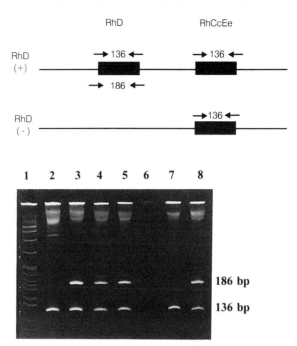

**Fig. 4.11   The Rhesus genes and use of PCR to determine blood
type.**
The genomic structure of the two Rh (D, C/E) genes is shown.
Amplification primers are selected to amplify a common region of
both genes (136 bp in size) and a region unique to the RhD gene
(186 bp in size) (from Bennett et al 1993). The gel shows DNA
amplified with these primers. Lane 1 has size markers; lanes 2 and 7
are from RhD(–/–) individuals and lanes 3–5 are from RhD(+)
individuals. This method does not distinguish RhD(+/–) from RhD
(+/+) but this is of no consequence in the prenatal detection
scenario since only the fetus with an RhD(–/–) genotype is no longer
at risk.

vaccination they still occur because of limitations in the
vaccination programmes and lack of compliance with
them. In these circumstances, pregnancy is frequently
terminated unless accurate diagnostic tests are available to
exclude fetal infection. Diagnosis of intrauterine infection
depends on culture of the virus or measurement of a
specific immunoglobulin M (IgM) response in fetal blood.
An important disadvantage of the latter test is the need to
wait until the 2nd trimester since the fetus does not fully
develop IgM responses until about 22 weeks of gestation.
Viral infections per se can inhibit the fetal immunological
response and so false-negative results are possible in these
circumstances. Failure to culture the virus does not exclude
infection.

Rubella-specific RNA sequences can be sought in the
fetus (through blood sampling or chorionic villus biopsy) by
Southern blotting or the more sensitive DNA amplification
by PCR. Trials are presently under way to assess the utility
of nucleic acid-based tests in rubella as well as a number
of other intrauterine infections such as toxoplasmosis,
cytomegalovirus and HIV.

Toxoplasmosis
Congenital toxoplasmosis results from transplacental infection with the protozoan *Toxoplasma gondii*. The affected
fetus can develop serious complications such as chorioret-
initis, hepatitis, hydrocephalus and pneumonitis. Mortality
is high and survivors may develop long-term neurological
or intellectual deficits. The fetus is mostly at risk from
complications if maternal infection occurs during early
pregnancy. Milder forms of toxoplasmosis in the fetus may

only be detected years after in the form of cerebral
calcification, neurological and intellectual impairment.

Treatment of maternal infection can reduce the risk to
the fetus. However, screening pregnant women for
toxoplasmosis is a debatable issue. There are few data to
indicate the percentage of women who are at risk, which
varies in each community. For example, the risk is low in
countries such as the USA, UK and Australia but higher in
France and Austria. Serological testing for toxoplasma-specific IgM in the mother is unreliable. The transmission rate
from mother to fetus is considered to be low so the finding
of a high (or rising) IgM titre in maternal serum does not
necessarily indicate infection in the fetus. To confirm the
latter requires culture of amniotic fluid or fetal blood in
fibroblast cell cultures. This may provide a diagnosis within
a week but only half the cases will be positive. Inoculation
of mice is more sensitive although it involves a longer
culture time (3–6 weeks). The necessary expertise for the
culture systems described would be available in few
laboratories. Testing for toxoplasma-specific IgM in fetal
blood has the same difficulties as described above for
rubella.

To overcome these problems DNA oligonucleotide primers directed at toxoplasma-specific DNA sequences have been designed. Results obtained to date are as sensitive as the mouse culture technique with the added advantage that the time required for diagnosis is reduced to 1 or 2 days. This would have practical implications if treatment of the affected fetus is being considered. Alternatively, termination of pregnancy would be an option if congenital infection could be confirmed.

A pregnant woman who is shown to have a recent toxoplasma infection is placed in a dilemma, particularly late in pregnancy when there is no guarantee that her fetus has been infected but there is little time to wait for culture results. In these circumstances, termination of pregnancy might be avoided if a rapid detection system, such as is possible with PCR, could be used to exclude infection in the fetus.

Cytomegalovirus
Cytomegalovirus is the most common cause of intrauterine infection. It can follow a primary infection or reactivation of latent infection in the mother. Similar dilemmas to those described above are found with congenital cytomegalovirus infection which can produce mental retardation and deafness. In some developed communities over 50% of pregnant women are at risk, i.e. they are seronegative. The number of at-risk women is lower in the less-developed countries. Fetal complications are more likely to result following primary infection in the mother.

Confirmation of active infection requires either detection of viral antigens or viral isolation by tissue culture. The latter is more accurate but time-consuming. Therefore, diagnosis of intrauterine infection is difficult. Effective treatment is not available. A further complexity arises since our knowledge of the effects of cytomegalovirus infection remains incomplete. For example, the majority of congenitally affected children show no long-term sequelae. This presents ethical problems in relation to termination of pregnancy even if cytomegalovirus infection can be detected during pregnancy. DNA technology for cytomegalovirus detection is still in its early days but promises an additional approach both in terms of diagnosis and research into the effects of the cytomegalovirus on the fetus.

Maternal screening

In many communities, pregnant women who are aged 35 or over have access to amniocentesis and cytogenetic analysis. This is directed predominantly at detecting Down syndrome although other cytogenetic abnormalities will also be found. Problems with this approach include the costs involved and the fact that most cases of Down syndrome occur in women who are younger than 35.

A more recent approach to screening for abnormalities in the fetus involves maternal serum testing. A number of biochemical markers are measured in the maternal blood, e.g. α-fetoprotein, human chorionic gonadotrophin and unconjugated oestriol. Results based on these studies identify pregnancies at risk for Down syndrome and neural tube defects such as spina bifida. Cytogenetic analysis or ultrasound will confirm the screening tests. The maternal screening methods mentioned are undertaken in the 2nd trimester. Biochemical markers which could be tested for in the 1st trimester of pregnancy are now being sought. DNA technology may also have a role to play if markers specific for Down syndrome or neural tube defect are identified. FISH could be used to look for common chromosomal abnormalities.

Future directions

Preimplantation genetic diagnosis
Two approaches to prenatal diagnosis have been described in the circumstances where there is a positive family history for a genetic disorder but termination of pregnancy is unacceptable or the couple have had difficulty conceiving and proceed to in vitro fertilisation. Prenatal diagnosis following in vitro fertilisation could be considered an unacceptable risk after the steps taken to conceive.

Preimplantation biopsy of the fertilised oocyte as a source of tissue to detect genetic defects is possible since blastomeres which make up the early pre-embryo remain undifferentiated. Thus, isolation of 1–2 cells from the 6–10 cell blastomere by microdissection will provide a source of DNA for PCR. If a genetic defect is excluded from the biopsied cells, the remaining blastomeres can be safely implanted into the mother since they will develop normally (Fig. 4.12).

Preimplantation diagnosis has been successfully used for sexing to identify male pre-embryos in situations where the mother is a carrier for an X-linked disorder. Preimplantation diagnoses in couples at risk of having a fetus with cystic fibrosis, Tay–Sachs disease, myotonic dystrophy, haemophilia or α_1-antitrypsin deficiency have been undertaken and the results confirmed to be correct following delivery.

Another strategy involves biopsy of the first polar body in the *preconception* oocyte. The first polar body following meiosis 1 divides away from the unfertilised oocyte but remains under the zona pellucida. Using microdissection techniques similar to those required for blastomere biopsy, it is *sometimes* possible to isolate the first polar body after removing the zona pellucida. Finding that the first polar body contains the mutant gene would indicate that the oocyte itself has the normal allele and vice versa. At the laboratory level, the preconception genetic diagnosis approach has been shown to be feasible in diseases such as cystic fibrosis and the haemoglobinopathies.

Further studies are required to determine the sensitiv-

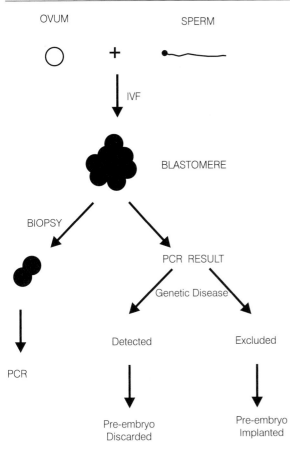

Fig. 4.12 Preimplantation diagnosis of genetic disease.
Preimplantation biopsy of an in vitro fertilised oocyte enables a few blastomeres from the pre-embryo to be removed before differentiation. Amplification of DNA from these blastomeres allows screening for possible genetic disease without the additional risks involved in prenatal screening methods or the necessity for a termination of pregnancy.

Table 4.3 Some comparisons between early, non-invasive (i.e. maternal blood sampling) approaches for prenatal diagnosis.

Target cell	Isolation procedures	Applicability
Syncytio-trophoblasts	Microscopy (physical appearance), MoAb	Not that useful, cell numbers can vary between pregnancies but overall remain small
Lymphocytes	MoAb or Y-gene markers or genes (e.g. HLA) which are paternal-specific	The concern that lymphocytes may persist in the maternal circulation, perhaps between pregnancies, has limited the use of lymphocytes
Nucleated red blood cells	MoAb or Y-gene markers or genes (e.g. HLA) which are paternal-specific	At present the best candidate, if it becomes possible to obtain sufficient numbers without maternal contamination

cules produced by a single cell compared to the amount of DNA present.

Maternal blood sampling

A non-invasive source of fetal tissue will become available if it is possible to identify and then isolate the occasional fetal cells present in the maternal circulation during pregnancy. These cells are syncytiotrophoblasts, lymphocytes and erythroblasts (nucleated red blood cells). Monoclonal antibodies have been described which are reported to be specific for fetal syncytiotrophoblast (the outer layer of the chorion frondosum). These cells can be isolated by flow cytometry or by using antibody-coated beads. The beads are mixed with 10–25 ml of maternal blood and are then separated from the blood on the basis of their centrifugation or magnetic properties. After washing, the mixture of beads plus fetal cells becomes the substrate for DNA analysis. Very few cells are isolated but PCR can be utilised to overcome the volume problem.

Lymphocytes are the second source of fetal cells. They are also few in number, e.g. one estimate would suggest that there is ~0.05 ml of fetal blood in the maternal circulation. However, on the basis of different maternal and paternal HLA antigen types, it would be possible to utilise antibodies to select fetal-specific lymphocytes, i.e. cells which carry a paternal HLA antigen type not found in the mother. These cells could then be tested as described above. A third and potentially very promising source of fetal tissue is the erythroblast. At present, a major problem with this cell is a lack of antibodies which can identify unique determinants, i.e. present on fetal but not maternal cells.

A variant of the maternal blood sampling approach relies on the detection of paternal-specific DNA sequences in the fetus. This avoids the requirement for an

ity and specificity of PCR in pre-embryo diagnosis before this option for prenatal diagnosis becomes more widely accepted. Particular worries include the potential for contaminating material to cause error which is increased since only few cells form the template for PCR. Other PCR-based artefacts which occur because of the very limited template include preferential amplification of one allele (either normal or mutant) and the inability to repeat the test for verification. One way to increase the yield of DNA is to biopsy the *trophectoderm*, a component of the ~200 cell blastocyst. This is now being investigated to determine if there are effects on the embryo's subsequent development and how representative are the cells obtained. A recent approach which is claimed will avoid differential amplification of alleles utilises mRNA as the template for amplification. This has proven successful in a pilot study of the fibrillin gene associated with Marfan syndrome. The rationale behind this strategy is that there are many mRNA mole-

antibody to isolate fetal cells. For example, Y-specific DNA sequences have been sought in maternal blood with PCR. Total maternal DNA can be tested without the necessity to fractionate fetal cells or DNA since the mother does not have a Y chromosome. A number of studies have reported promising results with accurate prediction of male fetuses whose sex was confirmed following delivery.

Many questions need to be resolved before maternal blood sampling becomes acceptable for prenatal diagnosis. For example, how specific are the antibodies for fetal cells? What is the likelihood of maternal contamination being present in the sample? This is relevant since there is little fetal tissue available and so PCR becomes mandatory. In this circumstance, amplification of maternal DNA could easily occur. PCR itself becomes technically more demanding since it is essential to include appropriate controls to ensure that fetal cells were in fact present and the DNA amplification had occurred appropriately (Table 4.3). Approaches which avoid the use of antibodies to select for fetal cells are limited by the necessity to identify a paternal-specific HLA or DNA component which is not present in the mother.

NEWBORN SCREENING

Current strategies

Newborn screening programmes are available in many communities. They are directed at diseases which satisfy the following criteria: (1) the disease occurs at reasonable frequency and is serious in nature, (2) effective treatment is available, (3) there is an advantage in early treatment rather than waiting until clinical features develop, (4) laboratory tests are available, e.g. there is a suitable screening test with appropriate *sensitivity* (the proportion of affected neonates which have an abnormal screening test) and *specificity* (the proportion of all normal neonates which have a normal screening test) and a confirmatory test can be undertaken if required, and (5) the programme has a clear cost-benefit.

The first example of this type of mass screening (see Box 9.2 for a summary of the various types of screening approaches) occurred in the 1950s and involved the testing of newborns' urine for phenylketonuria. In the 1960s, the impetus and scope for newborn screening increased with the availability of *blood spots*. These were obtained from heel pricks with the blood being spotted onto filter papers. Newborn blood spots were initially tested for phenylketonuria using the Guthrie bacterial inhibition assay. The blood spots, or 'Guthrie spots' as they are now frequently called, can be utilised for a number of diseases. Screening will depend on what at-risk conditions are present in particular communities. A common combination of tests involves the 'Guthrie spot' being screened for phenylketonuria, hypothyroidism, galactosaemia and cystic fibrosis (see Table 4.4 for a list of potential candidates for screening in the newborn).

Future directions

Genetic disorders

The cystic fibrosis newborn screening programme involving the assay of 'Guthrie spots' for immunoreactive trypsin has been successful if this is measured by the number of affected newborns which are detected. There are also some data to suggest that early detection of affected children will improve prognosis. The availability of newborn screening has reduced parental consternation since children with cystic fibrosis are likely to undergo repeated hospital and medical contacts before the diagnosis is established. However, a problem with the immunoreactive trypsin screening assay is its low *positive predictive value* (the proportion of neonates with abnormal screening tests which are affected). Therefore, a large number of false-positives will result. In these circumstances ~1% of infants need to be recalled, retested and if again positive the diagnosis is confirmed with a sweat test. This is a time-consuming process. Apart from economic considerations, recall will provoke anxiety in the families concerned.

Data are now emerging to suggest that the problems relating to false-positive results can be reduced by combining the immunoreactive trypsin with DNA testing for the ΔF508 mutation in the *original* blood spot. The finding of a homozygote for ΔF508 will confirm the diagnosis of cystic fibrosis without the necessity for further tests. A heterozygote for the ΔF508 mutation will either be a carrier of cystic fibrosis or have a second (unknown) mutation associated with the disease. The two will be differentiated by a sweat

Table 4.4 Metabolic, genetic and infectious disorders which could be screened for in the newborn (from Wilcken 1995).

Phenylketonuria*
Congenital hypothyroidism*
Galactosaemia
Cystic fibrosis
Maple syrup urine disease
Sickle-cell disease
Homocystinuria
Congenital adrenal hyperplasia
Biotinidase deficiency
Tyrosinaemia
Human immunodeficiency virus

* These are usually included in most screening programmes.

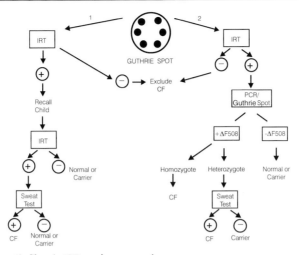

Fig. 4.13 Alternative protocol for cystic fibrosis (CF) newborn screening.
This method combines the immunoreactive trypsin (IRT) test and DNA amplification to detect the ΔF508 mutation. **(1)** the conventional screening approach, **(2)** the addition of DNA testing.

test. The exclusion of ΔF508 (which will be the usual outcome of DNA testing) will make it more likely that the immunoreactive trypsin result was a false-positive, particularly in those communities where the ΔF508 mutation accounts for over 70% of the cystic fibrosis defects. Using this strategy, some newborns with cystic fibrosis will be missed since both their cystic fibrosis mutations will be non-ΔF508 but these will be few. Apart from the inadvertant identification of newborns as cystic fibrosis carriers, the combination of the traditional biochemical screen and confirmation by DNA analysis is now considered to be a useful development in newborn screening. Studies are in place to compare the traditional approach with DNA testing (Fig. 4.13). The potential consequences of detecting asymptomatic carriers for cystic fibrosis is discussed further in Chapter 9.

The applications of PCR on newborn blood spots are in many ways unlimited. Therefore, decisions which involve the testing of DNA from blood spots need to be taken with care and foresight (see Ch. 9). Factors to be considered when determining the types of screening tests undertaken in newborns are: cost-effectiveness, community knowledge, the availability of counselling facilities and whether there is DNA sequence information to synthesise oligonucleotide probes for DNA amplification. Of the four, the last is the least limiting at present and will become even less so as more genes are sequenced (see the Human Genome Project discussion in Ch. 10). Therefore, conflicts are bound to arise. These reflect an increasing pool of DNA information but limited (or even diminishing) resources required to ensure this knowledge is utilised appropriately (see Chs 9, 10).

Infectious disorders

The value of PCR in the identification of intrauterine infections has already been discussed. Controversies about the significance of infections in different communities may be resolved with better epidemiological data which could be obtained from PCR-based newborn screening studies. Cytomegalovirus is an important congenital infection in humans. A rapid and inexpensive test for cytomegalovirus infection may prove useful in a newborn screening programme. DNA amplification has already been shown to detect accurately cytomegalovirus-specific sequences in urine from newborns. Infected newborns would benefit from early diagnosis which would enable special developmental follow-up and prompt diagnosis of hearing impairment. Prior to complications becoming established, therapeutic interventions might offer some assistance in the future.

Although a sensitive issue, DNA testing for HIV in newborn blood spots is no more technically difficult than what has been described for toxoplasmosis or cytomegalovirus. IgG to HIV can also be measured in newborn blood. In an attempt to gain information on the frequency of HIV-positive newborns or pregnant women, studies have been undertaken based on a blind approach to ensure anonymity. Newborn testing for HIV does not fulfil all the criteria mentioned earlier, e.g. no effective treatment is available. Nevertheless, it is likely that HIV testing will become more routine in the newborn particularly in those communities where AIDS is becoming a major public health issue.

THERAPY

Fetal therapy

In utero fetal therapy is available in a very limited number of situations. Haemolytic disease of the newborn, usually secondary to rhesus immunisation (Rh disease), can be treated by intrauterine blood transfusions via cordocentesis until the fetus is considered to have reached an age where the risk of further transfusion is greater than the complications associated with prematurity. The ability to obtain pure fetal blood samples by cordocentesis means that the clinical progress of an affected fetus can be monitored more accurately through serial haemoglobin estimations rather than the less precise bilirubin levels in amniotic fluid.

In utero surgery to correct abnormalities such as obstructive uropathy, abdominal wall defects, congenital hydrocephalus, neural tube defects and a number of other conditions is available in very few centres. Fetal surgery is still in its early days, so it remains an experimental form of treatment. One of the important problems to overcome before fetal surgery becomes a realistic option is the control of premature labour, a major limitation to this form of therapy. Nevertheless, a start has been made which enables the fetus in utero to undergo surgical manipulation (Box 4.2).

Box 4.2 In utero fetal surgery for spina bifida

Spina bifida (one form of which is known as myelomeningocoele) results from failure of the vertebral arches to fuse. An 'open' myelomeningocoele affects approximately 1 in 2000 births. The spinal cord is exposed and this leads to paraplegia, hydrocephalus, incontinence, sexual dysfunction and skeletal deformities. Further evidence for the suggestion that the long-term consequences of spina bifida result from exposure of the spinal cord to amniotic fluid rather than the primary defect has come from a sheep animal model in which spina bifida was created surgically at 75 days of gestation. The affected animals developed the expected complications. However, in utero surgery which covered the defect at 100 days gestation so that it was no longer exposed to the amniotic fluid enabled most neurological function to be preserved. The authors of this report correctly noted that the sheep model was not identical to the human disease which starts at a much earlier stage of development. Perhaps this will be resolved if the genetic components to this multifactorial disorder (see Ch. 3 for further discussion of multifactorial disorders) are able to be identified and a more suitable animal model is created (see Ch. 10). In the meantime, there is some evidence that early in utero intervention might be beneficial for what is a relatively common fetal abnormality (from Meuli et al 1995).

Treating the neonate

The placenta, which would normally be discarded following childbirth, is now proving to be an important source of haematopoietic stem cells. Cord blood is rich in CD34-positive cells (see Ch. 7) and so provides an alternative to bone marrow for transplantation. There is also some preliminary evidence that cord blood is: (1) less immunogenic, thereby reducing the risk of transplant rejection, and (2) immunologically less active and so the frequency of the second problem related to transplantation (graft versus host disease) is also reduced.

Cord blood cell transplanations in the neonate have proven to be effective in the treatment of some genetic disorders, e.g. β thalassaemia. A natural progression from this has been the early studies in which cord blood taken at birth from a newborn with the genetic disorder ADA (adenosine deaminase deficiency) has been transduced in vitro with a normal gene and the cord blood then returned to the newborn within a few days. Early results from these studies suggest that there is long-term expression from the transduced cells, i.e. DNA has been transferred into stem cells.

A HLA-identical match for bone marrow transplantation is difficult to find even with a number of potential siblings as donors. An alternative in this situation is cord blood which is easy to obtain and store. It also provides a broad ethnic representation which can be a problem with conventional sources for bone marrow or organ donation. The potential advantage of cord blood for transplantation has led to 'cord blood banks' being established. A disadvantage of cord blood as a source of marrow for transplantation is the small volume involved so that this form of treatment is predominantly used in children. However, the volume problem might be overcome by using recombinant DNA-derived growth factors to increase the number of stem cells circulating in the cord blood.

Future directions

In utero transplantation, particularly that involving bone marrow, would be useful to correct genetic defects by somatic cell gene therapy, i.e. transfer of DNA is made into non-germline cells (see Ch. 7). Prenatal correction might be required to minimise end-organ damage which would develop once the fetus was born. Fetal tissues for transplantation might also provide a better source of stem cells into which could be inserted normal genes. The underdeveloped immunological system in the fetus would be useful in situations where transplantation from a genetically dissimilar donor will induce both graft rejection and graft versus host disease. Implicit in the scenarios described above is an early detection system (i.e. DNA analysis) for the underlying genetic defect.

Germ cell therapy in the human is prohibited. Nevertheless, there is now a method available to biopsy blastomeres or the first polar body to determine genetic status in the pre-embryo or unfertilised oocyte. If the problems inherent in germline therapy can be overcome, it is possible in the future that genetic defects identified in the pre-embryo could be corrected prior to implantation in the mother (see Ch. 10).

FURTHER READING

General

D'Alton M E, DeCherney A H 1993 Prenatal diagnosis. New England Journal of Medicine 328: 114–120

Prenatal diagnosis

Adinolfi M 1992 Breaking the blood barrier. Nature Genetics 1: 316–318

Anderson J C 1995 Amniocentesis, chorionic villus sampling and fetal blood sampling. In: Trent R J (ed) Handbook of prenatal diagnosis. Cambridge University Press, Cambridge, p 58–79

Bennett P R, Le Van Kim C, Colin Y et al 1993 Prenatal determination of fetal RhD type by DNA amplification. New England Journal of Medicine 329: 607–610

Eldadah Z A, Grifo J A, Dietz H C 1995 Marfan syndrome as a paradigm for transcript-targeted preimplantation diagnosis of heterozygous mutations. Nature Medicine 1: 798–803

Geifman-Holtzman O, Holtzman E J, Vadnais T J, Phillips V E, Capeless E L, Bianchi D W 1995 Detection of fetal HLA-DQα sequences in maternal blood: a gender-independent technique of fetal cell identification. Prenatal Diagnosis 15: 261–268

Gilbert G L 1995 Diagnosis, prevention and management of infectious diseases in the fetus and neonate. In: Trent R J (ed) Handbook of prenatal diagnosis. Cambridge University Press, Cambridge, p 142–171

Hohlfeld P, Daffos F, Costa J-M, Thulliez P, Forestier F, Vidaud M 1994 Prenatal diagnosis of congenital toxoplasmosis with a polymerase-chain-reaction test on amniotic fluid. New England Journal of Medicine 331: 695–699

Suthers G, Haan E 1995 Maternal serum screening and prenatal diagnosis for birth defects. In: Trent R J (ed) Handbook of prenatal diagnosis. Cambridge University Press, Cambridge, p 4–27

Thomas M R, Tutschek B, Frost A et al 1995 The time of appearance and disappearance of fetal DNA from the maternal circulation. Prenatal Diagnosis 15: 641–646

Trent R J 1995 DNA technology and prenatal diagnosis. In: Trent R J (ed) Handbook of prenatal diagnosis. Cambridge University Press, Cambridge, p 110–141

Newborn screening

Belnaves M E, Bonacquisto L, Francis I, Glazner J, Forrest S 1995 The impact of newborn screening on cystic fibrosis testing in Victoria, Australia. Journal of Medical Genetics 32: 537–542

Wilcken B 1995 Screening of the neonate. In: Trent R J (ed) Handbook of prenatal diagnosis. Cambridge University Press, Cambridge, p 248–269

Therapy

Flake A W, Harrison M R 1995 Fetal surgery. Annual Review of Medicine 46: 67–78

Issaragrisil S, Visuthisakchai S, Suvatte V et al 1995 Transplantation of cord-blood stem cells into a patient with severe thalassemia. New England Journal of Medicine 332: 367–369

Kohn D B, Weinberg K I, Nolta J A et al 1995 Engraftment of gene-modified umbilical cord blood cells in neonates with adenosine deaminase deficiency. Nature Medicine 1: 1017–1023

Meuli M, Meuli-Simmen C, Hutchins G M et al 1995 In utero surgery rescues neurological function at birth in sheep with spina bifida. Nature Medicine 1: 342–347

5

MEDICAL MICROBIOLOGY

INTRODUCTION

Traditional or *phenotypic*-based methods for the detection of pathogens include staining and visual identification under the microscope, culture, growth and biochemical characteristics of an organism, and immunological responses, i.e. recognition of antigenic determinants related to an organism and the host's specific immune responses to it (production of antibodies). However, direct visualisation or culture is not always possible. These techniques can also be time-consuming and technically difficult. Phenotypic variation can occur during a pathogen's life-cycle, e.g. eggs, larvae and adult forms of a species may alter depending on the stage of development, the associated host/vector and whether the organism is free-living. Thus, antibodies (polyclonal or monoclonal) or isoenzyme techniques for detection may become limiting and dependent on certain stages in the life-cycle. Host immune responses can be delayed or conversely they remain persistent even after resolution of a previous infection. Cross-reacting antibodies acquired from natural infections or vaccination can produce false-positive results. Phenotypic methods can be used to discriminate between isolates, genera and species. However, the approaches are less effective when it comes to distinguishing differences within species.

In the long term, many of the conventional approaches described will be complemented or even replaced by *genotypic* analysis, i.e. detection of DNA/RNA-specific sequences through nucleic acid hybridisation techniques or DNA amplification by PCR (Box 5.1). An understanding of a pathogen's life-cycle and the host's responses to the infectious agent will be enhanced by characterisation of its genome using molecular technology. The spread of epidemics or hospital-acquired (nosocomial) infections will be followed and characterised with more accuracy by the identification of unique DNA fingerprints for individual pathogens.

Box 5.1   Diagnostic approaches in infectious diseases

Conventional
- Direct visualisation (light, electron microscopy)
- Culture (isolation, characteristics, biochemistry)
- Antibody/antigen reaction (enzyme-linked, fluorescence, agglutination)

Molecular (nucleic acid detection)
- DNA amplification (polymerase chain reaction – PCR)
- Hybridisation (Southern blots, dot blots, pulsed field gel electrophoresis)
- In situ hybridisation (DNA probes, PCR)

LABORATORY DETECTION

DNA probes

To obtain an appropriate DNA probe it is necessary to clone and sequence a microorganism's DNA to identify species-specific regions. From these, DNA probes can be isolated for nucleic acid hybridisation or primers can be synthesised for DNA amplification by PCR. A useful target for a probe is *repetitive* DNA, an example of which is ribosomal RNA (rRNA). rRNA as well as the rRNA genes in chromosomal DNA provide naturally derived amplified products which enhance their hybridisation potential. In these circumstances, non-P^{32}-labelled DNA probes become feasible since the signal-to-noise ratio is increased because of the amplified target sequence (Fig. 5.1 and Table 5.1). rRNA probes have another useful property, i.e. their ability to distinguish different levels of relatedness between organisms. In other words, whether organisms are bacteria or more closely related to another kingdom such as the fungi can be determined by their rRNA hybridisation patterns. A genus-specific probe against rRNA will hybridise to many species of the organism in the one reaction. On the other hand, a species-specific probe will identify only a single species. This method of differentiating between bacteria is called *ribotyping* and relies on the variable distribution of rRNA genes on bacterial chromosomes. Thus, hybridisation patterns will be different, which is comparable to the DNA polymorphisms found in eukaryotic DNA (see Ch. 2).

Clinical examples

Intestinal infections

Intestinal infections involve a range of pathogens and occur in an environment which may contain the same organisms as part of the normal flora. Thus, an important consideration in these organisms, for example, *E. coli*, is the differentiation of pathogenic from non-pathogenic strains.

They can also be fastidious growers, such as the *Campylobacter* species, which will give a delayed diagnosis if detection is sought through traditional culture means. In these circumstances, recombinant DNA (rDNA) methods will increasingly become more helpful.

DNA probes specific for toxins produced by enterotoxigenic *E. coli* or enteroinvasive organisms such as *E. coli* or *Shigella* species are available. *Staphylococcus aureus* is associated with a wide spectrum of clinically important infections (Box 5.2). *S. aureus*-related toxins can be identified and distinguished by DNA testing. *Clostridium difficile* is a normal component of the gut flora which will flourish if the growth of other gut bacteria is inhibited by antibiotics. Spread by the faecal-to-oral route can also occur. Since many strains do not produce toxins, the microbiology laboratory must first detect the microorganism and then follow this by identifying a toxin before the bacteria can be considered pathogenic. Traditional biochemical or immunological ways to look for toxins are time-consuming and costly. A PCR-based assay would provide a rapid way to identify the toxin.

Detection, as well as subclassification, of *Campylobacter* strains, and identification of virulent *Yersinia enterocolitica*, *Helicobacter pylori* and *Clostridium perfringens* may now be undertaken through rDNA approaches (Table 5.2).

Neurological infections

Bacterial meningitis is a treatable disorder but delays or ineffective treatment can lead to serious consequences. Infections with the bacterium *Neisseria meningitidis* (meningococcus) can develop rapidly into a septicaemia or meningitis leading to disseminated intravascular coagulation, shock and death. Rapid diagnosis and treatment are essential in this circumstance. Detection of the meningococcus by Gram stain of cerebrospinal fluid and/or culture is simple and able to be undertaken in most laboratories. However, culture of cerebrospinal fluid will not give a

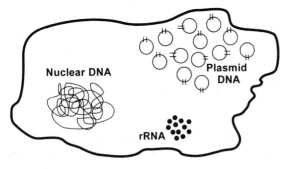

Fig. 5.1 Potential targets useful for nucleic acid (DNA/RNA) probes in a microorganism.
Nucleic acid (DNA or RNA) probes can be species- or genus-specific depending on the reason for their use.

Table 5.1 Comparison of nucleic acids as potential probes.

Useful properties of a DNA probe	Nuclear DNA	Ribosomal RNA and DNA	Plasmid DNA
Ubiquitous and stable	Yes	Yes	No
Species-specific ± repetitive regions identifiable	Yes	Yes	Identification of plasmid not necessarily species-specific
Abundance	May need amplification to detect	Abundant, ~10^5 copies/cell	Multiple copies may be present

Table 5.3 *Herpesviridae* family (from Mims et al 1993).

Virus	Abbreviations	Clinical features
Herpes simplex type 1	HSV1 (HHV1)	Cold sores, corneal infections, necrotising encephalitis
Herpes simplex type 2	HSV2 (HHV2)	Genital herpes, encephalitis, meningitis
Varicella zoster virus	VZV (HHV3)	Varicella (chicken pox), herpes zoster (shingles)
Epstein–Barr virus	EBV (HHV4)	Infectious mononucleosis, Burkitt lymphoma
Cytomegalovirus	CMV (HHV5)	Mononucleosis-like, hepatitis, pneumonitis, congenital infection
Human herpes virus-6	HHV6	Exanthem subitum, encephalitis
–	HHV7	Still in search of a disease
Kaposi sarcoma associated herpesvirus*	HHV8	Associated with Kaposi sarcoma although cause and effect not proven

*Detected by molecular means as still not cultured or shown conclusively by electronmicroscopy.

positive result for 24 hours. Additional delays in making a diagnosis can arise if antibiotics had been used earlier on in the illness or the meningococcus did not survive transportation of the specimen (e.g. cerebrospinal fluid) from the referring centre. Tests directed at identification of meningococcal-specific antigens are available but these have the disadvantages of low sensitivity and false-positives. DNA oligonucleotide primers specific for the meningococcus have now been described. They are likely to provide a rapid DNA amplification-based detection system for this microorganism.

Serotyping and serogrouping based on *protein antigens* are useful in epidemiological studies but do not differenti-

Table 5.2 Bacterial infections involving the gastrointestinal tract for which diagnosis by the DNA approach is useful.

Organism	Utility of DNA test
E. coli	Distinguishes the pathogenic strains by looking for virulence factors and invasiveness. Resistance to antibiotics sought by detecting the relevant plasmids
Shigella spp.	Invasiveness plasmid can be identified
C. difficile	Time-consuming isolation and toxin identification can be avoided
Campylobacter spp.	Fastidious growing; DNA probes useful for strain classification
Y. enterocolitica	Special culture conditions avoided; serological changes delayed
H. pylori	DNA tests compare favourably with urease tests
C. perfringens	DNA strain classification useful in food poisoning outbreaks

ate between outbreak strains. Meningococci from the nasopharynx are often difficult to type or group. Therefore, an alternative approach utilising *DNA typing* by Southern analysis to identify strain similarities becomes an option (see p. 99 for a further discussion on this use of rDNA technology in tracing the sources of infections).

Herpes simplex is one of a number of viruses belonging to the family *Herpesviridae*. Distinguishing features of this family include a double-stranded DNA structure and a tendency to chronic persistence often in latent form which can be reactivated (Table 5.3). Encephalitis following infection with herpes simplex virus type 1 is associated with significant mortality unless treatment is prompt. However, this type of viral encephalitis is difficult to diagnose clinically. Non-invasive investigations such as CT scans and EEGs are non-specific and may be normal in the acute phase of the infection. Antibodies to herpes simplex virus in peripheral blood or CSF are unreliable indicators of infection because of a delayed response and cross-reactivity with type 2 herpes simplex virus. Searching for herpes simplex virus-specific antigen in CSF may enable an earlier diagnosis to be made although this approach is relatively insensitive. To obtain a specimen for direct detection and culture of herpes simplex virus requires brain biopsy, a procedure which has morbidity of its own.

The difficulties described above have meant that empirical treatment with an antiviral agent such as acyclovir is usually started before a definitive diagnosis of herpes

Box 5.3 Detection of herpes simplex DNA is a simple and sensitive technique for the early diagnosis of herpes encephalitis

A male of 65 years presented to the hospital's emergency department with confusion. He had a temperature of 38.5°C but no focal neurological signs. Examination of his CSF showed 66 lymphocytes/μl. A CT scan was normal and herpes simplex DNA was detected in the CSF. He was commenced on the drug acyclovir. It was only on the sixth day of his illness that a magnetic resonance imaging scan showed the characteristic bilateral temporal lobe changes found in this infection. PCR products specific for the conserved glycoprotein D gene of herpes simplex virus types I and II were sought. The ability to detect viral-specific DNA within a few hours was critical for early treatment. Case courtesy of Dr N Kappagoda, Department of Microbiology, Royal Prince Alfred Hospital, Camperdown NSW.

simplex virus encephalitis is made. This diagnostic dilemma will be resolved with the increasing availability of DNA amplification tests for the herpes simplex virus (Box 5.3). rDNA-based tests for a range of neurological infections have now been described. A summary is given in Table 5.4. With time, DNA tests will be used more frequently and the range of organisms detectable will expand considerably.

Mycobacterial infections

Mycobacterium tuberculosis illustrates some of the difficulties which can be encountered in the diagnosis of bacterial infections. Although searching for mycobacteria by microscopy (on the basis of their acid-fast staining characteristics) is rapid, it does not allow *M. tuberculosis* to be distinguished from the atypical mycobacteria (e.g. *M. avium*, *M. intracellulare*). Large numbers of organisms ($>10^4$/ml) must also be present to be detected reliably. The specificity of the acid-fast bacilli approach is close to 100% in experienced hands but the sensitivity is variable with a wide range from 10–60% being described. Further discussion on the

Table 5.4 Neurological infections in which diagnosis by the DNA approach is or will be useful.

Infection	Microorganisms detectable
Meningitis	*N. meningitidis, Mycobacterium tuberculosis, Borrelia burgdorferi*, enteroviruses, herpes simplex virus 2, parvovirus B19
Encephalitis/ encephalopathy	Human immunodeficiency virus (HIV), herpes simplex virus, herpes zoster virus 1, *Toxoplasma gondii*, enteroviruses, JC virus, human herpes virus-6, *Plasmodium falciparum*
Slow-virus infection	Measles (subacute sclerosing panencephalitis), Creutzfeldt–Jakob (prion) disease*

*See Box 10.2 for further discussion of prion diseases.

implications of *specificity*, *sensitivity* and related measurement parameters in diagnostic assays is found in Box 5.4 and the section on human immunodeficiency virus (HIV) (p. 95).

Culturing of mycobacteria to determine antibiotic sensitivity or to distinguish *M. tuberculosis* from atypical forms is an additional problem since it will take a number of weeks to get sufficient organisms to grow. This has practical significance since the atypical mycobacteria are unresponsive to conventional therapy. The increasing emergence of *M. tuberculosis*, as well as the atypical mycobacteria, in immunocompromised hosts, particularly AIDS, has en-

Box 5.4 Laboratory parameters to assess utility of new technology

There are a number of parameters which are useful when assessing the value of a particular diagnostic test. *Sensitivity* refers to the percentage of true positives that will be detected as being positive by the test (or the probability that the test will be positive when the disease is present). *Specificity* refers to the percentage of true negatives that will be detected as being negative by the test (or the probability that the test will be negative when the disease is not present). The above parameters are largely a product of the test and do not alter according to the population tested. However, they do not directly demonstrate a test's usefulness. The latter is better assessed by the predictive value of positive and negative results coming from the test. *Positive predictive value (PPV)* refers to the percentage of all positive test results that are truly positive (or the probability that the disease is present when the test is positive). PPV is influenced by the specificity but more importantly by the prevalence of what is being tested for in that population. The higher the prevalence, the higher will be the PPV since there will be less chance of false-positive results in populations which have fewer true negatives. The *negative predictive value (NPV)* refers to the percentage of all negative test results that are truly negative (or the probability that the disease is absent when the test is negative). NPV is influenced by test sensitivity as well as prevalence. Thus, a test can be made to have a high PPV if performed in a population with a high prevalence for a particular infection. Conversely a high NPV will be easier to obtain if the infection has a low prevalence in a population. *Test accuracy* is defined as the percentage of positive and negative results obtained with a test that are correct. *Tests with high PPVs* are required for conditions in which a false diagnosis will have significant consequences, e.g. treatment regimens are potentially toxic or there are medical or psychological stigmata associated with a positive test such as might occur in the case of sexually transmitted diseases. *Tests with high NPVs* are required when it is essential that positives are not missed, e.g. screening tests or treatable infections with fatal outcomes if missed (Balows et al 1991).

couraged exploration for alternative and more rapid means of detection.

DNA probes distinguishing the various mycobacteria through hybridisation procedures are now commercially available. Kits utilise either radioisotopically labelled DNA probes or non-radioactive labelling methods. The latter avoid the potential hazards of radioactivity, the probe has a longer shelf-life and there is no radioactive disposal problem. These tests are becoming more popular. As indicated previously, DNA probes are often designed to hybridise to repetitive sequences (e.g. rRNA) in order to provide sufficient target DNA/RNA to identify microorganisms *directly* in clinical specimens. This is possible with some organisms, e.g. *Neisseria gonorrhoeae* or *Chlamydia* spp. However, in the case of the mycobacteria, even DNA probes to rRNA are not sensitive enough if clinical specimens are used for direct detection. An additional step, which involves a short-term culture of the sample, can be undertaken to overcome this problem.

Special liquid media are now available which allow rapid culture and from this an aliquot is taken for testing with the DNA probes. In practical terms, hybridisation with two sets of probes (one for *M. tuberculosis* and the second for the atypical mycobacteria) provides a sensitive and relatively rapid method to diagnose mycobacterial infection as well as distinguish *M. tuberculosis* from the atypical mycobacteria. In a number of laboratories, the DNA approach described above is now being preferred over the time-consuming and slower conventional methods.

An alternative DNA strategy involving amplification by PCR has been successful in limited trials in which DNA from clinical samples (sputum, gastric aspirate, biopsy, blood and abscess material) was amplified with primers derived from different mycobacterial antigens including repetitive regions and rRNA. The amplified fragments were hybridised to species-specific oligonucleotides for *M. tuberculosis*, *M. avium* and *M. intracellulare*. Results were obtainable within 24–36 hours. This approach showed greater sensitivity than direct examination and was more rapid than culture. In some patients with AIDS and proven mycobacterial infections it was possible to identify mycobacterial-specific DNA in their peripheral blood. The PCR strategy would be particularly valuable in the case of immunocompromised patients in whom diagnosis needs to be both rapid and sufficiently specific to distinguish the atypical forms. Commercial PCR-based kits for *M. tuberculosis* detection are now available. Their sensitivity and specificity will need to be assessed since a recent comparative study of this approach to clinical detection has shown that sensitivity and false-positive results remain a problem with PCR. *M. tuberculosis* as a cause of meningitis is discussed in Box 5.5.

M. leprae cannot be cultured. Furthermore, the demonstration of an acid-fast bacillus from tissue scrapings does not necessarily indicate *M. leprae*. Serological assays and skin tests are unsatisfactory since they lack sensitivity and

***

Box 5.5 Tuberculous meningitis

This form of meningitis occurs in ~10% of patients with tuberculosis in developing countries. Despite the availability of effective chemotherapeutic drugs, mortality and morbidity remain high – which partly reflects the delay in diagnosis. In one clinical study comparing three diagnostic approaches – culture, PCR and enzyme immunoassay for cerebrospinal fluid antibodies – it was shown that the overall sensitivities of the three methods were 12%, 65% and 44% respectively. Although false-positives were found with the enzyme immunoassay test (6%), the PCR technique was associated with even greater number of false-positives (12%). However, the PCR result represented cross contamination of specimens since repeat analyses on uncontaminated cerebrospinal fluid samples were shown to be negative. This study confirmed both the advantages and potential problems of PCR for diagnosis of tuberculous meningitis (Shankar et al 1991). At present, the PCR approach remains useful within the confines of a specialised laboratory. Further trials to determine sensitivity and specificity and to ensure that the potential for cross contamination can be avoided are awaited. As the technology is simplified and/or kits are produced, DNA testing will become attractive to a wider range of diagnostic laboratories.

***

specificity. Even if positive serology is obtained, it does not define current bacteriological status but may simply reflect a past infection. DNA probes specific for *M. leprae* are now available and these allow detection of the organism in biopsy material from patients with lepromatous leprosy. If there are few bacilli, such as might be found in tuberculoid leprosy or in a subclinical infection, the DNA hybridisation approaches are less sensitive. In these circumstances, PCR is more appropriate. Oligonucleotide primers specific for *M. leprae* have been described and they enable the detection of few organisms. The utility of the DNA approaches for diagnosis will take time to assess since there is no satisfactory 'gold standard' with which to compare results from DNA tests. Nevertheless, the inadequate options currently available make it essential to find an alternative approach to diagnosis of this mycobacterial infection.

Chlamydial infections

There are three major species of the very small bacteria known as chlamydiae: *Chlamydia trachomatis*, *Chlamydia psittaci*, and *Chlamydia pneumoniae*. All are obligate intracellular parasites which cause disease in humans and a wide variety of animals (Box 5.6).

Conventional tests

Although chlamydiae are bacteria they are difficult to culture for the following reasons: (1) *C. pneumoniae* has low virulence for mammalian cells, (2) *C. trachomatis*

Box 5.6 Clinical manifestations of chlamydial infections

C. trachomatis is a major cause of sexually transmitted diseases. In women, the bacterium is an important contributor to infertility. Asymptomatic carriers are found in 25–50% of cases, which is a contributing factor to the spread and prevalence of this infection. *C. pneumoniae* is a recently discovered respiratory pathogen in humans. *C. psittaci* is an example of a zoonotic infection with birds being the main reservoir of infection for humans. Clinical consequences of the chlamydiae include:

- *Ocular*: neonatal opthalmia, inclusion conjunctivitis, trachoma (blindness), opthalmia conjunctivitis
- *Genitourinary*: urethritis, cystitis, epididymitis, prostatitis, proctitis, cervicitis, lymphogranuloma venereum
- *Respiratory*: pneumonia, adult and neonatal respiratory infections, psittacosis
- *Reiter syndrome*: urethritis, arthritis, conjunctivitis, colitis

Box 5.7 The spread of AIDS

Factors associated with the worldwide AIDS pandemic include the independence of former African colonies, the greater availability of travel, changes in sexual behaviour and the increasing frequency of intravenous drug addiction. An additional factor has been the improved access to blood or blood-derived products. In the West, AIDS has affected mostly homosexual men and intravenous drug users. Today, particularly in the developing countries, intravenous drug addicts, heterosexual spread (particularly in terms of prostitution) and mother-to-infant transmission are increasing concerns since the numbers infected in these groups is increasing. Although the risk from blood products has diminished with comprehensive screening programmes underway in developed countries (p. 97), blood and its products remain a significant source of infection in the developing countries. It is estimated that by the turn of the twenty-first century there will be more than one million new cases of AIDS each year and it will be predominantly a heterosexual disease, as is already found in Africa and other developing countries.

requires the availability of cell culture facilities, and (3) *C. psittaci* is very infectious and places laboratory workers at risk during culture. Thus, only a limited number of laboratories are willing to provide a diagnostic service for these infections, which reach high prevalences in some regions. Transport of specimens also requires care since the organism is fragile and this increases costs for diagnosis as well as being a source of false-negative results if organisms do not survive transportation.

Commercial kits are now available for detecting chlamydia. These rely on direct immunofluorescence or monoclonal antibody-based enzyme immunoassays. Antibodies can be directed against the outer membrane protein which is species-specific (e.g. *C. trachomatis*) or the genus-specific lipopolysaccharide (*C. trachomatis*, *C. pneumoniae*, *C. psittaci*). Direct immunofluorescence has the added advantage that a rapid diagnosis is possible and a small number of specimens can be managed (the situation which would be found in non-reference laboratories). One drawback of this technique is the potential for false-positives which partly reflects the subjectivity of the operator. Enzyme immunoassay can overcome this problem through automation. Nevertheless, antibody-based assays have a lower sensitivity than culture and there is the possibility of cross-reactivity to other bacterial antigens, particularly if rectal specimens are used. Antibody responses following a primary infection are also variable both in their development and decline. For example, in diagnosis of *C. pneumoniae* both IgG and IgM may need to be measured and to detect a rise in titre will require a second serum specimen which is taken a number of weeks following the initial infection. None of the conventional diagnostic approaches is effective if urine is tested, which would be useful in the case of asymptomatic infected males.

DNA tests
Commercial diagnostic kits based on DNA hybridisation or PCR have been produced. They are user-friendly in the setting of a routine diagnostic laboratory. The PCR chlamydia kit has the additional advantage that *C. trachomatis* can be detected directly in clinical specimens including urine from males. PCR-based kits promise high specificity and high sensitivity even in low prevalence populations.

Human immunodeficiency virus (HIV)
The emergence of AIDS following infection with HIV has only become apparent in the past 20 years or so. Regions affected include Africa, Haiti, the USA, South America, Europe, Australia and more recently spread of the infection has been rapid in South East Asia (Box 5.7). There are two viral types (HIV-1, HIV-2). HIV-1 was first isolated in 1984 (Fig. 5.2). A year later, a similar virus (HIV-2) was found in Paris from West African patients with AIDS but seronegative for HIV-1. Fortunately, HIV-2 has not spread as rapidly although it has been reported in the USA, Europe and more recently in Australia. The identification of HIV viruses has required extensive changes to be made in blood screening protocols (see p. 97).

Over 20 million people are estimated to be infected with HIV. Nearly three-quarters are to be found in developing countries. The urgency and potential implications of the AIDS pandemic has resulted in many research programmes directed to better diagnostic tests, more effective therapeutic regimens and a greater understanding of the virus's

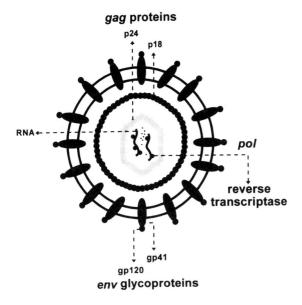

gag proteins

env glycoproteins

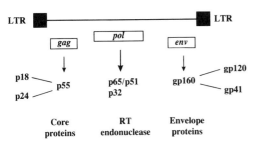

Core proteins   RT endonuclease   Envelope proteins

Fig. 5.2 Human immunodeficiency virus (HIV).
(**Top**) The HIV genome comprises two identical strands of RNA within a core of viral proteins. Two key viral enzymes are reverse transcriptase and integrase. The core is surrounded by a protective envelope derived from viral glycoprotein and membrane from the previous host cells. (**Bottom**) A number of genes make up HIV. *Gag* (group specific antigens) codes for a protein p55 which is cleaved to give core structural proteins including p18 and p24. *Pol* codes for reverse transcriptase (p65/p51) and an endonuclease (p32). *Env* (envelope) codes for a glycoprotein precursor gp160 which is cleaved to give envelope structural proteins gp120 and gp41. Although an RNA virus, HIV is able to produce DNA through its reverse transcriptase enzyme once it infects a cell. The viral DNA then integrates (via the enzyme integrase) into double-stranded host DNA to produce the proviral form. Not shown in this diagram are two regulatory genes (*rev, tat*) and four accessory genes (*nef, vpr, vpu* and *vif*) which give HIV a more complex structure compared to the animal retroviruses. The HIV regulatory and accessory genes are discussed further in Chapter 7. LTR: long terminal repeats of the retrovirus; RT: reverse transcriptase.

biology. In all these areas, rDNA technology is playing a major role. In the context of the present theme of laboratory detection, molecular medicine has an important function to play since opportunistic infections are the chief cause of death in AIDS. All who have AIDS will develop infections at some time in their illness. The pathogens involved include both those that normally do not produce

Box 5.8 Laboratory diagnosis of HIV-1

There are essentially two types of screening tests available for HIV-1. The first identifies viral-specific *antibodies* in the serum, the second looks for viral *antigens*. Antibody-based tests are the most frequently used, with the enzyme immunoassay (EIA) preferred for routine testing because it can be automated. With this test, components of the viral antigen(s), the quality of which has improved dramatically with the availability of recombinant DNA-derived products, are coated onto a solid surface (e.g. beads) and allowed to react with the patient's serum. HIV-specific antibodies in the serum will bind to viral antigen(s) on the beads. The beads are then exposed to a second antibody (conjugated to an enzyme) which attaches to human immunoglobulin. This reaction will produce a colour change which indirectly indicates the presence of HIV antibody. Samples positive by EIA are then confirmed by western blotting (also called immunoblotting), an assay which has higher specificity. In this test various combinations of HIV antigens are transferred by electroblotting onto a strip of nitrocellulose (this is the western blot). Sera are reacted with these strips. Just as for the EIA, antibody binding is detected with a second antibody which is labelled with a colour-producing enzyme. In one commercial immunoblotting kit, results are grouped into positive, negative and indeterminate. A *positive* result requires reactivity to one *env* component together with three other viral-specific proteins for the *gag* or *pol* components. *Negative* results have no reactivity with any of the HIV-specific bands. *Indeterminate* results are somewhere in between and require follow-up and retesting since they can represent early infection, individuals who cannot mount an adequate immune response, infected infants and normals (see also Figs 5.2, 5.3).

overt disease, such as *Candida albicans*, and the more exotic organisms that will not usually be seen in clinical practice, e.g. JC virus, which gives rise to a demyelinating neurological disorder called progressive multifocal leukoencephalopathy. In these circumstances, current microbiological detection methods often prove ineffective.

Conventional diagnostic tests
HIV can be cultured, but this procedure is: (1) time-consuming, i.e. it may take 1–4 weeks, (2) costly, (3) dependent on the numbers and types of mononuclear cells available from the patient, and (4) potentially hazardous to laboratory staff. There are many commercial enzyme immunoassay kits which detect *antibodies* to HIV-1 and HIV-2. These are used to screen those who are at risk and blood donors. The assays are well standardised with good *sensitivities* and *specificities*. A high sensitivity (i.e. the ability to detect antibodies in sera that contain antibodies) will mean that few false-negatives should occur. Specificity

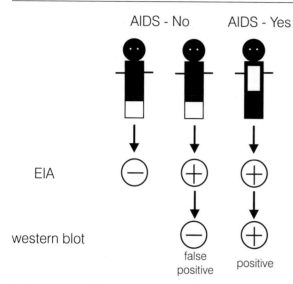

EIA

western blot

Fig. 5.3 A schematic representation of serological testing for HIV.

refers to the ability to identify as negative those sera which do not have antibodies, i.e. the number of false-positives will be reduced if the specificity is high (see also Box 5.4). Screening tests must have a high sensitivity so that positives will not be missed. False-positives are then excluded by using a second assay with a high specificity. In the case of HIV infection, positives from the initial enzyme immunoassay are confirmed by a second test, e.g. a western blot (Box 5.8, Fig. 5.3).

The present serological assays have proven to be very effective in identifying those who are infected. However, there remain a number of unanswered questions:

- Are potentially infectious individuals not being identified because of a delay in the development of a detectable antibody titre? Seroconversion following infection can take several weeks to 6 months. Therefore, in terms of testing high-risk individuals and screening blood donors, this window period as well as the possibility of a donor acquiring HIV infection around the time of donation must be considered.
- What is the HIV-1 status of infants born to seropositive mothers? This is a significant issue since a number of trials have shown that the rate of vertical transmission from an infected mother to her unborn child is ~15–35%. Most infected infants will develop AIDS within a few years and about 25% die in the first year. However, serological testing will not identify the HIV status of the newborn during the first 6 months or more of life because there are maternal antibodies in the newborn's sera and the relatively immature newborn immune system means that neonatal IgG is either not yet present or present at very low levels. In this circumstance, an alternative diagnostic test is required.

- What is the prognosis for those who are seronegative for HIV but positive by DNA amplification? The latter is extremely sensitive and can theoretically detect 1–2 copies of the viral genome. The biological significance of this finding remains to be established. It will require longer follow-up studies.

DNA detection

HIV-1 testing by DNA analysis can be directed towards the *gag*, *env* or *pol* genes of the virus (Fig. 5.2). Laboratory techniques and detection strategies ('algorithms') utilising these sequences have been constructed to optimise sensitivity and specificity and minimise false-positive or false-negative results. It has been estimated that one HIV-1 proviral copy can be detected against a background of 10^5 mononuclear cells in peripheral blood. Virus isolation is the only other method with comparable sensitivity although it has a number of disadvantages which have already been mentioned.

Variations of the PCR technique including 'hot-start' PCR, 'nested' PCR (see Ch. 2) and various commercially produced safeguards can be used to reduce the false-positivity rate which is due, in most cases, to contamination of samples or reagents. On the other hand, false-negative results can reflect the presence of inhibitors in the specimen being tested or chemical substances such as anticoagulants. This problem will be identified if suitable controls are included in the PCR. Another source of a false-negative result with HIV occurs because of the considerable diversity found in the viral DNA sequence so that a mismatch between the usual primer pairs inhibits the subsequent PCR step. This difficulty can be avoided by use of primer pair *combinations* derived from segments of the viral genome which are conserved.

In one multicentred study comparing PCR in 105 HIV-1 seronegative and 99 HIV-1 seropositive/culture-positive specimens it was shown that the overall sensitivity rate for PCR was 99.0%, specificity 94.7% and there was a 3.2% (32 out of 1005 determinations) misclassification rate. The latter included 1.8% false-positives, 0.8% false-negatives and 1.9% in which a diagnosis could not be made. The above trial (see Sheppard et al 1991) illustrates the potential of PCR in HIV-1 testing. However, serological testing for HIV-1 and HIV-2 will remain the preferred screening method until there is better standardisation of DNA sample preparation and more stringent and uniform PCR protocols become available (see also the section on transfusion-related infections, p. 97).

The first generation of commercial PCR kits is now coming into the marketplace. Kits tend to be more 'user-friendly' and hopefully the quality control within and between laboratories will be more rigidly maintained. The claim from one company's HIV kit is impressive, i.e. a sensitivity of 97–100% and a specificity of 100%. A PCR-based method which would allow DNA to be *quantitated* has long been sought. This is essential if therapeutic responses to treatment in AIDS are to be objectively assessed. Monitoring HIV-1 seropositive pregnant women would also be assisted if accurate viral quantitation were possible since there is evidence that the risk for neonatal infection is related to the viral load in maternal blood. In situ PCR is yet another variation which allows PCR to be performed within intact cells. Thus, a very sensitive method is available to identify individual cells which have been infected with HIV.

Cytomegalovirus

Cytomegalovirus belongs to the viral family *Herpesviridae* (Table 5.3) and is a ubiquitous human pathogen. Infections with cytomegalovirus are usually asymptomatic and often give rise to undetected latent infections or reinfections. Transmission of the virus occurs via transplanted organs, blood or its products, placenta, urine, breast milk and other body fluids. Cytomegalovirus infection is significant in the following clinical situations:

- Immunosuppressed patient in whom disseminated infection can be fatal.
- Transplanted organ, tissue or bone marrow where rejection or disseminated infections can occur secondary to cytomegalovirus being transmitted with the transplant or blood products given during transplantation.
- Intrauterine infection, which can produce mental retardation and deafness (discussed further in Ch. 4).

The viral status of immunosuppressed patients (AIDS, transplant recipients) should be checked, as should blood products and donors in these circumstances. With improved processing and preservation of human tissues, there has been an increase in the number and range of transplants involving solid organs (e.g. kidney, heart, heart/lung, pancreas) and tissues (e.g. bone, semen, cornea, skin). Tissues can be preserved for months whilst solid organs need transplantation within hours of collection. In the context of transplantation, it should be noted that conventional serological assays for detecting antibodies to cytomegalovirus do not readily distinguish primary from reactivated infections. Active infection can be confirmed by viral culture but this takes time, is expensive and will not detect latent virus. In these circumstances, products will be unnecessarily excluded from use because of the potential for cytomegalovirus infection. For example, in some communities, 80% of normal blood donors are seropositive for cytomegalovirus.

DNA tests

DNA amplification tests for cytomegalovirus are now available and will become increasingly useful for screening donors, transplants, blood products and recipients. However, further refinements of the DNA-based tests for cytomegalovirus are required. Cytomegalovirus DNA which has been amplified is detectable earlier than an-

tigenaemia or viraemia although in this situation the patient is usually asymptomatic. Does the DNA finding indicate imminent infection or latent (non-replicating) virus? Quantitation of amplified DNA is difficult but would be helpful in this clinical setting. The ability to detect viral DNA in the case of possible fetal infection is discussed further in Chapter 4.

Cytomegalovirus is not the only infectious agent which can be transmitted during blood transfusion or transplantation. A wide range of bacteria, fungi and viruses has been implicated (Table 5.5). In many cases serological testing will identify active infection in the donor. In other situations this will not be possible or the necessity for transplantation to be undertaken quickly excludes many of the more conventional diagnostic approaches. DNA amplification to detect infectious agents in the donor or transplant will prove very useful in these circumstances.

Transfusion-related infections

Blood transfusion services screen donors for a range of infectious agents, e.g. hepatitis B virus, hepatitis C virus, HIV-1, HIV-2 (and in some regions human T cell lymphotropic virus types I and II [HTLV-I, HTLV-II]), syphilis and cytomegalovirus. Cytomegalovirus testing is undertaken in selected cases (e.g. blood required for an immunosuppressed recipient, fetus or neonate). Enzyme immunoassay-based assays are available for hepatitis B and C viruses and the retroviruses (HIV-1, HIV-2, HTLV-I, HTLV-II). These are indirect measures of infection. Positive results require additional investigations for confirmation or assessment. Serological testing is also used for cytomegalovirus and syphilis.

The development of post-transfusion HIV infection has become a rare event following screening of donors for risk factors and donor sera for HIV-related antibodies. Nevertheless, cases have been reported. The actual risks for transfusion-associated HIV remain unclear. One retrospective survey in a US military hospital has shown that conventional enzyme immunoassays failed to detect an infected donor although stored sera from this individual was found to be positive by western blots, PCR and a p24 antigen-based assay for the virus. Although 700 000

Table 5.5 Infections associated with transplantation of tissues.

Transplanted tissue/organ	Transmitted infections
Cornea	Rabies, Creutzfeldt–Jakob
Semen	*N. gonorrhoeae, T. vaginalis*, HIV, hepatitis B virus
Human heart valves	*M. tuberculosis*, various bacteria, fungi, yeasts, *T. gondii*
Bone marrow	HIV, cytomegalovirus, bacterial infections
Skin	HIV

Box 5.9 Comparison of different viral assay methods

Current HBsAg (hepatitis B surface antigen) enzyme immunoassays can detect ~3×10^7 hepatitis B virus (HBV) particles per ml of serum, i.e. 100–200 pg per ml of HBsAg. These tests will not detect individuals with low levels of circulating antigens/infectious virions. DNA hybridisation tests are more sensitive, identifying approximately 10^5 HBV genome equivalents per ml (0.1–1 pg). Even better is DNA amplification by PCR which can detect as little as 10 HBV DNA copies, i.e. 10^4 times more sensitive. The sensitivity of PCR can be improved further by hybridising amplified products with labelled probes or using nested primer techniques. Apart from the potential for contamination with PCR, the biological significance of such small numbers of virions awaits confirmation. However, experimentally infected chimpanzees demonstrate HBV-specific DNA by PCR before HBsAg is detected. HBsAg-negative (serological) but PCR-positive blood has also been shown to produce acute hepatitis if transfused into chimpanzees (Jackson 1991).

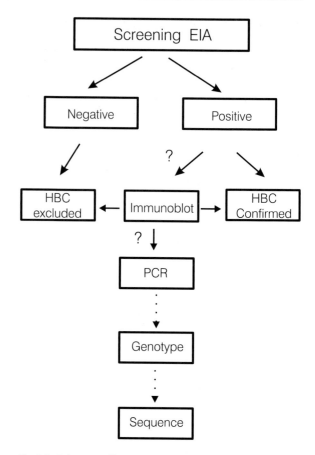

Fig. 5.4 Laboratory diagnostic strategies for the hepatitis C virus. A simplified version of steps taken in testing for the hepatitis C virus is shown. To exclude a false-positive result following the primary screen, it may be necessary to utilise immunoblots (e.g. nitrocellulose filters which have a combination of viral antigens to which the patient's serum must react) as a more definitive test. Just as for HIV, cases can remain indeterminate even after the immunoblot (western blot). In these circumstances, the results from PCR can be helpful. In terms of epidemiological studies, additional steps are required once a diagnosis is confirmed. Genotyping of the viral isolates provides useful information, but ultimately it may be necessary to sequence regions of the viral genome to provide definitive identity. Testing protocols vary and these are often determined by local expertise, the availability of kits and economic considerations. The latter is not trivial since commercial kits are relatively expensive.

donors' units have been screened in US Army blood donor centres since HIV-1 testing was started in 1985, the one donor missed resulted in two individuals becoming infected. The study by Roberts et al (1994) highlights the necessity for better screening assays particularly as the potential for spread from the heterosexual community increases.

PCR-based technology is starting to be included in blood transfusion screening protocols. This partly reflects the potential of DNA amplification to become a more efficient and sensitive test. Data are also accumulating to suggest that individuals who are negative by conventional screening, e.g. surface antigen for the hepatitis B virus, can still harbour the viral genome as demonstrated by PCR. This is not surprising given the exquisite sensitivity of DNA amplification over both immunoassays and DNA hybridisation tests (Box 5.9). Another advantage of utilising viral DNA/RNA as the target for screening is in overcoming the problem of a window period, i.e. a false-negative serological result which can occur if there is a delay before the host mounts an immune response. The availability of kits or screening protocols which can be automated will further facilitate the move from indirect, antibody-based assays to the detection of specific viral nucleic acid sequences.

Post-transfusion hepatitis is caused by hepatitis B virus in up to 10% of cases. Serological assays for hepatitis B virus antigens (surface or e antigens) and antibodies (directed towards the surface, e or core antigens) are well established. They are discussed further in the section on virulence factors (p. 104), and will continue to play a role in screening for some time. On the other hand, hepatitis C virus is the cause of what used to be known as 'non-A, non-B' post-transfusion hepatitis. Despite many attempts to culture the hepatitis C virus, success has been difficult to achieve. Therefore, in terms of viral detection systems there remain those that are based on antibody screening or identification of viral nucleic acid.

Since the single-stranded ~10 kb RNA sequence of the hepatitis C virus was reported in 1989, it has been possible to use an enzyme immunoassay to detect antibodies against a number of viral antigens in patient testing or blood donor screening. However, the first-generation immunoassays gave rise to many false-positives. This problem with serological testing has been overcome, to some extent, by the availability of more sophisticated second and third generation enzyme immunoassays. Similar to the HIV story, resolution of false-positive results is undertaken by an additional assay such as an immunoblot which incorporates multiple viral-specific antigens. Viral antigens in many of the modern tests have been prepared by using DNA expression systems such as *E. coli* or yeast (see Chs 2, 7). Indeterminate or equivocal results can be further evaluated with the use of PCR-based assays if these are available (Fig. 5.4). Screening of blood donors with the more recently available immunoassays has virtually eliminated post-transfusion hepatitis.

Nevertheless, in terms of screening a relatively low-risk population, such as blood donors, the sensitivity of current assays, which approach 90%, may not be enough. There is also the concern about the time required for seroconversion or the 'window period' between infection and development of an antibody response which, like HIV, can be as long as 6 months. DNA/RNA-based assays may help resolve these difficulties. The long-term implications of hepatitis C infection are considerable and will be discussed further (see p. 110). There is also the suggestion that at least one other virus may be implicated in non-A, non-B hepatitis (see p. 111).

EPIDEMIOLOGY

Conventional typing of pathogens based on their phenotypes (phage susceptibility, biochemistry, antigen profiles, antibiotic resistance, immune response, fimbriation) is not always successful in epidemiological studies. A changing spectrum of infectious agents, particularly in the immunocompromised host and in hospital outbreaks, has meant that newer epidemiological approaches are required to complement or replace the more traditional methods. Five strategies based on characterisation of pathogen-derived DNA are possible in these circumstances:

- Nucleic acid hybridisation
- Plasmid identification
- Chromosomal DNA banding patterns
- PCR
- Sequencing.

Nosocomial infections

It is important in nosocomial (hospital-acquired) infections to have methods to type pathogens to enable their source(s) and mode of spread to be identified. A composite of DNA restriction fragments will provide a unique DNA pattern for an individual person. DNA fingerprinting, as it is popularly known, is now well established in forensic practice (see Ch. 8). The DNA/RNA profile of infectious agents is similarly being exploited for diagnostic or epidemiological purposes. One approach to fingerprinting microorganisms is called chromosomal fingerprinting or RFLP (*restriction fragment length polymorphism*) analysis. Digestion of genomic DNA with one or more restriction enzymes and then hybridisation to a DNA probe will produce a number of fragments. Probes can be selected from a conserved region of the microbial genome if less discrimination is required and so species or even subspecies within an organism will be detected. On the other hand, choosing a DNA probe from a highly variable (i.e. polymorphic) region of the genome will allow discrimination between closely related organisms.

Not surprisingly, another DNA approach utilises PCR to amplify DNA. With this technique the identity or relatedness of organisms is based on DNA sequence or variations in single or small segments of DNA. DNA sequencing is now possible by a number of methods and the technology will only improve as the Human Genome Project progresses. The term 'human' is in fact a misnomer since components of this project involve the sequencing of model organisms such as yeast, *E. coli* and *M. tuberculosis* (see Ch. 10). Once a segment of microbial DNA is amplified, it is also possible to use it for detection purposes or for comparison with amplified fragments from other organisms by using techniques such as SSCP or CCM (see Ch. 2). The latter has been applied successfully in distinguishing RNA from different dengue virus isolates. Some of the molecular approaches applied in tracing infections in the clinical or public health setting will be illustrated by reference to the *Legionella* spp., *Pseudomonas aeruginosa* and *S. aureus*.

Legionella

Legionella spp. are ubiquitous bacteria present in domestic and industrial water systems, tanks and other sources of pooled or collected water. More than 30 different species of the genus *Legionella* have been described. The bacteria cause Legionnaire disease, a proportion of which is considered to be nosocomial in origin. Both sporadic cases and outbreaks of this respiratory disorder have been reported. The organism (e.g. *L. pneumophila*) is difficult to grow and it is frequently necessary to obtain bronchial aspirates or lung biopsies for direct detection. Serological testing is the mainstay of diagnosis although a number of DNA-derived tests are now becoming available. These frequently involve DNA to DNA/RNA hybridisation with probes which are often rRNA specific. Typing of *Legionella* spp. for epidemiological purposes follows similar strategies (Fig. 5.5).

More recently, another rDNA approach to fingerprinting

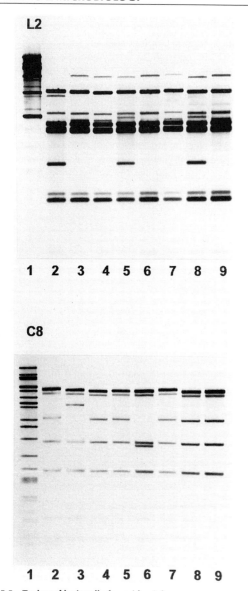

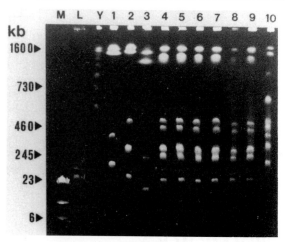

Fig. 5.6 Analysis of a legionella outbreak in an intensive care ward.
PFGE map of *Legionella pneumophila*. Bacterial-derived DNA has been digested with the rare cutting restriction enzyme *Not*I. Tracks M, L and Y are molecular weight size markers. Tracks 1–3 and 10 are control samples of *L. pneumophila* (track 10 is DNA from bacteria obtained from a patient who acquired Legionella pneumonia outside a hospital environment). Tracks 4–6 are bacterial isolates taken from three individuals who developed legionella pneumonia whilst in the intensive care unit of a hospital. Tracks 7–9 are bacterial isolates taken from the water supply in the same intensive care unit. It can be seen that legionella-specific DNA in tracks 4–9 are identical (compared to the different DNA patterns seen in tracks 1–3 and 10). This identifies the source of the *L. pneumophila* in the three patients (PFGE map courtesy of Dr M Ott, Wurzburg, Germany).

Fig. 5.5 Typing of legionella for epidemiology.
The DNA/RNA profile of infectious agents can be used to 'fingerprint' them, thereby enabling their source and mode of spread to be identified. This is particularly important in hospital-acquired infections. Chromosomal fingerprints (RFLPs) of *Legionella longbeachae* serogroup 1 strains, which are usually less common causes of human infection, are shown in this figure. Microbial DNA was obtained from various sources (human infections, environment – in this case commercial potting mixes) and digested with two restriction enzymes (*Hind*III and *Bam*HI). The two DNA probes used for hybridisation were obtained from random clones of the organism's DNA in a λ phage (L2) or a cosmid (C8). Lanes 1 are DNA size markers; the remaining lanes are bacterial isolates in identical order for each gel. Considerable variability between the eight samples of the one serogroup can be seen (Southern blot courtesy of Dr J Lanser, Institute of Medical and Veterinary Science, Adelaide).

bacterial genomes has become available. The DNA strategy is called PFGE (see Ch. 2). The feature of this technique is that it allows *large segments of DNA* to be separated by using restriction enzymes such as *Not*I and *Sfi*I which digest DNA very infrequently. The upper limit for resolution in conventional DNA gel electrophoresis is ~30 kb whereas for PFGE it is approaching 10 Mb (see Ch. 2 for further discussion). PFGE is particularly attractive in microbiological work because the genomes of infectious agents are relatively small. For example, total DNA from *Legionella* spp. can be cleaved into a limited number of fragments (5–10 with *Not*I depending on the strain) and these fragments or 'fingerprints' can be directly visualised from an ethidium bromide-stained gel thereby avoiding the requirement for a DNA probe and a hybridisation step.

The value of PFGE is illustrated by a study which confirmed that an outbreak of *Legionella* in patients admitted to an intensive care ward was caused by *L. pneumophila* which originated from the water taps in the ward. This was possible since restriction fragment patterns from isolates of *L. pneumophila* obtained from the patients and the ward's water taps were identical, thereby proving the source of infection. *Legionella* isolates from other geographical regions were distinguished by their different PFGE patterns (Fig. 5.6). In a similar way, profiles of

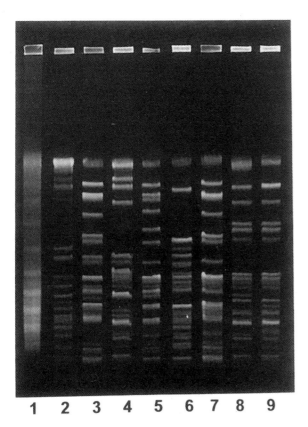

Fig. 5.7 Monitoring hospital infections by DNA analysis.
A PFGE analysis of chromosomal DNAs from *Pseudomonas aeruginosa* strains that were digested with the restriction enzyme *Spe*I. Isolates were recovered from children in a neonatal intensive care unit over a period of several years. From left to right: lane 1 is λ DNA concatamers used as molecular weight standards, lanes 2–9 are *P. aeruginosa* isolates from neonates. Isolates in lanes 2–8 show highly distinctive patterns that are probably not related to one another, while isolates in lanes 8 and 9 are very similar but not identical. These strains are most likely related to one another (PFGE courtesy of Ms Zaiga Johnson and Professor M. L Vasil, University of Colorado Health Sciences Center, Denver).

bacterial DNA obtained by PFGE were used to monitor *P. aeruginosa* infections in the hospital setting (Fig. 5.7).

Methicillin-resistant *Staphylococcus aureus*

Antibiotic-resistant *S. aureus* is an important and frequent cause of nosocomial infections in hospitals. Infection control protocols must first identify that an outbreak has occurred. The source(s) of the methicillin-resistant *S. aureus* is next sought to determine if there has been a breakdown in infection control practices so that more effective preventative measures can be implemented in future. For the above to occur, the different bacterial subtypes need to be distinguished. Phage typing has, for over 30 years, been the traditional tool in epidemiological studies involving *S. aureus*. However, this method has poor reproducibility, is not able to type all isolates and is

Box 5.10 Utility of DNA subtyping for infection control

A nosocomial methicillin-resistant *S. aureus* (MRSA) infection was detected in a hospital. DNA fingerprint profiles of MRSA-derived plasmid DNA identified 10 distinct subtypes from 24 patients infected or colonised with MRSA. Nine of 24 had a single subtype (A2). The nine came from the surgical service and eight had been hospitalised in the surgical intensive care unit. None of the health care workers in that unit were colonised by MRSA. On the basis of this, it was concluded that the MRSA outbreak represented a breakdown in infection control procedures within that unit so that nosocomial transmission of a single MRSA subtype (A2) had occurred from patient to patient (Pfaller et al 1991).

impractical for most laboratories since it requires a large number of phage stocks to be maintained. Phage typing has now been complemented, and in some cases replaced, by DNA-based tests. DNA plasmid analysis to identify polymorphisms has been very useful in deriving a fingerprint for the various methicillin-resistant subtypes since it is possible to use a number of restriction enzymes to produce different restriction fragment patterns (Box 5.10). More recently, PCR analysis and PFGE have been used to distinguish various isolates. It is possible to show genetic variability by PCR despite homogeneity based on phage typing. The Center for Disease Control and Prevention (CDC) in Atlanta has now converted from phage typing to PFGE. Guidelines have been published by the CDC which will enable PFGE to be used more effectively and in a wider range of laboratories to type isolates of *S. aureus* (Bannerman et al 1995).

Zoonoses

In the past, the emergence of infectious agents reflected changed patterns of human movements which disrupted traditional geographical boundaries. For example, yellow fever is thought to have emerged in the New World as a result of the African slave trade which brought the mosquito *Aedes aegypti* in ships' water containers. More recently, *Aedes albopictus*, a potential vector for dengue virus, has become established in the USA following its conveyance from South East Asia in old car tyres. With this, the threat of dengue in the North American continent has become real. Most emergent viruses are zoonotic, i.e. they are acquired from animals which are reservoirs for infection. Thus, completely new strains are less likely than the appearance of a virus following a change in animal reservoirs. This is particularly relevant to the modern world where the consequences of easy migration, deforestation, agricultural practices, dam building and urbanisation are having and will continue to have a major impact on the ecology of animals. At the same time, humans have increasingly populated rural areas as well as taking part in more outdoor recreational activities (see also p. 109).

Lyme disease (Lyme borreliosis)

Lyme disease was first recognised as a distinct clinical entity in the USA in 1975. Since then it has been reported in a number of European countries, Canada, Asia and Australia. Lyme disease is a tick-transmitted disorder shown only in the early 1980s to be due to a spirochaete *Borrelia burgdorferi* which was found in patients and ticks of the *Ixodes* spp. *B. burgdorferi* has now been divided into three distinct species on the basis of DNA typing (*B. burgdorferi* sensu stricto, *B. garinii* and *B. afzelii*). Although a rare condition on a worldwide basis, it is associated with considerable morbidity. In the USA it is the most common vector-related disease and has spread rapidly with more than 9600 human infections reported during 1992. Accurate diagnosis and a knowledge of the organism's life-cycle are essential to initiate early treatment and to attempt preventative measures when it first appears.

Clinical and laboratory detection

There are a number of problems associated with current diagnostic approaches to Lyme disease. A history of tick bite may not be obtained and a characteristic skin rash need not be present. Symptoms associated with the early stages of the disease, e.g. malaise and fever, are non-specific. Potentially debilitating sequelae of Lyme disease such as arthritis and neurological and cardiac problems are not diagnostic. Direct detection of the spirochaete by microscopy rarely gives a positive result since there are few organisms present. Culture is difficult because it is not widely available, special conditions are required for this fastidious organism and results are slow to obtain. The mainstay of diagnosis is serological with immunofluorescence or enzyme immunoassay methods to detect IgG or IgM antibodies. Serological testing can lead to confusion rather than clarity in some circumstances because of: (1) technical problems, (2) slow development of antibodies, and (3) false-positive results from cross-reactivity with other spirochaetes, e.g. *Treponema pallidum* and *B. hermsii* (another borrelia but not a cause of Lyme disease).

rDNA technology

Characterisation of DNA from *B. burgdorferi* has shown that there are two distinct types: (1) low molecular weight, multiple copy, plasmid DNA, and (2) a high molecular weight chromosomal DNA. In one study, a sequence of chromosomal DNA was identified which was specific to *B. burgdorferi* but not the closely related *B. hermsii*. Chromosomal DNA rather than plasmid DNA, was selected since the latter can be unstable. From the DNA sequence, oligonucleotide primers for DNA amplification by PCR were constructed. A diagnostic test based on PCR could now be used for *B. burgdorferi*. The test was highly sensitive and capable of detecting few organisms, although studies

> **Box 5.11 Lyme arthritis and detection of organisms by PCR**
>
> Various forms of arthritis can accompany Lyme disease. Accurate clinical and laboratory diagnosis is difficult in this circumstance. An adequate course of penicillin or related antibiotics is usually effective in treating arthritis. However, some patients continue to have a persistent and recurrent joint involvement. Those with a certain HLA type, e.g. HLA DR4, are thought to be particularly at risk. The underlying mechanism for this complication remained unclear until PCR was used to look for *B. burgdorferi* DNA in synovial fluid. The study showed both the effectiveness of antibiotics in removing borrelia from the synovial fluid and the absence of bacterial-specific DNA in those who had chronic relapsing arthritis. This would support an immunological basis for the persistent arthritis. Another interesting observation to emerge was the use of PCR primers which were specific for plasmid DNA. These proved to be more sensitive than primers for chromosomal DNA. The reasons proposed included the likelihood that there would be multiple copies of the plasmid, or plasmid was all that remained in the synovial fluid (Nocton et al 1994).

were based on *culture-derived* organisms. Clinical specimens will be more difficult to assay because of inhibitory factors which can be present and the greater potential for contamination (Box 5.11).

Epidemiology

The specificity of PCR in the diagnosis of Lyme disease is high although a European-derived isolate could not be detected in one study. In this circumstance a reassessment of the DNA sequence from various borrelias enabled alternative DNA amplification primers to be constructed. These were then able to detect both European and US isolates (see Rosa et al 1991). As further knowledge is gained about the DNA structure of *B. burgdorferi* it will be possible to design better primers. This is important since clinical heterogeneity is geographically related. For example, late clinical features in Europe are predominantly neurological and dermatological, whilst arthritis is the most frequent complication in the USA. A further clue to the clinical heterogeneity came from a study which characterised isolates from North America and Eurasia on the basis of their rRNA patterns (Liveris et al 1995). This work showed that American isolates were *B. burgdorferi* sensu stricto whereas the Eurasian ones comprised all three species mentioned earlier.

The few species of *Ixodes* ticks implicated in transmission of the disease in the USA and Europe have not been found in relatively isolated regions such as Australia. Identifying the insect vector in these areas will be facilitated with a test such as PCR. This will be an important preliminary step in planning effective preventative measures. The value of PCR is further illustrated by the finding of *B. burgdorferi*-specific

DNA in museum specimens of *Ixodes* ticks. From this it was possible to conclude that a number of ticks from different geographical locations had been infected with the spirochaete many years before Lyme disease was described.

Leishmaniasis

Leishmaniasis illustrates some of the difficulties facing epidemiologists in their study of infectious diseases. Conventional identification of leishmania involves biochemical isoenzyme characterisation or detection by monoclonal antibodies. The former approach is difficult since a minimum of 10^7 cells is required and the cells need to be grown in experimental animals or in culture. Monoclonal antibodies will identify leishmania but are not satisfactory in distinguishing the various species. To find the insect vectors for leishmania, it may be necessary to dissect thousands of sandflies looking for the organism. Even when found, it is difficult to be absolutely sure that the leishmania is human-specific and has not come from other animals such as birds and reptiles. In view of the above, it is understandable that prevalence figures, particularly in tropical and subtropical countries where leishmaniasis is an important public health problem, are inadequate.

Alternative diagnostic strategies based on DNA identification will help in overcoming the above problems. Parasites obtained from cultures or within insects can be smeared onto microscope slides. DNA is gently denatured with alkali to make it single-stranded but at the same time retain the cell's morphology. In situ hybridisation with DNA probes will allow this protozoan's detection within its potential vectors. Any leishmania found can be discriminated to the subspecies level if necessary. The potential for a PCR-based DNA strategy to identify both hosts and vectors in bacterial infections has been discussed above in reference to Lyme disease and would apply equally to the leishmania example.

DISEASE PATHOGENESIS

Pathogenesis of many infections, particularly viral ones, has been deduced from experimental strategies based on light and electron microscopy, cell culture and immunoassay. To these research tools can now be added nucleic acid (DNA, RNA) probes for in situ hybridisation. Advantages provided by this technique include the ability to detect latent (non-replicating) viruses and to localise their genomes to nuclear or cytoplasmic regions within cells. Tissue integrity remains preserved during in situ hybridisation and so histological evaluation can also be undertaken. Nucleic acid probes or the hybridisation conditions can be manipulated to enable a broad spectrum of serotypes to be detectable. This is particularly valuable in those emerging infections where the underlying serotypes are unknown. The utility of in situ hybridisation to detect viral DNA/RNA sequences is illustrated by animal and human studies of enterovirus-induced cardiomyopathy (Box 5.12).

Box 5.12 In situ hybridisation and acute dilated cardiomyopathy

Enteroviruses of the group B coxsackievirus (types 1–5) are considered to be the most common causes of viral myocarditis. Infections by these viruses can produce life-threatening arrhythmias or acute dilated cardiomyopathy. Treatment options for cardiomyopathy include cardiac transplantation. Molecular cloning and characterisation of the single-stranded RNA from the cardiotropic B3 coxsackievirus has enabled cDNA to be made. With this as a probe, it has been possible to show by in situ hybridisation in a mouse model for acute dilated cardiomyopathy, that B3 coxsackievirus could be detected in muscle cells and small interstitial cells thought to be fibroblasts in a multifocal and random distribution in heart muscle. Progression of the infection could also be seen from areas with myocardial fibrosis to as yet uninfected myoctes, indicating possible cell-to-cell spread of the virus. In human patients with acute dilated cardiomyopathy, the incidence of enterovirus infection has been estimated to be ~30% using in situ hybridisation on cardiac biopsies (Kandolf and Hofschneider 1989).

Virulence factors

Toxins

As mentioned earlier in the section on intestinal infections (p. 90), a number of bacterial toxins are identifiable through rDNA testing and so their clinical effects can be predicted. Molecular technology will also assist in determining the toxin's mode of action. For example, toxigenic strains of *Clostridium difficile* are responsible for pseudomembranous colitis which follows the use of broad-spectrum antibiotics. The organism produces two toxins: A (known as 'Enterotoxin') and B ('Cytotoxin'). How the two toxins exert their effects is unclear and so rDNA technology is being used to investigate this further. First, the genes for both toxins have been identified. Next, the *C. difficile* toxin A gene has been cloned into *E. coli* and expressed. From this it can be shown that the known toxin A effects (haemagglutination, cytotoxicity, enterotoxicity and lethality) are able to be elicited by the cloned toxin A gene. The gene has now being subcloned into defined segments and the various toxin effects are being analysed in more detail.

Hepatitis B virus and liver disease

The hepatitis B virus was first identified in 1963. Today, over 400–500 million people are chronically infected despite the availability of an effective vaccine. A wide spectrum of liver disease is associated with infection: (1) acute self-limited hepatitis, (2) chronic hepatitis which can progress to cirrhosis including liver failure or primary liver cancer, (3) an asymptomatic carrier state, and (4) fulminant hepatitis. The latter is a rare complication of hepatitis B infection but leads to high mortality and will be discussed further. The role of the hepatitis B virus in primary liver cancer is well proven and will be examined in Chapter 6.

The hepatitis B virus is a DNA virus. The intact virus particle (virion) is also called a Dane particle. Three different antigens are associated with this virus: HBsAg (hepatitis B surface antigen); HBcAg (hepatitis B core antigen), and HBeAg (hepatitis B e antigen). Following infection with the hepatitis B virus, antibodies to the various antigens are formed. These assist with recovery from infection and the maintenance of immunity for several years thereafter. Occasionally, antibodies do not develop but antigenaemia persists. In these circumstances, the infected individual becomes a chronic carrier. The finding of hepatitis B e antigen in acute cases or the chronic carrier state also correlates with infectivity.

There are a number of controversial issues surrounding the hepatitis B virus. For example, what is the role played by the virus in those who are hepatitis B surface antigen-negative but have chronic liver disease? Studies based on serological and DNA techniques have shown conflicting data, e.g. viral-specific DNA detected in the blood and liver cells despite absence of the surface antigen. Individuals who are persistently positive for hepatitis B surface antigen may also be positive for hepatitis B e antigen and/or develop antibodies to the e antigen. However, the significance of hepatitis B e antibodies has at times been unclear. One suggestion was that these antibodies indicated loss of viraemia. Recent analysis of hepatitis B virus DNA has shown how serological assays can confuse the picture, e.g. it is possible to have persistent viral replication leading to very severe liver disease with inconsistent or negative serological findings for the hepatitis B e component. Molecular characterisation of the hepatitis B virus in cases of fulminant hepatitis has provided an explanation for the incongruous serological observations as well as a potential mechanism for pathogenesis in this potentially lethal form of hepatitis (Box 5.13).

Vaccine-acquired poliomyelitis

Although rare, poliomyelitis may occur following administration of the oral attenuated poliovirus (Sabin) vaccine, particularly types 2 and 3. It is known that these two Sabin type viruses revert to the wild-type (neurovirulent) phenotype on passage through the human gastrointestinal tract. To understand further the change in virulence of the poliovirus, its RNA was sequenced and a point mutation

---

Box 5.13 Mutation of hepatitis B virus and virulence

The hepatitis B virus e protein comprises both the hepatitis B core (capsid) protein and a pre-core component. The latter contains a signal peptide which enables secretion of the hepatitis B virus e protein into the serum. The function of the e protein in patients with severe liver disease who were hepatitis B e antigen-negative and anti-e antibody-positive was studied by PCR to isolate hepatitis B virus DNA from their sera. Sequencing the DNA identified a mutation which produced a premature stop codon at the end of the pre-core region. Thus, synthesis of the hepatitis B e protein could not be completed. A similar type of mutation in the pre-core region has now been detected in a number of other studies associated with fatal hepatitis. Recently, DNA analyses of the hepatitis B virus involved in Japanese cases of fulminant hepatitis have identified a second region which is mutated. This involves the viral core promotor sequence and so has the potential to interfere with DNA transcription. How these mutations in viral DNA lead to a more severe form of liver disease is presently being investigated, particularly in relation to possible immunological sequelae which might result from the body's failure to develop tolerance to the e antigen (because it is not produced) rather than a direct cytopathic effect of the virus (Sato et al 1995).

---

uracil (vaccine) to cytosine (wild-type) was detected at position 472 in the non-coding region of the viral genome. This was a consistent finding and the significance was further reinforced when faecal samples were progressively examined following oral intake of the Sabin vaccine. In a study by Evans et al (1985), polio virus in faecal samples at 24 hours after vaccination demonstrated uracil at position 472. At 35 hours, a mixture of viruses with either uracil or cytosine at the 472 locus was present. From 48 hours, all viral isolates had cytosine at this position. Two questions remain unresolved. Why should there be selection in the gut for 472-cytosine bearing polioviruses? What is the molecular basis for the increase in neurovirulence which alone does not produce the complete poliomyelitis phenotype? A clue to the latter may come from molecular analysis which suggests that the 5' non-coding region at position 472 has functional significance since comparative studies show this area to be highly conserved between the various polio serotypes and a rhinovirus.

Resistance

Antimicrobial drugs

Resistance to antibiotics can occur by a number of mechanisms: (1) reduced uptake by a cell, e.g. chloramphenicol, (2) efflux from a cell, e.g. tetracycline, (3) reduced binding to a target, e.g. erythromycin, (4) inactivation, e.g. penicillin, (5) protein binding, e.g. fucidic acid, (6) metabolic bypass, e.g. sulphonamides, trimethoprim, and (7) overproduction of antibiotic target, e.g. sulphonamides, trimethoprim (reviewed in Davies 1994). The genes involved in these processes are gradually being characterised. DNA-based testing protocols will make it possible to detect the presence or development of resistance early on in an infection. Information from DNA studies will enable drugs to be modified or will enable the development of treatment regimens to overcome this problem.

Rifampicin occupies a pivotal place in the WHO multidrug resistance programme for treating leprosy. The development of resistance would have major therapeutic and public health implications. Since *M. leprae* grows slowly it is essential to identify the development of drug resistance by alternative means to the traditional culture approach. There are two potential mechanisms for resistance to rifampicin. Either there is a mutation in the target for the drug (DNA-dependent RNA polymerase) or there is reduced drug intake by the cell. A gene which codes for a component of the RNA polymerase is now suspected to play a role in rifampicin resistance. This gene is called *rpoB*. DNA characterisation of *rpoB* has identified a number of conserved regions which have been shown to contain missense mutations in association with drug resistance. PCR-based assays could be developed to detect these DNA changes. Further understanding of how *rpoB* works in its native and mutated forms may indicate ways in which to bypass rifampicin resistance.

M. tuberculosis

This organism illustrates another problem with the worrying emergence of multidrug resistance. Until recently, the use of various drug combinations selected from isoniazid, rifampicin, ethambutol, streptomycin and pyrazinamide, as well as improved living standards played an important role in the reduced incidence of tuberculosis. However, this is now changing. The number of new cases is increasing. Traditional dogma that active cases usually arise from infections acquired years previously needs to be reviewed since in some US cities DNA studies have shown that approximately one-third of active cases are the result of person-to-person transmission. Contributing factors to the increase in tuberculosis and the finding of multidrug resistant strains are HIV infection, intravenous drug use and the decline in living standards resulting from political changes or civil war (see also the section on re-emerging infections, p. 110). At the molecular level, the genes associated with resistance to isoniazid and streptomycin are being defined. This, as well as data coming from study of rifampicin/DNA, will be of value in identifying resistant strains early. To be more effective than programmes presently underway, modern global strategies to deal with tuberculosis will require laboratory facilities which utilise state-of-the-art technology to provide health professionals

Table 5.6 DNA changes in the dihydrofolate reductase (DHFR) gene leading to drug resistance in malaria (Wellems 1991).
Point mutations produce single amino acid changes which are thought to exert their effect by inhibiting drug binding to the enzyme's active site.

Drug	Mutation in DHFR Residue No.	Amino acid change
Pyrimethamine	108	Serine → asparagine
Increased resistance to pyrimethamine	108	Serine → asparagine
	51	Asparagine → isoleucine
	59	Cystine → arginine
Proguanil	108	Serine → asparagine
	16	Alanine → valine
Pyrimethamine and proguanil	108	Serine → asparagine
	164	Isoleucine → leucine
	59	Cystine → arginine

with accurate information in an efficient and short time frame.

Malaria

Two major classes of drugs used to treat malaria are the cinchona alkaloids (e.g. chloroquine and its derivatives) and the antifolates (sulphonamides or the inhibitors of dihydrofolate reductase such as pyrimethamine or proguanil). The spread of drug-resistant *Plasmodium falciparum* and, to a lesser extent, *Plasmodium vivax* has made the prophylaxis and treatment of malaria a global public health issue. Chloroquine-resistant *P. falciparum* is present in every major region where malaria is endemic. Air travel means that drug-resistant strains can appear in any city. Resistance is thought to have spread since the 1950s from a limited number of foci in South East Asia and South America. The epidemiology is consistent with multigenic effects, i.e. rare events occurring initially in South East Asia or South America and then spreading slowly worldwide. In contrast, resistance to pyrimethamine and proguanil has arisen independently in many different regions where these drugs have been used, which would suggest a single gene is involved.

The molecular basis for resistance in malaria is complex. Resistance to the dihydrofolate reductase inhibitors pyrimethamine and proguanil occurs on the basis of point mutations in the parasite's dihydrofolate reductase inhibitor gene. These point mutations are thought to exert their effect by inhibiting binding of the antimalarial drugs to the enzyme's active site. The inhibition can interfere with binding by both the drugs or can be selective for one alone (Table 5.6). As the molecular defects become more comprehensively characterised it will be possible to attempt manipulation of the dihydrofolate reductase inhibitors to overcome the effects of the point mutations on binding, e.g. data in Table 5.6 would suggest that combinations of pyrimethamine and proguanil are better than either drug given alone as the mutations which lead to resistance are different.

Less is known about chloroquine resistance although the protein product from the normal *pfmdr1* gene isolated from the parasite can increase the level of chloroquine in cells. This is relevant since chloroquine-sensitive strains of *P. falciparum* are able to accumulate this drug to a very high level compared to resistant strains. Molecular characterisation of the *pfmdr1* gene shows that it is related to the ATP-binding cassette transporter family. Other genes in this group include *CFTR* (cystic fibrosis, see Ch. 3) and the mammalian P-glycoprotein. The latter is implicated in the multiple drug resistance phenotype involving mammalian cells (see Ch. 7 for further discussion of P-glycoprotein). The finding of other genes and further characterisation of *pfmdr1* will provide valuable information on chloroquine resistance. Since chloroquine is safe and effective, its reintroduction would be highly desirable if resistance mechanisms can be bypassed.

Host factors

The host's response to an infection involves a complex mix of genetic and environmental factors. In humans, evidence for a genetic component contributing to the outcome of an infectious disease comes from the observation that some racial groups are more resistant to infections, whilst others appear to have an increased susceptibility, e.g. resistance to malaria in Black Africans and chronic carrier state for hepatitis B virus in Chinese.

Malaria

Protection from malaria can result from single-gene effects, i.e. inheritance of the sickle-cell gene, thalassaemia or the Duffy-negative blood group. Recently, an intriguing observation was made relating to genetic susceptibility and cerebral malaria in the Gambia (McGuire et al 1994). Since tumour necrosis factor-α (TNF-α) is considered to play a role in the pathogenesis of severe infections including cerebral malaria, a study was undertaken to compare genetic markers in the TNF-α promotor gene region with the development and complications of cerebral malaria. The results showed that children who were homozygous for a particular TNF-α DNA polymorphism had a relative risk of 4.0 for cerebral malaria and 7.7 for death or complications following cerebral malaria. Although the TNF-α gene is located near the HLA class I and class II gene, the researchers were able to exclude HLA as the critical factor in this association. One mechanism proposed for this genetic link was an increased propensity to produce TNF-α by the gene with the above DNA polymorphism. Since TNF-α stimulates the expression of adhesion molecules by the host, there would be a greater risk for parasites to bind to vascular endothelium in those homozygous for the 'susceptibility' TNF-α polymorphism.

Hepatitis B

Unlike the malaria example, the explanation for a predisposition to the hepatitis B carrier state is more difficult to explain and may involve the effects of a number of genes and/or viral-specific factors. For example, it is considered that the hepatitis B virus does not have a direct cytopathic effect on liver cells but damage results from the host's cellular immune responses. Following an acute infection, a vigorous T-cell response against the various viral-specific components (core, envelope and polymerase) limits the infection. In contrast, those with a poor cellular immune response are more likely to develop a chronic carrier state. In this scenario, the MHC could be predicted to play a role since class II molecules from the HLA complex are critical for the T cells to function. Evidence for a role of the MHC in hepatitis B infection is starting to appear, e.g. a study comparing HLA types with recovery from hepatitis B virus in the same population from Gambia as mentioned above for malaria has shown that one particular HLA group (DRB1*1302) appears to protect against developing the chronic carrier state (Thursz et al 1995).

Human immunodeficiency virus

Following primary infection with HIV, there is a viraemia associated with high viral titres in blood. This is followed by reductions in titres and the load of provirus in peripheral blood mononuclear cells which are presumably related to host immune responses. This situation can persist in asymptomatic individuals for 10–15 years. However, to date there have been no reports of HIV-1 being completely cleared once a human has been infected.

It has been mentioned earlier that up to 35% of infants of HIV-infected mothers will themselves become infected (see p. 96). However, it remains a puzzle why the great majority escape infection. Reports, often incomplete, describe HIV-positive infants who become seronegative. Recently, an interesting case was documented of an infant infected with HIV-1 who completely cleared his infection (serologically, by culture and PCR) at 5 years of age. Both PCR and DNA sequence analysis were used to exclude the possibility of contamination or error in blood collection (an alternative explanation which has yet to be excluded is that there were circulating maternal cells in the infant's blood and these were the source of the HIV) (Bryson et al 1995). Confirmation that the infant did have HIV and clarification of the mechanisms enabling this infant to remove the virus are important in further understanding HIV infection and perhaps the development of alternative therapeutic measures.

Animal models

Polygenic traits contributing to susceptibility in humans would be difficult to separate into their individual components because of genetic variability in the human genome. However, this can be overcome by using differ-

Table 5.7 Some mouse-specific loci and genes associated with host resistance to infection (from Malo and Skamene 1994).

Infection	Consequence	Mouse chromosome	Locus (gene)
Influenza	Lethal infection	16	Mx1 (Mx1,Mx2)
Cytomegalovirus	Unrestricted splenic replication	6	Cmv1
Moloney virus	Leukaemia	17 (H-2)	Rmv1-3
Mycobacterium spp.	Unrestricted growth in pre-immune phase	1	Bcg(Nramp)[a]
Salmonella typhimurium	Unrestricted growth in pre-immune phase	1	Ity[b](Nramp)
Legionella pneumophila	Deficient macrophage inflammatory response	15	Lgn
Leishmania donovani	Unrestricted growth in pre-immune phase	1	Lsh[b](Nramp)

[a]Nramp – natural-resistance-associated macrophage protein, [b]same locus as Bcg.

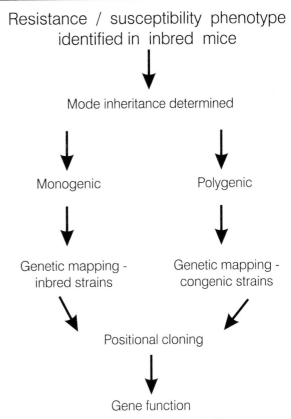

Fig. 5.8 Identifying genetic factors associated with resistance or susceptibility to infections in mice.
Inbred mice are studied to identify strains which show resistance or susceptibility to certain phases of different infections, e.g. macrophage inflammatory response. The mode of inheritance of the trait is followed along several mouse generations. A 'yes or no' phenomenon might suggest a single-gene effect whilst a more complex pattern of inheritance is consistent with a number of genes being involved. Inbred mice (i.e. genetically identical except for sexual differences) are useful to investigate further *single-gene* effects. Congenic strains (i.e. genetic differences between inbred mice involve a small chromosomal segment) are more appropriate in the case of the *polygenic effects*. Genetic mapping in both cases enables chromosomal loci to be identified. From these, the usual positional cloning approaches such as linkage analysis or a search for candidate genes will lead to isolation of the relevant genes (see Fig. 2.14 and Ch. 3 for further discussion on positional cloning. Information on how resistance or susceptibility has arisen will come from characterisation of the genes. (Modified from Malo and Skamene 1994.)

ent strains of animals, particularly mice. Since chromosomal segments in the mouse can be traced back to their homologous, i.e. syntenic, regions in humans, it becomes possible to identify a genetic locus or gene in a mouse and then go back in the corresponding region of the human genome to identify the human counterpart. The potential to alter the genetic background in mice helps to identify either polygenic or single-gene loci which play a role in the host's response to an infection. Loci identified so far involve H-2 (the mouse equivalent of the MHC) on mouse chromosome 17 as well as others on different chromosomes (Table 5.7). To date a limited number of genes have been isolated from these loci by positional cloning (Fig. 5.8, see also Ch. 10 for further discussion on animal models of human disorders).

Cancer

Both DNA and RNA viruses can cause tumours in animals. The DNA papillomaviruses are of medical interest because they are associated with tumours which involve the anogenital and respiratory tracts. During the 1970s, developments in rDNA technology were particularly significant in our subsequent understanding of these viruses and their relationship to cancer.

Papillomavirus

There are experimental and epidemiological data to implicate the sexually transmitted human papillomaviruses and, to a much lesser extent, herpes simplex virus type 2 in the pathogenesis of cervical cancer and its precursor, cervical intraepithelial neoplasia. Cervical cancer is one of the most common neoplasms in women. Over half diagnosed with this malignancy will die as a result. The cancer begins as a pre-invasive neoplastic change in cells (histologically described as CIN – cervical intraepithelial neoplasia). This

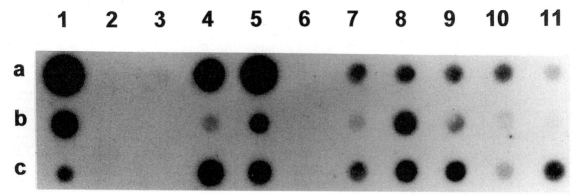

Fig. 5.9   DNA testing by dot blotting for the detection of the human papillomavirus type 11, i.e. low risk virus.
Detection of papillomavirus by conventional techniques is difficult and this has hampered epidemiological studies of its role in cervical
carcinoma. Lane 1 is a positive control DNA in three concentrations, a, b and c, 125 pg, 12.5 pg and 1.25 pg. Lanes 2 and 3 are negative
controls. Lanes 4–11 demonstrate 24 different clinical specimens tested for papillomavirus (dot blot courtesy of Professor Y Cossart,
Department of Infectious Diseases, University of Sydney).

may regress, remain unchanged or progress to an invasive
malignancy. Human papillomavirus has been implicated in
the various stages of cervical cancer with different types
correlating with the histological findings (see below).
Human papillomavirus has more recently been implicated
in anal cancer (particularly in those who are HIV-positive),
penile, laryngeal and oral cancers. Human papillomavirus-
related skin cancers can be found in the immunosuppressed.

Human papillomaviruses induce cellular transformation
and may also interact with other viruses, oncogenes or
carcinogens to bring about neoplastic changes. The virus
alone is not the sole factor in the pathogenesis of cervical
cancer since some tumours do not have viral DNA
detectable and the virus's transforming proteins (E7, E6,
E5) alone are not sufficient for tumourigenicity. Transformation is considered to involve a multistep process
requiring additional factors besides the papillomaviruses.
At present, there are over 70 types of human papillomaviruses which can be distinguished on the basis of
their DNA sequences. Comparisons between histological
findings and viral types has enabled classification into
three groups: (1) *low risk* (genital condylomas but not
invasive cancer), types 6, 11, 42, 43, 44; (2) *intermediate
risk* (intraepithelial neoplasia but less with invasive
cancer), types 30, 31, 33, 34, 35, 39, 40, 49, 51, 52,
57, 58, 63 64; and (3) *high risk* (intraepithelial neoplasia
and invasive cancer), 16, 18, 45, 56 (from Gross Fisher
1994).

Conventional diagnosis
Detection of the human papillomavirus by conventional
techniques is difficult since there are no suitable serological responses to measure and these viruses cannot be
cultured. By immunohistochemical and electron micro-

scopic means it is possible to identify viral-specific antigens but these may only be present at certain stages of
infection. Histological examination of tissue biopsies will
detect the characteristic dysplastic changes resulting from
infection with human papillomavirus but the different
types cannot be distinguished and changes may be
patchy in distribution. Thus, it is not surprising that early
epidemiological studies to determine the oncogenic potential of human papillomaviruses were significantly
constrained.

DNA studies
To date all standard DNA diagnostic approaches have been
tried, i.e. DNA Southern blotting, dot blotting, in situ
hybridisation and PCR (Fig. 5.9). The 'gold standard' is
considered to be Southern blotting. Commercial kits are
now on the market. One kit makes use of probe combinations in a non-radioactive DNA hybridisation test which
will identify the papillomavirus and then go on to determine its risk group. Results within a few hours are
guaranteed and the test is considered to be more sensitive
than PCR. Used in conjunction with the pap smear, the
papillomavirus DNA screen will be helpful if there is a
suspicious result. Papillomavirus DNA typing will also assist
in histopathological examination of cervical biopsies.

Questions will remain for some time about the role of
human papillomavirus in cervical cancer and why some
cases of non-invasive neoplasia regress and others progress
to an invasive form. The first step towards answering these
questions is in place now that there are DNA methods for
viral identification, typing and quantification. Assessment
of viral aggressiveness and/or interaction with other genetic
or environmental factors (e.g. cigarette smoking) will also
be more effectively studied through DNA technology (see
Ch. 6 for further discussion).

DIVERSITY

Antigenic variation

Influenza

The three RNA influenza viruses (A, B, C) are distinguished by their internal group-specific ribonucleoprotein. Influenza A causes epidemics and occasionally pandemics. It has an animal reservoir, particularly birds. The viral envelope has two important antigenic components – haemagglutinin (H, composed of 13 different types) and neuraminidase (N, 9 different types). Although the envelope antigens are capable of producing 117 different combinations, only three have been detected in humans to date (H1N1, H2N2, H3N2). As the influenza A virus passes through its hosts it undergoes genetic changes which are of two types. (1) *Genetic drift* – these result from small mutations involving the H and N antigens which are occurring continuously; for example, a change in the influenza A virus from (H3N2) A/Hong Kong/1/68 to (H3N2) A/England/42/72. On a local scale, the new subtype can avoid immune surveillance in those who have developed immunity. (2) *Genetic shift* – this results from major changes in antigenicity of the H and N components. The new viral type can now lead to a pandemic. For example, A/Japan/57/H2N2 changing to A/Hong Kong/68/H3N2.

Characterisation of the different viral types and their mode of development relies on analysis of the H- and N-specific amino acids and more recently the sequencing of the single-stranded RNA genome of this virus. The ability to identify the functional components of the viral ribonucleoprotein complex and then produce these by rDNA means will provide a potential source of antigens for immunisation programmes.

Trypanosomiasis

African trypanosomes, e.g. *Trypanosoma brucei*, are unicellular parasites that infect a variety of mammals. Infection is usually fatal if left untreated. The life-cycle of the trypanosome can be divided into two main stages: bloodstream forms in mammals and the procyclic form in the mid-gut of the vector *Glossina* (tsetse fly). Each form is covered by a coat composed of characteristic surface proteins: the variant surface glycoprotein in bloodstream forms and procyclin in procyclic forms. Variant surface glycoproteins are highly immunogenic and antibody-mediated destruction of trypanosomes is an important host response to infection. To evade the host's immune response *T. brucei* undergo continuous switching of their variant surface glycoproteins.

At the molecular level, the genetic events underlying variation in these glycoproteins is being defined. The gene for each variant surface glycoprotein is present in all trypanosomes of a stock regardless of whether the gene is expressed. Thus, antigenic variation is not produced by

Table 5.8 Some factors leading to the re-emergence of infectious diseases (from Lederberg et al 1992).

Factors	Components	Examples
Human demographics and behaviour	Population growth, density and distribution	Dengue
	Immunosuppression	TB, CMV*
	Sexual activity, substance abuse	Syphilis, HIV
Technology and industry	Modern medicine	Nosocomial infection, blood products
	Food processing/handling	Bacteria, helminths
	Water treatment	Bacteria, viruses, protozoans
Economic development and land use	Dam building	Rift valley fever
	Reforestation	Lyme disease
	Global warming	Viruses
International travel and commerce	Travel	Malaria, Lassa fever
	Commerce	Cholera, viruses
Microbial adaptation and change	Natural variation and mutation	Influenza A, HIV, bacteria
	Selective pressure and development of resistance	Antibiotics, antivirals, antimalarials, pesticides
Breakdown of public health measures	Inadequate sanitation	Cholera
	Complacency	Measles, pertussis
	War	Leishmaniasis, malaria, viruses

* TB, tuberculosis; CMV, cytomegalovirus.

recombination of coding segments as occurs with the immunoglobulin genes (see Ch. 6). From a repertoire of several hundred variant surface glycoprotein-specific sequences, only one is transcribed at any time. Antigenic variation occurs either through DNA rearrangements at the expressed site or through activation of a new expression site. Further work is required at the molecular level to unravel this parasite's interesting strategy to escape antibody responses generated by the host. Other, as yet unexplained, aspects of the trypanosome's life-cycle may

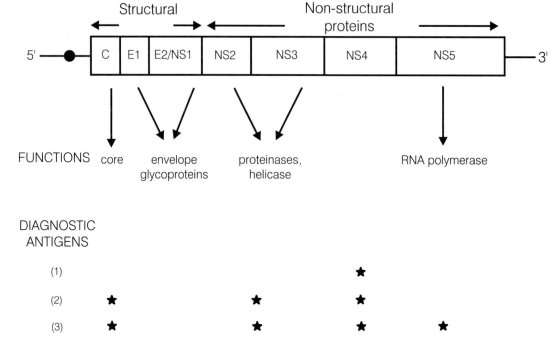

Fig. 5.10   Molecular structure of the hepatitis C virus.
Single-stranded RNA codes for a number of structural and non-structural proteins. Some functions for these proteins are listed. The 5′ untranslated region (●) demonstrates the most sequence conservation (>90%) between the various viral genotypes. On the other hand, the E2/NS1 segment has considerable variability in its nucleotide sequence. Corresponding variability in the same region of the other flaviviruses is considered a factor in escaping immune surveillance through selection of different mutants. A schematic representation of the evolution from first to third generation diagnostic assays shows how different antigens or combinations of antigens have been used in screening tests (★ = antigen component in enzyme immunoassay; 1, first generation; 2, second, and 3, third generation enzyme immunoassays).

also be studied through a molecular approach, e.g. infection can persist for months in an animal and show waves of serologically distinct parasites appearing at 7–10 day intervals.

Re-emerging infections

A number of previously controlled infections, e.g. malaria, tuberculosis, dengue fever, and more recently described microorganisms, e.g. HIV, *B. burgdorferi*, hepatitis C, are starting to pose major public health problems in terms of their frequency of infection or carrier state and the development of drug resistance. Complex social, environmental and other issues considered to underlie the emerging infections are listed in Table 5.8. To counter these infections will require modern and efficient surveillance methods so that appropriate responses can be mounted early. In this increasingly complex area, it will be necessary to supplement the conventional microbiological approaches with the potentially more rapid and specific nucleic acid-based technologies.

To monitor the appearances of new viruses requires a number of detection strategies. The 'gold standard' is viral culture. This is expensive and tedious since it is necessary to identify an appropriate host cell for growth and then

suitable markers (e.g. cytopathic effect, viral product, immunochemical characterisation) must be defined. Electron microscopy is an additional approach which has the advantage that it is rapid and can identify unknown species. The hepatitis A virus and the rotaviruses were found in this manner. Immunological detection systems rely on antibodies or antigens either host-derived or prepared in vitro. Although very useful in identifying viruses they are less helpful in understanding the agent's life-cycle and its pathogenic potential.

Nucleic acid-based strategies have similar uses to the immunological techniques, with the added potential that DNA amplification is rapid, it can be more specific than monoclonal antibody-based assays and can detect infectious particles in small numbers. Moreover, sequence data can be generated by PCR. This information on the virus's genome is useful in determining the relatedness of viruses, factors which may be implicated in pathogenesis and potential targets for therapy. In the long term, our deficient knowledge of viral evolution will be greatly enhanced through comparative studies of DNA/RNA sequences.

Hepatitis C virus

It took 13 years after the term 'non-A, non-B' hepatitis was first used for the causative organism to be found. A milestone in molecular virology came when cDNA for the hepatitis C virus was cloned in 1988, years before the virus had been visualised, cultured or even characterised. The structure of the virus is depicted in Figure 5.10. Based on its amino acid sequence and genomic organisation, hepatitis C virus has been classed in a separate genus in the viral family *Flaviviridae*. Modern serological screening tests utilise different combinations of structural and non-structural components of the hepatitis C virus. The conserved DNA sequences at the 5′ end of the virus are ideal when it comes to constructing DNA primers for PCR. Because the hepatitis C virus is an RNA virus it is essential to include an additional step (the making of complementary or cDNA) in the PCR, commonly known as RT-PCR (see Fig. 2.2). Digestion of cDNA with various combinations of restriction enzymes has enabled at least six genotypes (types 1–6) of this virus to be defined. These show both a geographical distribution and the potential to predict prognosis in terms of disease progression and response to treatment.

Infection with the hepatitis C virus is reported throughout the world. Transmission is essentially parenteral and can occur in a number of ways. The most frequently described is from blood or its products, particularly before screening programmes were implemented and viral inactivation steps were included in manufacturing. Another and potentially more serious mode of transmission is by contaminated needles from intravenous drug users. About 80% seropositivity has been reported in populations of US intravenous drug users. Trauma to the skin, e.g. needle-stick injury, acupuncture or tattooing is another way in which blood can be infectious. In contrast to HIV and hepatitis B, sexual transmission is uncommon. Mother-to-child transmission during childbirth has been reported. The single most important factor which determines infectivity of the hepatitis C virus appears to be its titre in the blood. At present, quantitation of the hepatitis C virus is only possible by a PCR-based assay. RNA sequence variation within segments of the viral genome as well as the potential for ongoing variation through selection of different mutants may explain how the immune system is unable to eradicate this virus. In the coming years the clinical consequences of infection with hepatitis C will be considerable (Box 5.14).

Other hepatitis-associated viruses
Hepatitis D, hepatitis E and hepatitis F viruses are also reported. Hepatitis D (delta) and hepatitis E viruses have single-stranded RNA genomes. The hepatitis D virus is infectious only in the presence of the hepatitis B virus. Transmission is similar to that described for hepatitis B and C. Hepatitis D infection produces a more serious disorder than is seen with hepatitis B alone. Hepatitis E is also called enteric non-A, non-B. Virus is spread by the oral–faecal route and is an important cause of hepatitis outbreaks in developing countries. Unlike other hepatitis viruses, hepa-

---

Box 5.14 Clinical consequences of infection with the hepatitis C virus

Clinical presentations of the hepatitis C virus range from an asymptomatic infection first detected in a blood donor who has normal liver function studies to an acute and often mild hepatitis or a chronic hepatitis. Conventional biochemical parameters used to measure progress in 'non-A, non-B' hepatitis initially provided a false sense of security since improving results suggested early resolution following infection. However, results from PCR and liver biopsies have demonstrated persistent hepatitis in these circumstances. The clinical consequences of the hepatitis C virus relate to persistence of infection which occurs in ~80% of cases. This produces a type of chronic hepatitis which can progress to cirrhosis (~20%), liver failure requiring transplantation or hepatocellular carcinoma. Risk factors which predispose to cirrhosis include alcohol-related liver damage and co-existent infection with hepatitis B or HIV. The long-term risk for hepatocellular carcinoma remains to be defined but the underlying mechanism is different from that following infection with hepatitis B since, unlike the latter, the hepatitis C virus does not integrate into the hepatocyte's genome. The reduction in blood transfusion-related hepatitis C following donor screening is gratifying but needs to be balanced by the growing pool of hepatitis C carriers related to intravenous drug use. These include both current drug users as well as those who were occasional users in the past. At present, the only form of therapy in hepatitis C infection is α-interferon, an antiviral and immunoregulatory drug, which is used in different circumstances depending on drug availability and cost considerations. The long-term benefit of this drug in hepatitis C infection remains to be determined.

---

titis E commonly produces intrauterine infection leading to considerable infant morbidity and mortality. A hepatitis G virus was also suspected since some cases of transfusion-acquired hepatitis were not caused by known viruses. Recent identification of this agent was another triumph for molecular technology.

Human herpes virus-6
Another of the new arrivals and a member of the family *Herpesviridae* is human herpes virus-6 (HHV-6, Table 5.3). This was first reported during 1986 in blood mononuclear cells from patients with lymphoproliferative diseases (lymphomas and leukaemias of lymphocyte origin). The virus is predominantly T-cell lymphotropic although it infects other cells including B lymphocytes and brain glial cells. Following viraemia there is frequent persistence of the viral genome in mononuclear cells. Seropositivity as high as 90% in the healthy adult population has been reported. Seroconversion appears to occur in early childhood.

Human herpes virus-6 produces exanthem subitum, an important cause of acute febrile illness in infants or young

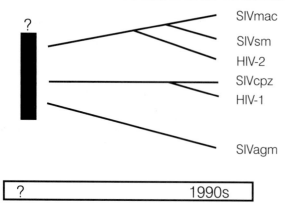

Fig. 5.11 The Lentiviridae family of viruses.
SIVmac – simian immunodeficiency virus in macaque (rhesus)
monkey, SIVsm – sooty mangabey monkey, SIVcpz – chimpanzee,
and SIVagm – African green monkey. The tree was constructed by
comparing DNA sequences (from Hirsch and Johnson 1994).

children. In a study by Pruksananonda et al (1992), 14%
of children in an emergency department had human
herpes virus-6 cultured from their peripheral blood and, at
the same time, viral-specific DNA identified by PCR. The
steps required to culture the virus would not be available
in most laboratories since they involve the isolation of
mononuclear cells, co-cultivation of these with cord blood
cells, the addition of a number of growth factors and finally
confirmation by cytopathic effects and immunofluorescent
staining. In contrast, detection of viral-specific DNA by
PCR is faster and technically less demanding. The latter is
as sensitive as culture in the acute phase and more sensitive
during the convalescence phase. Serology was only helpful
during convalescence with the majority of infected children demonstrating an elevation in their IgG antibody
titres. DNA characterisation of the human herpes virus-6
genome continues and will provide a better understanding
of its evolutionary origin and its apparent ability to remain
dormant. A DNA-derived test will be more rapid and
effective for what is proving to be a common infection in
children.

Human immunodeficiency viruses
HIV-1 and HIV-2 as recently emerged human pathogens
have been discussed earlier (p. 94). Epidemiological evidence would suggest that these viruses have involved
humans in the past 100 years or so and may have
come from Central African monkeys. HIV shares many
features with the simian immunodeficiency viruses (SIV).
These include a similar genetic organisation and the use
of the CD4 molecule on CD4+ T helper lymphocytes,
macrophages and, to a lesser extent, glial cells as a
receptor. Comparisons of protein, DNA and gene structure enable the viruses to be classified into related
groups (Fig. 5.11).

It is presently considered that HIV-2 is an example of a

zoonosis with infection in humans coming from apes. The
source of HIV-1 remains unclear. Its major similiarity is to
SIV-cpz but it is less obvious in which direction infection
has spread, i.e. ape to human or vice versa. Natural
infection with SIV has only been identified in the Old
World primates, e.g. African green monkeys, who commonly have antibodies but do not develop AIDS. On the
other hand, Asian (e.g. macaques) and New World primates are not naturally infected with these viruses and SIV
infection is rare. Although artificial, induction of AIDS by
infecting the macaque with various types of SIV provides a
model to study the evolution and potential treatment of
the human infection. The ability of some animals to adapt
to HIV and not develop AIDS may have important
implications in developing a strategy to control AIDS in
humans.

Taxonomy

Fungi
Fungi are eukaryotes that belong to a separate kingdom
from plants and animals. They are ubiquitous in the
environment and, in addition, have many commercial
applications. In humans, infections caused by fungi (mycoses) can be superficial, e.g. skin, hair, nails, or deep, e.g.
pneumonia, systemic infection and septicaemia. At particular risk are the immunocompromised in whom mycoses
can be life-threatening episodes which are not easily
treated.

The large diversity in fungal morphology, their ecologi-
cal habitats and wide spectrum of clinical consequences
have complicated diagnostic approaches and our understanding of infections caused by these organisms. An
important objective in studying the fungi is to deduce
evolutionary comparisons and from these determine relatedness. Fungal phylogenetics relies on parameters
described earlier such as morphology, biochemistry and
staining characteristics. In addition, life-cycles, e.g. morphological appearances, particularly of the sexual
reproductive structures, provide additional sources of
information for comparison. However, traditional typing
methods are not always helpful or sensitive enough and
even life-cycles can be uninformative, e.g. the presence
of asexual forms of Coccidioides immitis has made classification particularly difficult.

To demonstrate the utility of DNA sequencing in fungal
phylogenetics, one study has compared DNA sequences
for the 18S ribosomal DNA (riDNA) in a number of fungi
which produce superficial or deep mycoses. Other related
microorganisms were similarly sequenced. 18S riDNA was
chosen since it is highly conserved during evolution. By use
of various computer programs which allowed DNA sequences to be aligned along multiple regions, it became
possible to bypass morphological traits and produce an
evolutionary relatedness between various microorganisms
which was based on their DNA sequences (Fig. 5.12). In

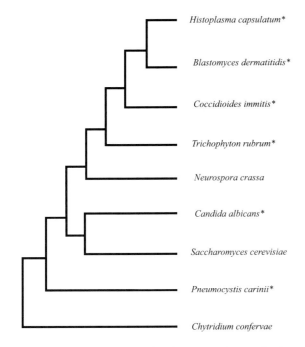

Histoplasma capsulatum*

Blastomyces dermatitidis*

Coccidioides immitis*

Trichophyton rubrum*

Neurospora crassa

Candida albicans*

Saccharomyces cerevisiae

Pneumocystis carinii*

Chytridium confervae

Fig. 5.12 Evolutionary relationships between pathogenic fungi and a number of other organisms.
A comparative tree based on a model of maximum parsimony. DNA sequences from various positions in 18S riDNA were used to create the tree. Pathogens are marked *; *Chytridium confervae* is a water mold which functions as a reference organism. The diagram illustrates the relationship between organisms as well as the fact that pathogens *Candida albicans* and *Pneumocystis carinii* are not closely related to each other or to other pathogens depicted. (Modified from Bowman et al 1992.)

Table 5.9 Clinical infections for which DNA or RNA diagnosis is presently of value.

Infections	Type rDNA test available	Advantages rDNA testing
Meningitis/ encephalitis (diagnosis)	PCR: *M. tuberculosis*, herpes simplex virus, rRNA: mycobacteria	Avoids long-term culture, likelihood negative microscopy (TB). Avoids brain biopsy, early detection system (HSV). Commercial rRNA kits: very promising results
Leprosy (diagnosis)	PCR: *M. leprae*	*M. leprae* cannot be cultured; few present in tuberculoid leprosy
Cytomegalovirus in the immuno-compromised (diagnosis)	PCR: CMV	Reactivation of latent infection/transfer CMV with organ transplants requires assessment of CMV status in transplantation
Sexually transmitted diseases (diagnosis)	PCR: rRNA: chlamydia	PCR still being evaluated to compare with immunofluorescence/enzyme immunoassay approaches. rRNA kits compare favourably with above methods (*C. trachomatis*)
Lyme borreliosis (diagnosis and epidemiology)	PCR: *B. burgdorferi*	Few organisms, difficult to culture, atypical clinical presentations, serological responses slow to develop
Legionella (diagnosis and epidemiology)	rRNA: PCR: PFGE: *L. pneumophila*, *L. longbeachae*	Fastidious grower; typing for epidemiological purposes
Intestinal infections (diagnosis and virulence)	rRNA; PCR: *Campylobacter* spp., PCR: rotavirus A,B,C	Fastidious growers (Campylobacter); avoidance immunoassay, electron microscopy or RNA characterisation (rotavirus)

TB, tuberculosis; HSV, herpes simplex virus; CMV, cytomegalovirus. PFGE, pulsed-field gel electrophoresis

this way four human pathogens (*Trichophyton rubrum*, *Histoplasma capsulatum*, *Blastomyces dermatitidis* and *Coccidioides immitis*) were confirmed to be closely related. Another opportunistic pathogen *Candida albicans* was shown to be closely related to the common yeast *Saccharomyces cerevisiae* but distinct from *Pneumocystis carinii*.

Pneumocystis carinii

This is a respiratory organism which produces a very common and serious infection in AIDS as well as in other immunocompromised patients (Box 5.15). Culture of human-derived *P. carinii* has been difficult. Its life-cycle and metabolic processes are poorly understood. The previous taxonomic classification as a protozoan is now considered to be incorrect since rRNA sequence comparisons would place this organism closer to the fungi or even in a distinct lineage. Analysis of the *P. carinii* genome is presently underway and from this it will be possible to appreciate better its evolution and biology.

Protozoans

A dilemma in classification similar to that described for the fungi applies to the protozoans, e.g. *Entamoeba histolytica*, *Giardia lamblia*, *Trichomonas vaginalis*, *Plasmodium falciparum* and *Toxoplasma gondii*. Many of the protozoan parasites infect humans and other animals. Because these are complex eukaryotes, investigation and culture has proven to be more difficult. In a number of cases the organisms' life-cycles remain to be fully understood. Not surprisingly, the same strategy described above with the fungi, i.e. comparisons of 18S riDNA sequences, has also proven to be beneficial in taxonomic classification of these organisms.

Box 5.15 *P. carinii*, its detection and the immunocompromised host

The natural history and epidemiology of *P. carinii* infections remain to be understood. Whether pneumocystis pneumonia in patients with AIDS or those immunocompromised for other reasons is the result of reactivation of a latent infection or an acquired event is not entirely clear. The ability to identify small numbers of organisms which can then be subtyped will be essential to answer these questions. Diagnosis of pneumocystis infection is difficult since for practical purposes the organisms will not grow in vitro. The present 'gold standard' for diagnosis in the clinical laboratory is immunofluorescent antibody staining. Tissues in which the organisms are sought include induced sputums and the more invasive bronchoalveolar lavage. A recent comparative study has shown that a PCR-based test is more sensitive than immunofluorescent antibody assays with induced sputum specimens which are likely to contain fewer microorganisms but are considerably easier to obtain (Cartwright et al 1994).

Box 5.16 PCR technology and the detection of microorganisms

DNA amplification by PCR is an important new laboratory diagnostic tool.

Advantages
- Speed, automation
- High degree of sensitivity and specificity
- Crude extract satisfactory
- Inoculum need not be viable
- Target need not be infectious (e.g. provirus)
- DNA or RNA suitable

Disadvantages
- DNA/RNA sequence required
- Contamination
- Quantification difficult
- Colonisation versus invasion can be difficult to distinguish
- Cultures are not available for further characterisation, e.g. typing

CONCLUSIONS

A number of considerations will influence the types of assays used in a microbiological laboratory. These include: cost, sensitivity, specificity, prevalence, demand, location as well as the equipment and expertise found in that laboratory. The exciting prospects offered by rDNA techniques have been quickly adopted by the research laboratories. On the other hand, traditional approaches remain firmly entrenched in many of the diagnostic laboratories. The reasons are varied. It is still cheaper and faster to culture many of the pathogens. Culture can also provide additional information such as antibiotic sensitivity, virulence factors, antigen status and strain variation. At present, DNA approaches are sought in situations for which conventional tests are unavailable, expensive, time-consuming or potentially hazardous (Table 5.9).

Nucleic acid hybridisation techniques show promise, particularly the use of the repetitive rRNA-derived probes. Disadvantages with hybridisation assays include the lack of automation and the requirement for skilled personnel. These are being resolved with the increasing availability of DNA hybridisation kits. DNA kits need not identify pathogens directly in clinical specimens but can also be used to confirm cultures more rapidly than would otherwise be possible. Although introduced only in 1985, PCR has emerged as a front runner for the identification and study of pathogens (Box 5.16). Automation is available and the technology is continually being simplified. Multiplex PCR will increase cost-effectiveness by enabling simultaneous detection of a number of microorganisms as well as their pathogenic features, e.g. virulence factors, antibiotic-resistance plasmids.

The potential role of rDNA in the areas of infection control and epidemiology has had a lower profile compared to its utility in diagnostic testing. Nevertheless, essential information to identify the source of an infection or the way it has spread is frequently incomplete after conventional approaches are exhausted. Knowledge of an organism's life-cycle, particularly newly evolved pathogens or re-emerging ones, may be obtained more easily through DNA characterisation. Other future contributions likely to come from molecular medicine will include the earlier identification of multidrug resistance and the ability to monitor more effectively the responses to therapy, particularly in the viral infections.

FURTHER READING

General

Balows A, Hausler W J, Herrmann K L, Isenberg H D, Shadomy H J (eds) 1991 Manual of Clinical Microbiology, 5th edn. American Society for Microbiology, Washington DC

Harper D R 1994 Molecular virology. BIOS Scientific Publishers Ltd, Oxford

Mims C A, Playfair J H L, Roitt I M, Wakelin D, Williams R, Anderson R M 1993 Medical Microbiology. Mosby, St Louis

Naber S P 1994 Molecular medicine. Molecular pathology – diagnosis of infectious disease. New England Journal of Medicine 331: 1212–1215 van Belkum A 1994 DNA fingerprinting of medically important microorganisms by use of PCR. Clinical Microbiology Reviews 7: 174–184

Wilson S M 1991 Nucleic acid techniques and the detection of parasitic diseases. Parasitology Today 7: 255–259

Laboratory detection

Becker Y, Darai G (eds) 1992 Diagnosis of human viruses by polymerase chain reaction technology. Springer-Verlag, Berlin

Char S, Farthing M J G 1991 DNA probes for diagnosis of intestinal infection. Gut 32: 1–3

Ferre F 1994 Polymerase chain reaction and HIV. In: Pomerantz R J (ed) Clinics in laboratory medicine: HIV/AIDS, vol 14. W B Saunders Co, Philadelphia, p 313–333

Hopkin J M, Wakefield A E 1990 DNA hybridization for the diagnosis of microbial disease. Quarterly Journal of Medicine 277: 415–421

Jackson J B 1991 Polymerase chain reaction assay for detection of hepatitis B virus. American Journal of Clinical Pathology 95: 442–444

Kakaiya R, Miller W V, Gudino M D 1991 Tissue transplant-transmitted infections. Transfusion 31: 277–284

Noordhoek G T, Kolk A H J, Bjune G et al 1994 Sensitivity and specificity of PCR for detection of Mycobacterium tuberculosis: a blind comparison study among seven laboratories. Journal of Clinical Microbiology 32: 277–284

Peckham C, Gibb D 1995 Mother-to-child transmission of the human immunodeficiency virus. New England Journal of Medicine 333: 298–302

Quinn T C 1996 Global burden of the HIV pandemic. Lancet 348: 99–106

Ratcliff R M, Goodwin A A, Lanser J A 1994 Use of gene amplification to detect Clostridium difficile in clinical specimens. Pathology 26: 477–479

Roberts C R, Longfield J N, Platte R C, Zielmanski K P, Wages J, Fowler A 1994 Transfusion-associated human immunodeficiency virus type 1 from screened antibody-negative blood donors. Archives of Pathology and Laboratory Medicine 118: 1188–1192

Shankar P, Manjunath N, Mohan K K, Prasad K, Behari Shriniwas M, Ahuja G K 1991 Rapid diagnosis of tuberculous meningitis by polymerase chain reaction. Lancet 337: 5–7

Sheppard H W, Ascher M S, Busch M P et al 1991 A multicenter proficiency trial of gene amplification (PCR) for the detection of HIV-1. Journal of Acquired Immune Deficiency Syndromes 4: 277–283

Smith K L, Dunstan R A 1993 PCR detection of cytomegalovirus: a review. British Journal of Haematology 84: 187–190

Tyler K L 1994 Polymerase chain reaction and the diagnosis of viral central nervous system diseases. Annals of Neurology 36: 809–811

Epidemiology

Bannerman T L, Hancock G A, Tenover F C, Miller J M 1995 Pulsed-field gel electrophoresis as a replacement for bacteriophage typing of Staphylococcus aureus. Journal of Clinical Microbiology 33: 551–555

Esteban J I, Gomez J, Martell M et al 1996 Transmission of hepatitis C virus by a cardiac surgeon. New England Journal of Medicine 334: 555–560

Krause R M 1992 The origin of plagues: old and new. Science 257: 1073–1078

Liveris D, Gazumyan A, Schwartz I 1995 Molecular typing of Borrelia burgdorferi sensu lato by PCR-restriction fragment length polymorphism analysis. Journal of Clinical Microbiology 33: 589–595

Nocton J J, Dressler F, Rutledge B J, Rys P N, Persing D H, Steere A C 1994 Detection of Borrelia burgdorferi DNA by polymerase chain reaction in synovial fluid from patients with Lyme arthritis. New England Journal of Medicine 330: 229–234

Pfaller M A, Wakefield D S, Hollis R, Fredrickson M, Evans E, Massanari R M 1991 The clinical microbiology laboratory as an aid in infection control. The application of molecular techniques in epidemiologic studies of methicillin-resistant Staphylococcus aureus. Diagnostic Microbiology and Infectious Disease 14: 209–217

Rosa P A, Hogan D, Schwan T G 1991 Polymerase chain reaction analyses identify two distinct classes of Borrelia burgdorferi. Journal of Clinical Microbiology 29: 524–532

Disease pathogenesis

Alland D, Kalkut G E, Moss A R et al 1994 Transmission of tuberculosis in New York city: an analysis by DNA fingerprinting and conventional epidemiologic methods. New England Journal of Medicine 330: 1710–1716

Bryson Y J, Pang S, Wei L S, Dickover R, Diagne A, Chen I S Y 1995 Clearance of HIV infection in a perinatally infected infant. New England Journal of Medicine 332: 833–838

Davies J 1994 Inactivation of antibiotics and the dissemination of resistance genes. Science 264: 375–382

Evans D M A, Dunn G, Minor P D et al 1985 Increased neurovirulence associated with a single nucleotide change in a noncoding region of the Sabin type 3 poliovaccine genome. Nature 314: 548–550

Gross Fisher S 1994 Epidemiology: a tool for the study of human papillomavirus-related carcinogenesis. Intervirology 37: 215–225

Heym B, Honore N, Truffot-Pernot C et al 1994 Implications of multidrug resistance for the future of short-course chemotherapy of tuberculosis: a molecular study. Lancet 344: 293–298

Honore N, Cole S T 1993 Molecular basis of rifampicin resistance in Mycobacterium leprae. Antimicrobial Agents and Chemotherapy 37: 414–418

Kandolf R, Hofschneider P H 1989 Enterovirus-induced cardiomyopathy. In: Notkins A L, Oldstone M B (eds) Concepts in viral pathogenesis III. Springer-Verlag, New York, p 282–290

Malo D, Skamene E 1994 Genetic control of host resistance to infection. Trends in Genetics 10: 365–371

McGuire W, Hill A V S, Allsopp C E M, Greenwood B M, Kwjatkowski D 1994 Variation in the TNF-α promotor region associated with susceptibility to cerebral malaria. Nature 371: 508–511

Moradpour D, Wands J R 1995 Understanding hepatitis B virus infection. New England Journal of Medicine 332: 1092–1093

Sato S, Suzuki K, Akahane Y et al 1995 Hepatitis B virus strains with mutations in the core promotor in patients with fulminant hepatitis. Annals of Internal Medicine 122: 241–248

Scorpio A, Zhang Y 1996 Mutations in *pncA*, a gene encoding pyrazinamidase/nicotinamidase, cause resistance to the antituberculous drug pyrazinamide in tubercle bacillus. Nature Medicine 2: 662–667

Thursz M R, Kwiatkowski D, Allsopp C E M, Greenwood B M, Thomas H C, Hill A V S 1995 Association between an MHC class II allele and clearance of hepatitis B virus in the Gambia. New England Journal of Medicine 332: 1065–1069

van Es H H G, Karcz S, Chu F, Cowman A F, Vidal S, Gros P, Schurr E 1994 Expression of the plasmodial *pfmdr1* gene in mammalian cells is associated with increased susceptibility to chloroquine. Molecular and Cellular Biology 14: 2419–2428

Wellems T E 1991 Molecular genetics of drug resistance in *Plasmodium falciparum* malaria. Parasitology Today 7: 110–112

Diversity

Bowman B H, Taylor J W, White T J 1992 Molecular evolution of the fungi: human pathogens. Molecular Biology and Evolution 9: 893–904

Cartwright C P, Nelson N A, Gill V J 1994 Development and evaluation of a rapid and simple procedure for detection of *Pneumocystis carinii* by PCR. Journal of Clinical Microbiology 32: 1634–1638

Cuthbert J A 1994 Hepatitis C: progress and problems. Clinical Microbiology Reviews 7: 505–532

Hirsch V M, Johnson P R 1994 Pathogenic diversity of simian immunodeficiency viruses. Virus Research 32: 183–203

Lederberg J, Shope R E, Oaks S C (eds) 1992 Emerging infections: microbial threats to health in the United States. National Academy Press, Washington

Le Guenno B 1995 Emerging viruses. Scientific American 273: 56–64

Morse S S, Schluederberg A 1990 Emerging viruses: the evolution of viruses and viral diseases. Journal of Infectious Diseases 162: 1–7

Pays E 1991 Genetics of antigenic variation in African trypanosomes. Research in Microbiology 142: 731–735

Pruksananonda P, Breese Hall C, Insel R A et al 1992 Primary human herpesvirus 6 infection in young children. New England Journal of Medicine 326: 1445–1450

Quinn T C 1994 Population migration and the spread of types 1 and 2 human immunodeficiency viruses. Proceedings of the National Academy of Sciences USA 91: 2407–2414

6

MEDICAL ONCOLOGY

INTRODUCTION

'Cancer is essentially a genetic disease at the cellular level' (W F Bodmer). Evidence for a genetic component in cancer includes:

- An increased risk for tumour development in those with a genetic defect in DNA repair
- Chemicals or physical agents which mutate DNA also elicit tumours in animals
- Some structural chromosomal rearrangements in the germline predispose to tumour development
- Somatically acquired mutations in the DNA of tumour cells can resemble those seen in familial cancers
- The malignant phenotype can be conferred on normal cells by gene transfer studies with oncogenes
- The malignant phenotype can be induced in vivo by gene manipulation in transgenic mice.

For over 30 years, tumourigenesis was hypothesised to represent a *multistep process*. However, it was only with the application of recombinant DNA techniques that evidence for this became available. It is now possible to identify molecular (DNA) changes which are responsible for the initiation, promotion and progression of cancers. Understanding the pathogenesis of cancer has become a realistic goal within the framework of molecular medicine. The ability to define mutations at the DNA level has enhanced the accuracy of diagnosis. Therapeutic options based on our knowledge of the DNA changes in cancer are now being attempted.

An individual's response to cancer is complex. It involves a number of factors such as the person's immunological state, his/her performance status, the extent of disease, its response to treatment and the development of drug resistance. In addition there are *genetic changes* which involve oncogenes, tumour suppressor genes, repair genes and perhaps epigenetic modifications. Genetic changes will be discussed in this chapter. The molecular basis of drug resistance will be reviewed in Chapter 7.

Since nomenclature in molecular oncology is cumbersome, a number of terms need definition. **Oncogenes** (*onkos* – the Greek word for mass or tumour) are genes associated with neoplastic proliferation following a mutation or perturbation in their expression. Oncogenes have normal counterparts in the genome. These antecedent genes or **proto-oncogenes** play an essential physiological role in normal cellular growth control. v-Onc or c-onc are abbreviations for oncogenes found in retroviruses (v) or their cellular equivalents (c). **Tumour suppressor genes** (also called recessive oncogenes, anti-oncogenes or growth suppressor genes) are genes which also play a role in normal cellular proliferation and differentiation. Loss or inactivation of tumour suppressor genes can lead to neoplastic changes. **Transformation** refers to the acquisition by normal cells of the neoplastic phenotype. **Transfection** refers to the procedure of gene transfer in which segments of DNA are incorporated into eukaryotic cells. **Transduction** describes the transmission of a gene from one cell to another by viral infection. In the context of the oncogenes, transduction has been used to describe the origin of retroviral oncogenes from cellular proto-oncogenes.

ONCOGENES

Retroviruses

The RNA tumour viruses (retroviruses) provided the first proof that genetic factors can play a role in carcinogenesis. Retroviruses have three core genes (*env, gag,* coding for structural proteins, and *pol*, coding for reverse transcriptase) (Fig. 6.1). As discussed in Chapters 1 and 2, reverse transcriptase is an enzyme which allows RNA to be converted into complementary DNA. In this way the retrovirus is able to make a DNA copy of its RNA which can then become incorporated into the host's genome (Fig. 6.2).

An additional gene gives retroviruses the ability to induce tumour growth in vivo or to transform cells in vitro. In the latter situation, cells lose their normal growth characteristics and acquire a neoplastic phenotype. An example of this would be loss of contact inhibition so that instead of growth in a single cell layer in vitro, there is unregulated proliferation into clumps. Retroviral DNA sequences which are responsible for transforming properties are called viral oncogenes (v-onc). Names are derived from the tumours in which they were first described. For example, v-*sis*, *s*imian *s*arcoma virus; v-*abl*, murine *Abelson leukaemia virus*; v-*mos*, *Moloney sarcoma virus*, v-*ras*, *rat sarcoma virus*.

Viral oncogenes were subsequently shown to have cellular homologues called cellular oncogenes (c-onc). The term proto-oncogene was coined to describe cellular oncogenes which do not have transforming potential to form neoplasms in their native state. Proto-oncogenes are

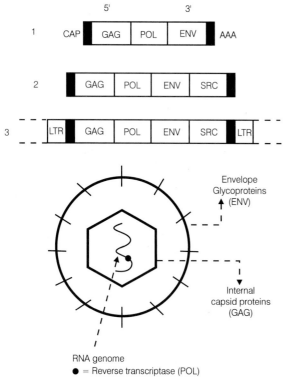

Fig. 6.1   The structure of a retrovirus.
(1) RNA tumour viruses (retroviruses) have an RNA genome. This RNA has two features of eukaryotic mRNA, i.e. a capped 5′ end and a poly-A tail at the 3′ end. Retroviral RNA codes for three viral proteins: (i) a structural capsid protein *(gag)* which associates with the RNA in the core; (ii) the enzyme reverse transcriptase *(pol)* and (iii) an envelope glycoprotein *(env)* which is associated with the lipoprotein envelope of the virus. (2) Transforming retroviruses have an oncogene. In the example here the oncogene is that of the Rous sarcoma virus *(src)*. (3) Retroviruses are so named because they have a RNA genome and are able to replicate through formation of an intermediate (provirus) which involves integration of the retroviral genome into that of host DNA. The provirus has LTRs (**l**ong **t**erminal **r**epeats) on either side of the RNA genes. The LTRs are several hundred base pairs in size and insert adjacent to smaller repeats derived from host DNA.

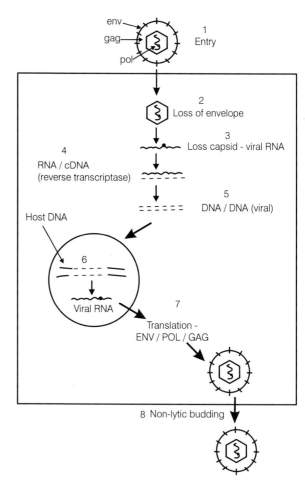

Fig. 6.2   Life-cycle of a retrovirus.
(1) The envelope protein enables the retrovirus to bind to the surface of host cells on infection. (2–5) Double-stranded DNA derived from viral RNA and the action of reverse transcriptase is required before the retroviral genome can be integrated into that of the host. (6–8) The provirus so formed replicates to produce mature viral particles which are extruded from the cell by non-lytic budding.

now known to play key roles in control of cell growth and will be discussed in more detail below.

The initial observation implicating viruses and cancer came in 1910 when Rous demonstrated that a filterable agent (virus) was capable of inducing cancers in chickens. Fifty-six years later, he was awarded a Nobel Prize for this work. Retroviruses have since been identified as the cause of cancers in many species, including mammals, although much less frequently in humans. Nevertheless, the direct relevance of the retroviruses to human cancers remained unclear until alternative lines of investigation became available. These included the availability of gene transfer studies and the potential to identify mutations affecting DNA or chromosomes.

Gene transfer

Gene transfer is illustrated in Figure 6.3. In brief, DNA derived from a malignant tumour is transfected or transferred into a cell line. In normal circumstances, cell lines grow as confluent monolayers with each cell remaining discrete from its neighbour. In the presence of a transforming oncogene, cells can form clumps. Additional evidence that the cells in clumps have acquired a neoplastic phenotype is obtainable by inoculating them into immunologically deficient 'nude' mice. Normal cells would not grow in these circumstances. Transformed cells will produce tumours.

Gene transfer studies have demonstrated the relevance of the oncogenes to human cancers such as bladder, colon and lung cancers and many others. New oncogenes have been found in this way. From the type of approach described above it is possible to fractionate DNA so that eventually the oncogene can be cloned. This approach was first successful with a human bladder cell line called T24. The oncogene isolated was H-*ras*. The in vivo equivalent of gene transfer is the transgenic mouse. Transgenic models can be produced with an oncogene linked to a tissue-specific promotor. In these circumstances, e.g. *ras*, *myc* transgenics, tumours develop but after a long latency.

Proto-oncogenes

There are many proto-oncogenes. Their roles in terms of cellular growth control are complex and may involve interactions between a number of the proto-oncogenes. The sites and modes of action of proto-oncogenes can be divided into: (1) growth factors, (2) growth factor receptors, (3) protein kinases, (4) signal transducers, and (5) nuclear proteins and transcription factors (Fig. 6.4, Table 6.1).

Growth factors

Growth factors are molecules which act via cell surface receptors to induce cellular division. The expression in a cell of a growth factor which normally does not function in that cell is one example of oncogene activation.

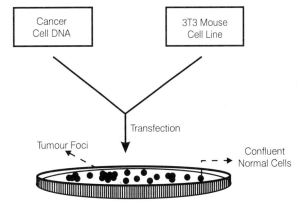

Fig. 6.3 Gene transfer to identify oncogenes.
DNA derived from cancer cells is transfected into a susceptible cell line such as the NIH 3T3 fibroblast mouse line. Transfection can be induced by co-precipitation of DNA with calcium phosphate. An alternative and more efficient method involves electroporation, i.e. DNA uptake occurs through permeabilising the cell to high molecular weight molecules by exposing it to an intense electrical field. Gene transfer studies enable new oncogenes to be identified and their role to be studied. Oncogenes can be cloned by fractionating DNA isolated from tumour foci.

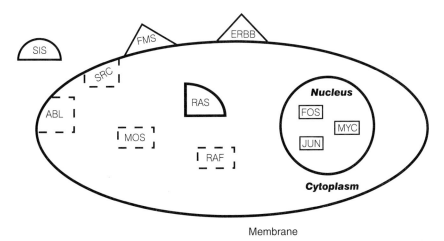

Fig. 6.4 The cellular localisation of proto-oncogene products (refer also to Table 6.1).

Table 6.1   Some proto-oncogenes and their modes of action.
Proto-oncogenes are normal genes involved in cellular growth control.

Proto-oncogene class	Example proto-oncogene	Function in the cell
Growth factors	*sis*	B chain of PDGF (platelet-derived growth factor)
Growth factor receptors	*erbB*	Epidermal growth factor receptor
	fms	CSF-1 receptor (colony stimulating factor)
Protein kinases	*abl, src*	Protein tyrosine kinases
	mos, raf	Protein serine kinases
Signal transduction (G proteins)	H-, K- and N-*ras*	GTP-binding/GTPase
Nuclear proteins	*myc*, N-*myc*, *myb, fos*, *jun*	DNA-binding proteins DNA-binding proteins Transcription factor

Table 6.2   DNA changes in proto-oncogenes and tumour development.
Some molecular rearrangements found in the proto-oncogenes and their neoplastic associations (from Bishop 1991).

Proto-oncogene	DNA abnormality	Human neoplasm(s)
abl	Translocation	Chronic granulocytic leukaemia
myc	Translocation	Burkitt lymphoma
erbB	Amplification	Squamous cell carcinoma, astrocytoma, carcinoma oesophagus
N-myc	Amplification	Neuroblastoma, small cell carcinoma of lung
K-ras	Point mutations	Carcinomas of colon, lung, pancreas, melanoma
N-ras	Point mutations	Acute myeloid and lymphoblastic leukaemias, carcinoma of thyroid, melanoma
H-ras	Point mutations	Carcinomas of genitourinary tract, thyroid

Growth factor receptors

Several proto-oncogene-derived proteins form components of growth factor receptors at the cell surface. Binding of growth factors and their respective membrane receptors is the first step in the delivery of mitogenic signals to the cell's interior to initiate cell division. Structural changes in these proto-oncogenes can initiate tumour formation.

Protein kinases

Protein products of these proto-oncogenes are associated with the inner surface of the plasma membrane or the cytoplasm. Structural changes increase the kinase activity in these genes which may then affect the membrane receptors or signal transduction.

Signal transduction

Once an extracellular growth factor has bound to its cell surface receptor it is able to transfer (transduce) its signal to the nucleus by a number of processes, some of which may involve 'second messengers'. GTP (**g**uanosine **trip**hosphate) and proteins that bind GTP (known as G proteins) play an important role in intracellular signal transduction. GTP is converted to GDP (**g**uanosine **dip**hosphate) by the GTPase activity of the G proteins.

The *ras* proto-oncogenes (namely H-*ras*-1, K-*ras*-2, N-*ras*) comprise a multigene family which is present in eukaryotic organisms as divergent as yeast and humans. This type of evolutionary conservation suggests that the *ras* genes have important functional significance. *Ras* proto-oncogenes encode related proteins of 21 kDa size called p21. Proteins derived from the *ras* proto-oncogenes have GTPase and GTP- and GDP-binding activities, i.e. they are related to G proteins. One physiological function of the *ras* proto-oncogenes is in modulation of cellular proliferation via the transduction of signals from the cell surface to the nucleus.

Wild-type *ras* p21 binds GTP and thence becomes inactive following conversion of GTP to GDP (see p. 121 for a continuation of the G protein story involving mutant *ras*).

Nuclear proteins and transcription factors

An increasing number of proto-oncogenes have been shown to encode nuclear-binding proteins. Some of these are previously described transcription factors which have an important role to play in the control of gene expression.

Oncogene activation

From molecular studies, a number of mechanisms have been defined in which the normal products of proto-oncogenes can be disrupted to produce uncontrolled cell division and hence neoplastic transformation (summarised in Table 6.2).

Chromosomal translocation

Changing the positions of oncogenes within the human genome can alter their function. A way to do this is through a translocation, i.e. the movement of a segment of a chromosome to another chromosome. Chromosomal breakpoints which produce the Philadelphia chromosome found in chronic granulocytic leukaemia, involve a translocation of the c-*abl* oncogene on chromosome 9 to a gene on chromosome 22 which is called the breakpoint cluster region (*bcr*). At the same time, the *sis* proto-oncogene on chromosome 22 is translocated to chromosome 9

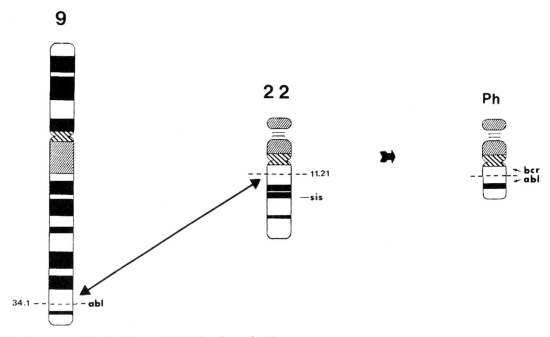

Fig. 6.5 Chromosomal translocation resulting in altered gene function.
A reciprocal translocation between chromosomes 9 and 22 produces the Philadelphia chromosome found in chronic granulocytic leukaemia.
The *sis* proto-oncogene is not considered to have a functional effect from this translocation because it is located at some distance
(22q12.3–q13.1) from the actual chromosome 22 breakpoint (22q11.21). – – – = breakpoints.

(Fig. 6.5). The hybrid *bcr-abl* transcript produces a novel protein (p210) with tyrosine kinase activity. Transgenic mice which express the *bcr-abl* fusion gene develop lymphoblastic leukaemia/lymphoma.

Gene amplification
Quantitation of gene numbers is possible by DNA mapping. Amplified *c-myc* sequences have been described in a number of human cancers. In the childhood tumour neuroblastoma, the N-*myc* gene may become amplified up to 300 times.

Point mutations
Single base changes in the DNA sequence of proto-oncogenes have been consistently observed in the different *ras* genes. A wide range of malignancies demonstrate *ras* mutations. How these mutations produce their effects remains unknown although it is noteworthy that the sites for mutations in *ras* are limited to codons 12, 13 and 61 which are located within the regions coding for binding of GTP/GDP. Therefore, failure by *ras* to convert the active complex (GTP-*ras*) to inactive GDP-*ras* would produce an excess of stimulatory activity leading to unregulated cellular proliferation.

Viral insertion
Another mode by which proto-oncogene function can be perturbed is via insertion of viral elements. An example of this is the hepatitis B virus which is discussed in more detail below (p. 135).

TUMOUR SUPPRESSOR GENES

The identification of proto-oncogenes and oncogenes in the pathogenesis of cancer was an exciting development in molecular medicine. However, results from in vitro and in vivo studies were not always consistent and only about 20% of human tumours demonstrated changes in these genes. Oncogenes were not abnormal in the inherited cancer syndromes. Thus, other molecular explanations were sought.

Perturbations in genes can produce stimulatory or inhibitory signals to cell growth. Examples of *stimulatory* effects are the proto-oncogenes described above. In these situations, mutations produce positive signals leading to uncontrolled proliferation. On the other hand, tumour formation can result through loss of *inhibitory* function which is associated with another group of genes known as the *tumour suppressor genes*. Evidence for these in the pathogenesis of cancer has come from three observations: (1) somatic cell hybrids and more recently transgenic mice produced by gene 'knock-out', (2) inherited cancer syndromes, and (3) loss of heterozygosity in tumours.

Experimental evidence

Somatic cell hybrids

In the late 1960s, murine cell hybrids formed by fusions between normal and tumour cells were found to revert to the normal phenotype. Subsequently, as the hybrid clones were propagated in culture, the tumour phenotype became re-established. This effect was seen in a wide range of tumour lines and was considered to indicate the influence of tumour suppressor genes derived from the normal cells. Subsequent loss of chromosomes, which occurred on serial passage of cell lines, enabled reversion to the neoplastic phenotype when the tumour suppressor genes were lost. Today, micro-cell transfers allow one or a few chromosomes from normal cells to be delivered within a reconstituted membrane to recipient tumour cells. In this way, it has been possible to identify tumour suppressor genes in a variety of chromosomes. Sophisticated molecular techniques such as gene 'knock-out' enable specific genes to be inactivated in

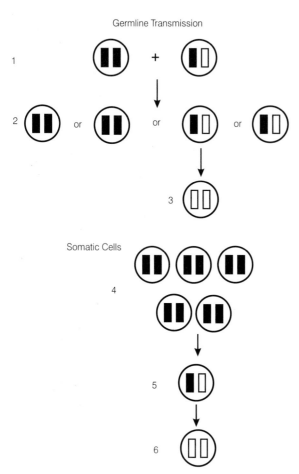

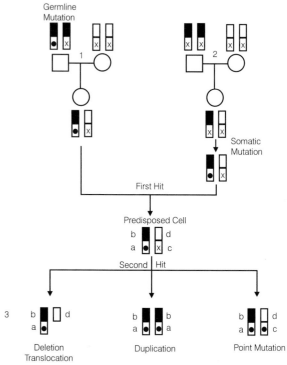

Fig. 6.6 Two-hit model for tumourigenesis at the gene level.
The first or predisposing event can be inherited through the germline or arise in somatic cells. A second event is required to inactivate the remaining normal allele. **(1)** First event occurring in the germline. The wild-type tumour suppressor gene for a particular locus is shown as ■. The first hit affects one of the germ cells to inactivate one tumour suppressor gene □. **(2)** Offspring have a 50% chance of inheriting the mutant tumour suppressor gene (dominant inheritance). **(3)** A second hit is required to inactivate the remaining tumour suppressor gene. Because the germ cells were initially involved all cells will be predisposed. Therefore, the second hit is more likely to happen in the appropriate cell, e.g. retinal cells in retinoblastoma. This produces multifocal tumours which present at an earlier age. Other members of the family can become affected. **(4)** Sporadic tumour formation requires two separate somatic events. **(5)** If the first hit affects a somatic cell it inactivates the tumour suppressor gene in that cell and its progeny. To get both tumour suppressor genes inactivated the second hit **(6)** must involve the predisposed line. This is less likely to occur than the situation in **(3)**. Tumour formation is delayed. If it occurs it is sporadic and unifocal.

Fig. 6.7 Molecular (DNA) defects which produce a second (somatic) loss at a tumour suppressor gene locus.
The mutation in the tumour suppressor gene is depicted by ●, the normal allele by X. **(1)** In the first family, there is inheritance of a germline mutation. **(2)** In the second family the germline is normal but a somatic mutation occurs. In both situations the end result is a predisposed cell. DNA markers for the two alleles in the predisposed cell are indicated by b/a (mutant) and d/c (normal). **(3)** Loss of the remaining (normal) tumour suppressor gene via a second hit can occur by a number of different mechanisms, e.g. from left to right: deletion or an unbalanced translocation, a duplication, or a more discrete abnormality such as a point mutation which occurs on the second (normal) allele. Loss of heterozygosity at the DNA level can be illustrated if it is presumed that there are four informative DNA markers which distinguish the two parental contributions and are situated at the tumour suppressor location. For simplicity, the polymorphic markers are depicted as 'a' (mutant); 'b, c and d' (normal) which would give two haplotypes of b,a/d,c. Loss of heterozygosity is seen as b,a/d– (deletion, translocation) and b,a/b,a (duplication). Loss of heterozygosity will not be apparent in the point mutation situation since the normal locus (identified by the 'c' marker) is present but non-functional since it has acquired a discrete mutation.

transgenic mice. This shows definitively that a particular gene functions as a tumour suppressor (see *P53* below and Ch. 10 for further discussion on gene 'knock-out').

Inherited cancer syndromes

Familial recurrence is seen in most tumours although the proportion is often low. One exception is the tumour retinoblastoma for which the inherited form comprises up to 40% of cases. The remainder are sporadic. Although rare, retinoblastoma is the most common intraocular malignancy in children with an incidence of about 1 in 14 000 live births. In familial cases, about half of the children can be affected, i.e. an autosomal dominant mode of transmission. In the early 1970s, epidemiological studies of both retinoblastoma and Wilms tumour led Knudson to propose his *two-hit model of tumourigenesis.*

Knudson's model required the tumour cells, in either the sporadic or genetic forms of retinoblastoma, to acquire two separate genetic changes in DNA before a tumour developed. The *first* or predisposing event could be inherited *either* through the germline (familial retinoblastoma) *or* it could arise de novo in somatic cells (sporadic retinoblastoma). The *second* event occurred in somatic cells. Thus, in sporadic retinoblastoma both events arose in the retinal (somatic) cells. In familial retinoblastoma the individual had

already inherited one mutant gene and required only a second hit affecting the remaining normal gene in the somatic cells. The frequency of somatic mutations was sufficiently high that those who had inherited the germline mutation were likely to develop one or more tumours (Fig. 6.6).

The mode of inheritance for retinoblastoma is dominant with incomplete penetrance. However, at the cellular level it is recessive since loss or inactivation of both alleles is required to change a cell's phenotype. A cell containing only one of its two normal alleles will usually produce enough tumour suppressor gene product to remain normal. Loss of the cell's remaining wild-type allele exposes it to

Table 6.3 Loss of heterozygosity in some human tumours.
Deleted loci will identify the potential location of tumour suppressor genes.

Tumour	Implicated chromosome(s)
Breast carcinoma	1p, 3p, 11p, 13q, 16q, 17p, 17q
Lung carcinoma	3p, 11p, 13q, 17p
Colorectal carcinoma	5q, 17p, 18q
Multiple endocrine neoplasia	11q, 1p, 10q
Brain tumours	9p, 10p, 10q, 17p
Oesophageal carcinoma	5q, 13q, 17p, 18q
Bladder carcinoma	11q, 13q
Renal carcinoma	3p, 14q, 6q, 8p
Neurofibromatosis type 1, (type 2)	17q, (22q)
Osteosarcoma	13q, 17p
Neuroblastoma	1p
Melanoma	1p, 9p, 17p
Hepatocellular carcinoma	11p, 13q, 4p, 5q, 16q, 17p
Prostate carcinoma	3p, 7q, 8p, 9q, 10p, 10q, 11p, 13q, 16q, 17p, 17q, 18q, Y

Box 6.1 The cell cycle

Growth and development of cells can be described in the form of a cell cycle which is divided into five phases. G_1 (G for gap) is a pause after stimulation when the cell is involved in considerable biochemical activity. S (synthesis) is the phase when the cell's DNA content is doubled in amount. G_2 is another pause and M is the phase involving mitosis when the cell divides into two. Cells can also stop cycling and enter a resting phase (G_0).

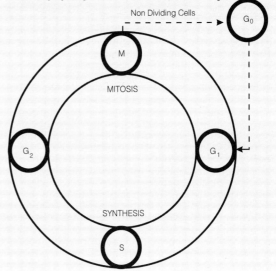

Control of this cycle is dependent on a number of interacting proteins. Stimulatory signals come from the *cyclins* which are a family of proteins named because their levels are cell-cycle dependent. Cyclins D and E are functional in the G_1 phase; cyclin A in the S phase and cyclin B in the G_2 and M phases. Cyclins are cofactors for the *cyclin-dependent kinases* (usually abbreviated to CDK) which comprise a family of protein kinases (CDK1 to CDK6) that phosphorylate critical target proteins involved in the regulation of the cell cycle. Inhibitory influences on the cell cycle can come from *tumour suppressor genes* such as *P53*, *RB1* and another class of compounds called the *cyclin-dependent kinase inhibitors.*

uncontrolled proliferation. Incomplete penetrance reflects the requirement for the second (somatic) mutational event and explains the 'skipping' of some generations. On the other hand, sporadic forms of the tumour involve two separate *somatic* events. The second hit must occur in the same cell lineage that has experienced the first or predisposing hit. The probability of this is relatively rare and so sporadic forms of the tumour occur later in life and have the additional features of being unifocal and unilateral (Fig. 6.6). The two hits proposed by Knudson were confirmed in the form of the two alleles for the retinoblastoma gene when it was finally cloned and characterised (see p. 128).

Loss of heterozygosity

In the inherited tumours, the somatic event that affects the second (normal) allele and so exposes the recessive mutation can involve chromosomal loss or molecular abnormalities such as a deletion. Chromosomal rearrangements, losses or aneuploidy (abnormal chromosome numbers) can be detected *cytogenetically* by studying constitutional cells (e.g. lymphocytes or fibroblasts) and comparing these to tumour cells. However, small DNA deletions (microdeletions) will be difficult to find. Cytogenetic analysis does not usually allow the parental origin for chromosomes to be determined. The importance of this will become evident in the section on imprinting (p. 137).

The availability of DNA polymorphic markers has meant that individual parental contributions in normal and tumour cells can now be distinguished (Fig. 6.7). DNA markers flanking a locus which contains a tumour suppressor gene will thus demonstrate loss of heterozygosity if that segment is lost. The same approach is useful in studying sporadic tumours with the base line or constitutional genotype determined from the affected individual's normal tissues and the two parental contributions from analysis of the pedigree. These can then be compared to DNA patterns in the tumour.

DNA studies to detect loss of heterozygosity in tumours have been used to identify putative tumour suppressor loci

and, from these, attempts to clone and characterise candidate genes have been made. Two tumour suppressor genes, *P53* on chromosome 17p and *DCC* (deleted in colon carcinoma) on chromosome 18q, have been detected in this way (see pp. 125, 129). Loss of heterozygosity has been described in many forms of cancer (Table 6.3). As is evident from the table, the same tumour suppressor regions may be involved in more than one malignancy.

Tumourigenesis

The various roles played by tumour suppressor genes in tumourigenesis are still being defined. Potential ways in which these genes can inhibit the development of cancer include: (1) inhibiting cell proliferation, (2) inducing differentiation or cell death, and (3) stimulating DNA repair.

Cell proliferation

The cell cycle ensures a regulated process so that each cell can complete DNA replication before cell division occurs. The cell responds to growth and environmental signals through the cell cycle. At the molecular level, recent observations based on studies from yeast and *Xenopus* have enabled the complex steps involved in the cell cycle to be better understood. Key components which have a stimulatory effect on the cell cycle include cyclins and cyclin-dependent kinases. Negative influences come from a series of check points which respond to external stimuli (Box 6.1). The tumour suppressor genes can reduce the potential for tumour formation by interfering with the progress of the cell cycle until damaged DNA is repaired.

Differentiation and cell death

Another way in which tumour suppressor genes can perform their role is by limiting a cell's proliferative capacity through inducing it to undergo differentiation. In this way, the relatively greater mitotic activity seen in the undifferentiated cell gives way to an end-cell which divides less frequently. As well as preventing the formation of tumours, this is a

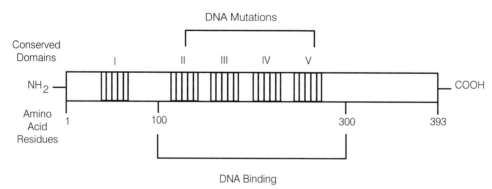

Fig. 6.8 The protein product of the *P53* gene.
The p53 protein has 393 amino acids. Five segments of the protein (I–V) have been strongly conserved during evolution and most DNA mutations are found in four of these conserved regions. DNA-binding sites associated with p53 are also indicated. The usual mutations in the *P53* gene involve single base changes. These will interfere with the *P53*'s transcriptional activity because DNA binding is impaired. Indirectly this will affect other genes influenced by *P53* (e.g. see Fig. 6.9).

way in which tumour suppressor genes can function during the development of an organism. Finally, tumour suppressor cells can reduce the risk of tumour formation by ensuring that a cell which has undergone a mutational event is induced to die by the process of apoptosis (see p. 126).

DNA repair

Although still preliminary, some evidence is now emerging that the tumour suppressor genes can have a direct stimulatory effect on DNA repair mechanisms. The various ways in which tumour suppressor genes can influence the development of cancer are well illustrated by *P53*, the paradigm for this type of tumour-related gene.

P53

P53 has been described as the most important cancer-related gene. This tumour suppressor gene is located on chromosome 17p13.1 and is implicated in both inherited and sporadic cancers. The gene has 10 functional exons and codes for an mRNA of 2.2–2.5 kb. The importance of this gene is suggested by the conservation that has occurred during evolution, e.g. mouse and human proteins are about 80% homologous. The gene is expressed in all cells but the highest levels of mRNA are in the thymus, spleen, testis and ovary. Mutations in *P53* are found in about 50% of cancers (Fig. 6.8). *P53* functions as a tumour suppressor gene since it inhibits the transformation of cells in culture by oncogenes and the formation of tumours in animals. Transgenic mice which have both *P53* genes inactivated by gene 'knock-out' studies are normal at birth, but by 6–9 months of age 100% develop a range of cancers. A key feature of tumour suppressor genes, i.e. loss through chromosomal or DNA rearrangement, is seen with the *P53* locus on chromosome 17p (Table 6.3). An important contribution to our understanding of cancer has also come from the study of the *P53* gene in the context of molecular epidemiology (p. 137).

Cancers in which there have been mutations affecting the *P53* gene include colon, lung, brain, breast and ovarian cancers, melanoma and chronic myeloid leukaemia in blast crisis. Defects observed lead to loss of both alleles in 75–80% of cases with one defect often being a deletion and the second a missense point mutation. The latter leads to production of an abnormal protein. Another way to interfere with *P53* is through the binding of exogenous viral antigens or cellular oncogenes to the normal p53 protein. *P53* plays a key role in inhibiting tumour development through the three mechanisms mentioned earlier: (1) checkpoint control of the cell cycle, (2) induction of apoptosis, and (3) stimulation of the DNA repair mechanism.

Cell cycle

Following exposure of the cell to DNA-damaging agents, e.g. irradiation or certain chemicals, the level of the p53 protein dramatisally increases. The 53 kDa protein encoded by *P53* is a transcription factor which can regulate a number of

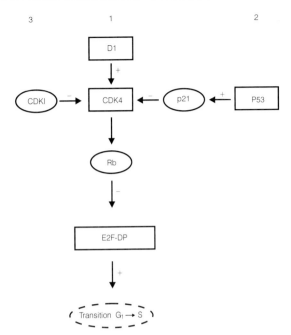

Fig. 6.9 A model illustrating the interactions between the various proteins which can influence the cell cycle.
Stimulatory or positive effects are shown as +; inhibitory or negative effects as −. The retinoblastoma protein (Rb) normally binds to and inhibits the transcription factor E2F-DP which is important for transition of the cell cycle from the G_1 to S phase. **(1)** Proceeding down in this segment of the pathway is cyclin D1 which binds to one of the cyclin-dependent kinases (CKD4). This leads to phosphorylation of Rb. Phosphorylated Rb cannot bind E2F-DP. The net effect is movement of the cell from the G_1 to the S phase of the cycle. **(2)** Next can be added the *P53* gene which responds to DNA damage by inducing a nearby gene known as *WAF1/Cip1* to produce the protein p21. p21 will inhibit CDK4 and so the shift now is towards a block in G_1 since the Rb protein will function without interference from CDK4. **(3)** Another way to inhibit CDK4 is through the more recently described cyclin-dependent kinase inhibitors (CDK1). In this scheme, D1 and CDK4 are demonstrating oncogene-like function while CDK1, Rb and *P53* have tumour suppressor gene effects. A CDK1 which has recently received considerable publicity is the *CDKN2* (or p16) gene. This relates to whether p16 is the tumour suppressor gene involved in one form of genetic melanoma and the potential that p16 may play a very important overall role in carcinogenesis since a mutation in p16 will remove one inhibitory effect on CDK4. Whether the final effect is movement from G_1 to S or arrest at G_1 will depend on the net effect of the above complex interactions.

genes at the DNA level. p53 blocks progression of the cell cycle in the G_1 phase (see Box 6.1). This allows DNA repair to occur prior to entry into the S phase. The cell cycle effect of p53 ensures that damaged DNA is not allowed to replicate and so this gene has been described as the 'guardian of the genome' (Fig. 6.9).

Mutant *P53* forms demonstrate altered growth regulatory properties and can also inactivate normal (wild-type) p53 protein. The latter phenomenon is called a 'dominant negative' effect since inactivation of one of the two tumour suppressor loci can produce what appears to be a dominant effect if the mutant protein inhibits the product from the remaining normal allele.

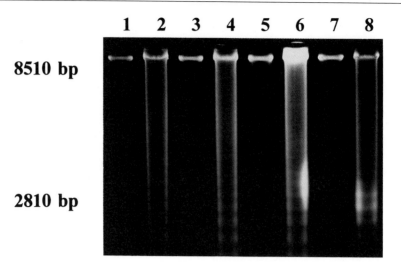

Fig. 6.10 Apoptosis.
Morphologic features of apoptosis include cell shrinkage, dense chromatin condensation, nuclear fragmentation and the cell can emit processes which contain pyknotic nuclear fragments. There is little or no swelling of intracellular organelles. At the molecular level, DNA is broken into segments, approximately 185 bp in size, which result from cleavage between nucleosomes. Pictured is an example of apoptosis induced in Burkitt lymphoma cells with a variety of damaging agents. Lanes 1, 3, 5 and 7 show DNA from untreated cells; lane 2 shows DNA from cells exposed to 10 Gy γ-irradiation; lane 4 shows DNA from cells treated with 60 μM etoposide; lane 6 shows DNA from cells exposed to 100 ng of the marine macrolide bistratene; lane 8 shows DNA from cells exposed to tetrandrine (an anti-inflammatory agent isolated from the plant *stephania tetrandra S*). Note the DNA ladders in lanes 2, 4, 6 and 8. DNA was separated on a 1% agarose gel and ranges in size from 8510 to 2810 bp. (Photo courtesy of M F Lavin and Q Song, Queensland Institute of Medical Research, Brisbane.)

Apoptosis

Cell death can occur in a number of ways. *Necrosis* is one form which has features that include an accidental initiation, e.g. severe injury, associated with degenerative and inflammatory responses. Nearby tissues or cells can be damaged and the final cellular debris is removed by phagocytes. In contrast *apoptosis* occurs in normal tissues and involves a type of cell suicide requiring the interaction of genetic events and producing a non-inflammatory demise of the cell (Fig. 6.10). At the DNA level, the internucleosomal degradation occurring in DNA can be seen in the form of small double-stranded fragments that migrate in a ladder pattern (composed of multiples of about 185 bp). Apoptosis has been described as a 'programmed cell death', although some researchers take exception to this term since in the normal course of events including development, some cells undergo a finite life span which could also be considered 'programmed' but the cells do not die by apoptosis.

At the molecular level, apoptosis requires the involvement of a number of genes. The best described are those from the nematode *Caenorhabditis elegans*. In the human, two genes are well characterised in respect to their effects on apoptosis. *Bcl-2*, which belongs to a multigene family, prevents some forms of apoptosis, i.e. it protects cells from death. Lymphoid cells which are exposed to an activated *bcl-2* gene following a chromosome 14:18 translocation eventually become malignant because spontaneous mutations which occur in these cells are unable to be contained (by the cell dying) and they accumulate. In contrast is the effect of the *P53* gene which can induce a damaged cell to undergo apoptosis and so remove a potential focus for tumour formation. The mechanism by which *P53* directs a cell into apoptosis remains to be defined although it is interesting that p53 can stimulate synthesis of the gene *Bax* which can bind and block *bcl-2*.

Cells damaged by chemotherapeutic agents will stimulate the production of *P53* and so undergo apoptosis. This additional antitumour effect may explain why cancers which have wild-type *P53* genes respond better to treatment. On the other hand, a mutant gene cannot function in this way and damage to the cancer cell produced by the chemotherapy will accumulate in cells which have not been directly killed by the treatment. These cells will then form a new clone of more malignant, treatment-resistant tumour cells. In this respect, *P53* may become useful as a prognostic marker.

DNA repair

GADD45 is a mammalian gene which is ubiquitously expressed. The gene is induced by DNA damage and can also be activated by *P53*. The role played by the gadd45 protein has not been fully defined but it appears to have an inhibitory effect on the cell cycle and interestingly it stimulates DNA excision repair (described in the following section). Thus, gadd45 has provided a mechanism by which the activation of *P53* following genotoxic stress can stimulate one of the DNA repair pathways.

Li Fraumeni syndrome

The Li Fraumeni syndrome is a rare autosomal dominant disorder which predisposes to a range of tumours including sarcoma, leukaemia and breast and brain tumours. Affected family members have one mutant *P53* gene in their germline. Malignant cells, on the other hand, have abnormalities affecting both alleles. These data reinforce the two-hit hypothesis for tumourigenesis and implicate the *P53* tumour suppressor gene as one component in the pathogenesis of the Li Fraumeni syndrome. What other genetic defects are involved in this disorder remain to be determined. Recently, it has been shown that germline *P53* mutations can occur without the classical Li Fraumeni syndrome. These are found in individuals who develop multiple malignancies.

DNA REPAIR GENES

Having described the multiple and complex events which are required for DNA to replicate itself *and* provide the backbone for the cells' proteins to be produced, it can only be considered amazing that the system works at all. A further source of damage to DNA occurs via the environment in the form of ultraviolet and ionising irradiation and exposure to chemicals. Not surprisingly, errors in DNA processes occur regularly and it is the group of enzymes required to monitor and repair these errors that have since 1994 formed a third distinct class of genes involved in carcinogenesis. A noteworthy feature of these cancer genes is that key discoveries were made using model organisms such as *E. coli* and the yeast *Saccharomyces cerevisiae*. The importance of studying model organisms at the DNA level will be discussed further in relation to the Human Genome Project.

Biology of repair

There are two major pathways for DNA repair: mismatch repair and nucleotide excision repair. *Mismatch repair* is the simpler of the two and involves three genes in prokaryotes and eukaryotes. In *E. coli* the genes are *mutS*, *mutL* and *mutH*, with the corresponding ones in humans being h*MSH2* (*mutS*), h*MLH1* and h*PMS1* (*mutL*). Mismatch repair is involved in errors which arise as DNA is being copied. It acts on single base mismatches as well as small displaced loops a few bases in size such as occur when repetitive regions (e.g. microsatellites) are replicated (Fig. 6.11). Mutations in mismatch repair genes in humans are associated with sporadic colon cancer and hereditary non-polyposis colon cancer (HNPCC), which will be discussed further on p. 128.

The *nucleotide excision repair* pathway is more complex, involving about 12 gene products in eukaryotes and a smaller number than this in prokaryotes. Excision repair does not deal effectively with the discrete defects described in mismatch repair but becomes involved in a broader range of DNA defects including those caused by chemicals and irradiation. In humans, the paradigm for a defect in excision repair is the genetic disorder xeroderma pigmentosa (see p. 128).

Two other features of the repair enzymes are noteworthy. The DNA excision repair pathway overlaps with that for transcription since some genes involved in repair also have the properties of transcription factors. Hence, different types of protein–protein interactions might explain the variable clinical phenotypes which emerge when there are mutations in these genes. In vitro studies have also shown that cells which are defective in DNA mismatch repair enzymes are more resistant to killing by anticancer drugs such as the alkylating agents. The mechanism has not been defined but the observation would provide another route for drug resistance and a way by which resistance might be overcome. Not surprisingly, the biological significance of the DNA repair enzymes was noted in the journal *Science*, which awarded the 1994 'Molecule of the Year' to this class of compounds (the 1993 award had gone to *P53*).

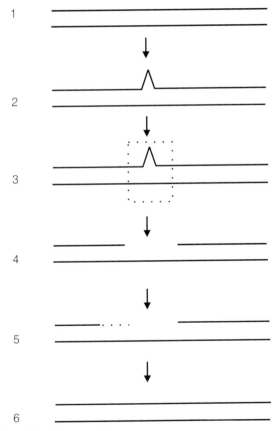

Fig. 6.11 DNA mismatch repair.
(1) A normal double-stranded segment of DNA. (2) During replication one of the strands incorporates an incorrect nucleotide base (indicated by the displacement in one strand). (3) The mismatch repair proteins (in bacteria these are *mutS*, *mutL* and *mutH*) recognise the error, bind to that site and produce a nick on one side of the mismatch. (4) The segment containing the mismatch is removed. (5) DNA polymerase and other enzymes can now start to repair the missing segment of DNA. (6) The normal double-stranded structure results.

DNA repair and cancer

Evidence for a role of DNA repair in the causation of cancer has come from several observations. (1) A number of rare human genetic disorders are associated with an increased predisposition to cancer and radiation sensitivity (Box 6.2). (2) In vitro studies using cells lines from the genetic disorders mentioned show chromosomal instability and increased sensitivity to a range of DNA-damaging agents. (3) Transgenic mice produced by mutating an excision repair gene die early in life and demonstrate aneuploidy and an increase in p53 protein (i.e. response to DNA damage). (4) Microsatellite instability is found in colon cancer tissue (both sporadic cancers and hereditary non-polyposis colon cancer).

Hereditary non-polyposis colon cancer (HNPCC)

HNPCC is the most common type of *hereditary* colon cancer and occurs in ~1–10% of colorectal cancers. The frequency of carriers for this autosomal dominant disorder is estimated to be between 1 in 200 to 1 in 2000. These figures are imprecise because diagnostic criteria for HNPCC are difficult to define. Clinical features of HNPCC include colonic polyps which are fewer than those in another genetic form of colon cancer known as familial adenomatous polyposis (discussed on p. 129) and a predominance for right-sided colonic involvement including the site for cancer development.

At the molecular level, the first clue that HNPCC was associated with a defect in DNA mismatch repair genes came with the observation in 1993 that microsatellites in tumour tissue were unstable, i.e. there were additional alleles of different sizes compared to what was present in normal tissues in the same individual. This was followed in the same year by the cloning of the gene h*MSH2*, a gene involved in DNA mismatch repair. In 1994, three more genes with similar function were isolated – h*MLH1*, h*PMS1* and h*PMS2*. The same changes in DNA have also been observed in about 15% of sporadic colon cancers. The exact role that these genes play in carcinogenesis remains to be determined and will be discussed further in the section on familial adenomatous polyposis. In practical terms, the finding of a mutation in DNA provides a means by which at-risk family members can be detected. This also will be considered more fully under familial adenomatous polyposis.

GENETIC MODELS AND CANCER

Retinoblastoma

Clinical features

Retinoblastoma is an important intraocular malignancy in children. The disease can be detected at birth although the usual time for detection is within the first 3 years of life. Since treatment with surgery or radiotherapy is potentially curative, it is essential to make an early diagnosis. With increasing survival of treated patients it has become apparent that those with the heritable form are at increased risk for other tumours, e.g. osteosarcoma.

An important breakthrough in understanding retinoblastoma came with the finding of a deletion involving chromosome 13q14 in the blood of patients with the heritable type and in the tumour cells of both the sporadic and heritable forms. Subsequently, DNA polymorphic markers showed that the *normal* allele on the chromosome 13 inherited from the unaffected parent was the one lost in the tumour cells, indicating that loss of heterozygosity produced its effect by uncovering a mutation in the germline. This was consistent with the two-hit model for cancer described earlier in this chapter.

Positional cloning

Using positional cloning, a candidate gene (*RB1*) at the retinoblastoma locus was identified. A 4.7 kb mRNA transcript which encoded for a 928 amino acid protein (called p105-RB) was isolated. Surprisingly, the protein was expressed in most tissues, and not only in the retina. Nevertheless, confirmation of *RB1* as a likely candidate gene came with the finding that the normal 4.7 kb mRNA transcript was not present in retinoblastoma cell lines. Mutations involving the gene, i.e. deletions and point mutations, were also detected in DNA from tumours. An apparent increased frequency of osteosarcoma in patients with retinoblastoma was strengthened when similar changes in *RB1* were found in osteosarcomas. Surprisingly, sporadic cases of osteosarcoma which had occurred unrelated to retinoblastoma also had mutations involving *RB1*.

DNA changes

The *RB1* tumour suppressor gene and its role in the cell cycle has previously been mentioned (Fig. 6.9). DNA probes are also available to identify those who are predisposed and this application of DNA technology will be discussed on p. 140. From DNA studies, it is estimated that in ~10% of cases the disease can arise following inheritance of a germline mutation from a carrier parent or through the occurrence of a de novo germline mutation (~30% of cases). This leads to bilateral heritable disease. Progeny of those with bilateral heritable disease have a 50% risk of developing retinoblastoma.

Familial adenomatous polyposis

Colon cancer has provided a model to study the evolution of a *solid* tumour. This is possible because there is a genetic form of colon cancer known as familial adenomatous polyposis that allows progression of the tumour to be followed from the premalignant stage to the advanced and then invasive (metastatic) cancer.

Clinical features

Colon cancer is one of the most common cancers in Western countries. Apart from the genetic variants present, the feature which distinguishes it from other frequently occurring malignancies is the well-defined precancerous state associated with the adenomatous polyp. *Familial adenomatous polyposis* is a rare form of colon cancer (~1% of all cases) and is inherited as an autosomal dominant disorder with high penetrance. It is characterised by many polyps in the colon with a probability close to 100% that one or more will become malignant. Because of this high risk, treatment involves prophylactic (preventative) removal of the colon in those who are predisposed.

Positional cloning

The various historical landmarks achieved in looking for the gene(s) causing familial adenomatous polyposis are summarised in Box 6.3. As has been seen in other examples, a key early finding was the observation of a deletion involving the long arm of chromosome 5. From linkage studies of affected pedigrees, the familial adenomatous polyposis locus was next localised to the 5q21-q22 region and DNA probes were soon isolated. During the next 3 years it became apparent that in both *familial* and *sporadic* colon cancers there were other frequently detected DNA abnormalities. These included the *P53* gene on chromosome 17p; a tumour suppressor locus on chromosome 18q and the *ras* proto-oncogene. In 1990, a tumour suppressor gene on chromosome 18q was isolated and called *DCC* (deleted in colon cancer). In the following year, the *MCC* gene (mutated in colon cancer) was isolated from the chromosome 5q locus which had been linked to familial

---

Box 6.3 Historical milestones in the familial adenomatous polyposis (FAP) story, including the developments in another genetic form of colon cancer known as hereditary non-polyposis colon cancer (HNPCC)

1986	Cytogenetic deletion on chromosome 5q observed in FAP
1987	Genetic linkage localises FAP defect to chromosome 5q21-q22
1987-9	Mutations/deletions in FAP involving chromosomes 17p (*P53* gene); 18q and the K-*ras* proto-oncogene
1990	DCC gene (chromosome 18q21.3) isolated in FAP
1991	MCC gene (chromosome 5q21) isolated in FAP
1991	APC gene (chromosome 5q21) isolated in FAP
1993	Microsatellite instability in colon cancer (sporadic and HNPCC)
1993	h*MSH2* mismatch repair gene for HNPCC isolated on chromosome 2p22
1994	h*MLH1*, h*PMS1*, h*PMS2* mismatch repair genes for HNPCC isolated on chromosomes 3p, 2q31-q33, 7p22

Abbreviations: FAP, familial adenomatous polyposis; *DCC*, deleted in colon cancer; *MCC*, mutated in colon cancer; *APC*, adenomatous polyposis coli; HNPCC, hereditary nonpolyposis colon cancer.

adenomatous polyposis. *MCC* thus became a good candidate for the familial genetic defect but a unique role was subsequently excluded when a number of affected families showed no defects in this gene. Further DNA searches then identified, not far from *MCC*, another gene called *APC* (adenomatous polyposis coli). *APC* has now been confirmed to be the correct gene for familial adenomatous polyposis, with DNA changes occurring early and persisting throughout the tumour's evolution. Knowledge of the underlying mutations in *APC* has been put to good use in terms of the screening and clinical management of those who are at risk. The implications of DNA testing for familial adenomatous polyposis will be discussed in the section on diagnostic applications (p. 139).

DNA changes

There are many DNA changes in familial adenomatous polyposis. From these, a number of observations can be made:

- Both proto-oncogenes and tumour suppressor genes are involved, with the latter having a major role.

- Multiple mutations are required: the numbers present correlate approximately with the stage of evolution, i.e. there are fewer changes in adenomatous polyps compared to carcinoma.

- The *APC* gene defect occurs early and persists. Other changes develop concurrently with the evolution of the tumour. No specific combination of mutations leads to a predictable phenotype.

- Over 90% of mutations in *APC* produce a truncated protein (discussed further on p. 139).

An important point to note is that DNA mutations similar to those found in the rare genetic form of colon cancer are also being detected in the *sporadic* cases. Thus, knowledge about a common disease is indirectly being obtained by studying genetic models.

Model for tumourigenesis

A number of hypotheses have been advanced to explain the clinical and genetic findings in familial adenomatous polyposis. One suggestion for the (precancerous) adenomatous polyp formation involves a dose effect from the

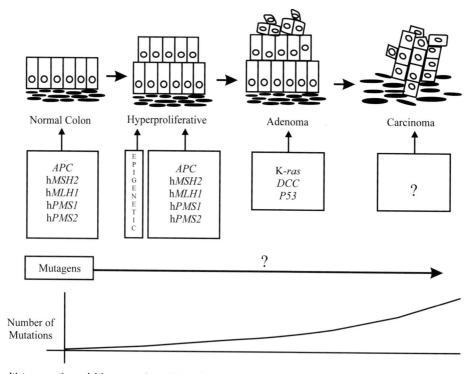

Fig. 6.12 A multistep genetic model for tumourigenesis in colon cancer.
An initial insult affecting the colonic tissue can involve any number of genes (*APC*, – adenomatous polyposis coli; h*MSH2*, h*MLH1*, h*PMS1*, h*PMS2* – DNA repair genes) and may have been inherited or acquired. This initiates the tumour pathway. Epigenetic factors such as hypomethylation (there is a loss of methyl groups on DNA early in the development of cancer) predispose to further DNA damage. At this stage the colonic epithelium is becoming overactive, i.e. hyperproliferative. Changes occurring in the genes mentioned may then affect others which had previously been normal, e.g. an initial h*MSH2* defect might predispose to *APC* mutations. The colonic epithelium with all these accumulating mutations develops a growth advantage over normal tissue. Additional mutations involving the proto-oncogene K-*ras* and tumour suppressor genes (*DCC* – deleted in colon cancer; *P53*) contribute to formation of an adenoma. Other, as yet unknown, genetic changes finally enable the early clone of tumour cells to become locally invasive and then metastatic. Throughout this process the environment, e.g. mutagens in food, can contribute to the DNA damage. The development of cancer from the first mutated cell relies on an accumulation of genetic defects until the appropriate 'combination' of oncogenes/tumour suppressor genes and DNA damage is present (from Vogelstein and Kinzler 1993; Cavenee and White 1995).

APC gene such that a single mutation is sufficient to initiate the growth of a polyp. This, as well as a number of additional genetic changes, enable the slow but steady progression to carcinoma. Important components allowing tumour progression include genetic instability and clonal growth advantage, e.g. loss of cell cycle control by the tumour clone. Invasion and metastasis require further genetic changes (Fig. 6.12). The net effect is multiple well-recognised mutations in key regions superimposed on which are other effects (genetic or environmental in origin). Clinical heterogeneity (e.g. invasiveness, the number of adenomatous polyps present) reflects the various genetic combinations or types of mutations present. Since similar changes in oncogene and tumour suppressor gene loci are seen in other malignancies, it is possible that there are common pathways in tumourigenesis.

Cancer of the breast

Clinical features

Breast cancer is the commonest cause of cancer death in women. By 75 years of age, nearly 1 in 10 women in the USA will develop this disease. Pathogenesis of breast cancer is complex, involving physiological, environmental and genetic factors. A positive family history is a significant risk factor which can exceed by 50% the lifetime risk if the affected are first degree relatives with early and bilateral breast involvement.

In contrast to the *inherited* cancer syndromes, such as retinoblastoma or familial adenomatous polyposis described above, *familial* cancers refer to neoplasms that cluster in families. However, because of inadequate markers to detect the predisposed phenotype, it is difficult to ascertain who are at risk. Many types of familial cancers have been reported but the sites most commonly involved are breast, ovary, melanoma, colon, blood and brain. Clinical features which would suggest a familial cancer include two or more close relatives affected, multiple or bilateral cancers in the same person, early age of onset and clustering (e.g. occurrence of cancer of both the breast and ovary).

The successful finding of the breast cancer-specific gene known as *BRCA1* was announced in the October 1994 issue of *Nature Genetics* as 'The glittering prize'. Subsequently, the intense efforts directed to this gene at the molecular level have emphasised a number of issues which have particular relevance to molecular medicine. These include: (1) the complex DNA changes which are likely to be involved in cancer; (2) the significance of DNA mutations in pathogenesis; (3) the resource-intensive efforts which will be required to utilise effectively information which comes from *BRCA1*; and (4) the additional ethical, social and educational issues which have appeared, in a very short time frame, in the management of a woman with breast cancer, particularly if there is a positive family history. The first two points will be dealt with in this chapter and the remaining two in Chapter 9.

Positional cloning

The following is a summary of historical developments in the breast cancer DNA story:

- During the late 1980s, loss of heterozygosity in breast cancer tissue was reported for a number of chromosomes (see Table 6.3).
- In December 1990, breast cancer was linked to chromosome 17q21.
- During 1990–1992, the chromosome 17q location was confirmed by the demonstration of loss of heterozygosity in both breast and ovarian cancer samples.
- In September 1994, the *BRCA1* gene was cloned and the BRCA2 gene locus on chromosome 13q12-q13 was identified. The potential for a third locus (BRCA3) was raised.
- By April 1995, it had been shown that cases of 'sporadic' ovarian cancers had mutations in the *BRCA1* gene but as yet no 'sporadic' breast cancers had abnormalities affecting this gene.
- In December 1995, the *BRCA2* gene was isolated.

At the molecular level, the breast cancer story is only just beginning. The *BRCA1* gene is large, with 24 exons spread over 5592 nucleotides in a genomic segment of DNA which is ~100 kb in size. The protein produced has 1863 amino acids. Comparisons between the DNA sequence of the *BRCA1* gene and entries in the DNA databases have shown that only the 5′ end has a match and from this it has been suggested that *BRCA1* might be involved in DNA transcription. Apart from this, nothing further is known about the functional significance of *BRCA1*.

Not unexpectedly the large numbers of laboratories and research funds available for work with *BRCA1* enabled mutation analysis to progress rapidly. By 1996, about 150 mutations had been described. The expected molecular heterogeneity was found with DNA changes involving point mutations (missense, nonsense, frameshifts), splicing defects and small deletions or insertions. Interestingly, over 80% of the mutations resulted from nonsense or frameshift changes so that truncated proteins were produced. Thus, the PTT (**p**rotein **t**runcation **t**est) which had earlier achieved considerable success with the *APC* gene was now also useful with *BRCA1*. Disappointingly, only few of the *BRCA1* defects were observed to recur in unrelated individuals. Thus, the potential for widespread screening in the general population was fast disappearing as the information on the underlying mutations accumulated. However, for individual families DNA testing for breast cancer predisposition became a realistic option (see p. 140).

Late in 1995, a group of scientists in the USA, UK and Europe developed an Internet-accessible database for *BRCA1*. This is available to all who are interested in the gene and provides information such as mutations detected, methodologies and a bulletin board. Information can be accessed through the World Wide Web address: http://www.nchgr.nih.gov/dir/lab_transfer/bic.

BRCA1 changes

The potential role of tumour suppressor genes in the pathogenesis of human breast cancer is complex, with many loci demonstrating loss of heterozygosity (Table 6.3). An interesting comparison can be made with the familial adenomatous polyposis story (Box 6.3), with the initial discovery of the *MCC* on the correct chromosomal location (5q21) subsequently giving way to the more relevant *APC* gene ~100 kb away. Perhaps a similar scenario will occur with *BRCA1* and a nearby gene will turn out to be significant. This would explain one disappointment in the *BRCA1* story, i.e. mutations detected in *BRCA1* have not to date been found in sporadic forms of breast cancer. This is unlike familial adenomatous polyposis and sporadic colon cancer, both of which show abnormalities in the same *APC* gene. The relationship between *BRCA1* and sporadic forms of cancer remains to be clarified, although it is noteworthy that loss of heterozygosity for sporadic breast cancer also involves the chromosome 17q locus.

An interesting case report has described a woman with mutations affecting both her *BRCA1* genes. Like her at-risk relatives, this individual developed breast cancer at an early age but her survival into adult life would suggest that *BRCA1* is not essential for normal development. It has also been proposed that *BRCA1* is an incorrect way to describe this gene since it appears to be more closely related to ovarian cancer (both familial and sporadic) than to breast cancer. The second or BRCA2 locus may be more relevant to breast cancer since it is associated with both females and males who develop this malignancy. Another suggested mechanism to explain anomalies in *BRCA1* relates to the position of the mutations so that exon-specific defects which have a direct effect on protein production lead to the more severe familial phenotype while mutations in the introns or 5′ or 3′ flanking regions are milder and perhaps only expressed in the form of sporadic cancer. A more recent study has suggested that germline mutations in the 3′ third of the *BRCA1* gene are associated with a lower proportion of

patients who develop ovarian cancer. These hypotheses and observations will be clarified as the molecular changes in breast cancer are documented further.

BRCA1 and the risks for breast cancer

To place some perspective on the *BRCA1* story, it should be noted that other genes are also important in the pathogenesis of breast cancer. *P53* and the Li-Fraumeni syndrome have been mentioned earlier. In 1995, the gene for ataxia-telangiectasia (known as *ATM*) located on chromosome 11q22-q23 was found. This gene, which is likely to be involved in tissue repair, is associated with a range of malignancies, e.g. leukaemia, lymphoma and breast cancer (Box 6.2). Another gene marker of significance in breast cancer involves the rare polymorphisms associated with the gene *HRAS1* on chromosome 11p15.5. A summary of the various risks associated with these genes and the development of breast cancer is found in Table 6.4.

An observation with potential clinical significance relates to *P53* and *ATM* and the fact that individuals with mutations in these genes are considered to have an increased sensitivity to ionising radiation. This is relevant since a reason for undertaking DNA mutation analysis is to enable

---

Table 6.4 Some statistics on breast cancer and at-risk genes or DNA markers (from multiple sources including Easton et al 1993).

Gene mutation	Risks	Women with this defect	% Breast cancer in the population
BRCA1[a]	~85% by age 70[b]	1 in 500 to 1 in 2000	8% < 40 years age 1% > 50 years age
ATM	~23% by age 70	1 in 80 to 1 in 200	up to 8%
P53	~30% by age 30	1 in 10 000	1% < 35 years age[c]
HRAS1	~30% by age 70	1 in 10 to 1 in 20	up to 8%

[a]*BRCA1* associated with female breast cancer and ovarian cancer; *BRCA2*, both male and female breast cancer. [b]Risk for ovarian cancer is less by age 70 but because of poorer survival *BRCA1* mutations have more serious consequences. [c]Women are unlikely to survive beyond this age.

> **Box 6.4 Germline and rearranged immunoglobulin and T cell receptor genes**
>
> Immunoglobulin (Ig) molecules consist of two identical heavy chain genes (either μ, δ, γ, ε, or α) and two identical light chain genes (κ or λ). The variable or antigen recognition portion of the Ig molecule (called Fab, *fragment antibody binding*) is located at the amino terminal end. At the carboxyl end is the constant region (called Fc, *fragment crystalline*) which defines the isotype or genetic subgroup of the Ig: IgM (μ), IgG (γ), IgE (ε), IgD (δ) or IgA (α) (Fig. 7.10). There are two types of T cell receptors (TCR): one is a heterodimer of α and β chains and the second involves $\gamma\beta$. Genes for the Ig and TCR loci are located on a number of different chromosomes and comprise various combinations of the following: variable (V), diversity (D), joining (J) and constant (C) genes (Table 6.5). Gene families encoding the Ig and TCR molecules are arranged in two distinct configurations: (1) a functionally inactive or germline state and (2) a functionally active or rearranged configuration (Fig. 6.13). During development of a stem cell into a B or T lymphocyte, there is rearrangement of its germline Ig or TCR genes. Where present, there is first a combination of a D (diversity) and a J (joining) segment (DJ rearrangement). The next step involves combination between the DJ and a V (variable) gene. This comprises the *variable* unit of the molecule. The fully rearranged Ig or TCR gene requires joining to a C (constant gene) (Fig. 6.13). Each Ig or TCR gene rearrangement is *unique*. The incredible diversity which can be generated explains in part the immunological repertoire which an organism can develop. Additional superimposed somatic mutations produce further diversity.

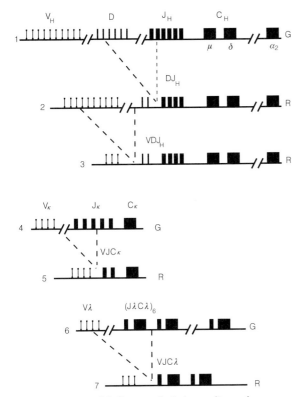

Fig. 6.13 Immunoglobulin genes in their germline and rearranged configurations.
During development of a stem cell into a B or T lymphocyte, there is rearrangement of the germline immunoglobulin or T cell receptor genes. This rearrangement generates the diversity in immune proteins necessary for effective antigen recognition. The different genes are: V = variable; D = diversity; J = joining; C = constant. A break in the baseline indicates that the full gene complement has not been given. G = germline; R = rearranged. **(1)** The immunoglobulin heavy chain locus. **(2)** The first recombination in the heavy chain locus involves a D to J step. **(3)** This is then followed by V to D–J recombination. **(4)** The germline κ light chain gene locus, which becomes rearranged in **(5)**. The germline λ light chain gene locus **(6)** and its rearrangement in **(7)**. See also Table 6.5 for a more detailed description of the above genes.

Table 6.5 Genes which comprise the immunoglobulin and T cell receptor loci.
A feature of the immunoglobulin and T cell receptor genes is the large numbers which are present.

Type	Chromosome	Component genes
Immunoglobulin heavy chains	14q32	V_H-D_H-J_{H1-6}-C_H (μ, δ, γ_3, γ_1, ε_2, α_1, γ, γ_2, γ_4, ε_1, α_2). There are ~200 V_H genes and >20 D_H genes
Immunoglobulin κ light chains	2p11	V_κ-$J_{\kappa 1-5}$-C_κ. There are ~100 V_κ genes
Immunoglobulin λ light chains	22q11	V_λ-$(J_\lambda C_\lambda)_6$. There are ~100 V_λ genes
T cell α, δ chains	14q11	V_α-$V_{\delta 1-3}$-$D_{\delta 1-3}$-$J_{\delta 1-3}$-C_δ-$V_{\delta 4}$-J_α-C_α. There are ~50 V_α and ~70 J_α genes
T cell β chains	7q34	V_β-$D_{\beta 1}$-$J_{\beta 1-6}$-$C_{\beta 1}$-$D_{\beta 2}$-$J_{\beta 7-12}$-$C_{\beta 2}$. There are ~75 V_β genes
T cell γ chains	7p15	V_γ-$J_{\gamma 1-3}$-$C_{\gamma 1}$-$J_{\gamma 4-5}$-$C_{\gamma 2}$. There are ~8 V_γ genes

V, variable; D, diversity; J, joining; C, constant. See also Figure 6.13.

six breast cancer families (two families with both female and male breast cancers; one with female breast and ovarian cancers). *BRCA2* shows no homology with entries in the DNA databases and so its function remains unknown. Further information on this gene will no doubt emerge in the next few years.

Leukaemias and lymphomas

Haematological malignancies either present in the first instance as an aggressive disorder or become more malignant during the course of their natural history. Access to abnormal cells in the peripheral blood or bone marrow is possible. Therefore, tumours of the haematopoietic cells have provided convenient models with which to study DNA changes during various stages of a malignancy.

Immunoglobulin and T cell receptor gene families
Lymphocytes are unique cells since they are able to undergo somatic rearrangements of their immunoglobulin or T cell receptor genes (Box 6.4). This is essential to generate molecules of sufficient diversity to enable recognition of the vast array of antigens to which an organism will be exposed. The process of gene rearrangements is error-prone. Therefore, it is possible that the immunoglobulin or T cell receptor genes can be accidentally spliced next to or with other genes including the proto-oncogenes. One way for this to occur is from a chromosomal translocation. Following this, the cells containing the rearranged immunoglobulin or T cell receptor genes are now driven by the associated

closer surveillance of at-risk women. Presently, the usual form of surveillance, apart from self-examination of the breasts, involves mammography, i.e. repeated doses of ionising radiation, but should this be recommended in those with changes in *P53* and *ATM*? It would seem that each interesting development in molecular medicine leads to new knowledge but at the same time raises more questions. For the researcher, this is a healthy situation. For the health professionals and affected families each bit of new information will hopefully lead to more effective preventative and therapeutic options.

BRCA2
The search for the second breast cancer-related gene was successful late in 1995, with the report that an ~10 kb RNA transcript isolated from the appropriate position on chromosome 13q12-q13 demonstrated germline mutations in

proto-oncogene and eventually a malignant clone can arise. Should a lymphoid cell form one such clone, all sister cells will carry the hallmark of its unique gene rearrangement. This can be utilised for the DNA investigation of patients with haematological malignancies (see p. 141).

Chronic granulocytic leukaemia

A preliminary description of the Philadelphia chromosome, the characteristic translocation in chronic granulocytic leukaemia, has already been given (p. 120). This leukaemia affects young adults and is a malignant clonal disorder involving a pluripotential haematopoietic stem cell. It usually presents in chronic phase and within 3 to 4 years develops into an accelerated or acute phase called blastic transformation. Over 95% of cases have the Philadelphia chromosome. The fusion gene product formed from the translocation between chromosomes contains *abl*, a proto-oncogene with tyrosine kinase activity. Following the translocation, the activity associated with *abl* is increased. This produces excessive stimulation of the cell cycle which

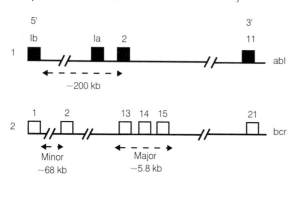

Fig. 6.14 Molecular (DNA) changes involving proto-oncogenes from the translocation which produces the Philadelphia chromosome.
(1) The *abl* gene (■) on chromosome 9 showing two alternatives for exon 1 (Ib and Ia) which are situated 5' to exon 11. The translocation breakpoint in chromosome 9 usually occurs 5' to exon 2. Thus, *abl* exons 2–11 are translocated onto chromosome 22. (2) There are two breakpoint regions in the *bcr* gene (□) which has 21 exons. Between exons 1 and 2 is the 'minor' region which is usually involved in the translocation found in Philadelphia chromosome-positive acute lymphoblastic leukaemia. The 'major' breakpoint cluster region is situated further 3' and is the site for the translocation break in most cases of chronic granulocytic leukaemia. (3) The normal *abl* mRNA transcript is 6–7 kb long depending on which exon 1 is utilised. (4) The *bcr*–*abl* fusion gene involving the major breakpoint region produces a transcript of 8.5 kb. (5) The *bcr*–*abl* fusion gene involving the minor region gives a smaller 7 kb transcript.

initiates the process leading to leukaemia (Figure 6.14). During development of blastic transformation, additional DNA changes are seen. These can involve *P53* or *RB1* (retinoblastoma). It is considered that mutations in tumour suppressor genes contribute more to progression than initiation in this disorder. The different fusion products described in Figure 6.14 both lead to the same type of leukaemia although the shorter product is considered to predispose to a more aggressive form, i.e. Philadelphia chromosome-positive acute lymphoblastic leukaemia in adults.

Acute granulocytic leukaemia

This is a leukaemia of primitive haematopoietic cells and, in contrast to chronic granulocytic leukaemia, it presents as an acute disorder which progresses rapidly without therapeutic intervention. *Ras* proto-oncogene abnormalities, predominantly N-*ras*, occur in 15–50% of cases. However, mutations in the N-*ras* proto-oncogene are not considered sufficient to explain the acute leukaemic phenotype. Additional DNA changes detected in this and other leukaemias occur in the *P53* tumour suppressor gene (~50%) and other proto-oncogenes, e.g. *myc* (Table 6.6).

Changes in the tumour suppressor genes *P53* and *RB1* also occur in the acute leukaemias but, as described for chronic granulocytic leukaemia, these are considered to contribute more to the progression of the leukaemia than

Table 6.6 Changes at the DNA level which are implicated in the pathogenesis of leukaemias and lymphomas (from Cline 1994). The study of chromosome translocations has been a fertile area for discovering genes which can produce haematological malignancies.

Property	Leukaemia/lymphoma (% with change)	Gene
Proto-oncogene, e.g. tyrosine kinase	Chronic granulocytic leukaemia (>95%)	*Bcr/abl* fusion[*]
Proto-oncogene, e.g. receptor	Acute promyelocytic leukaemia (100%)	PML/retinoic acid[*]
Proto-oncogene, e.g. signal transduction	Acute myeloid leukaemia (15-50%)	N-*ras*
Tissue differentiation	T cell acute lymphoblastic leukaemia (7%)	TCR/HOX-11[*]
Apoptosis	Follicular lymphoma (>75%)	Ig/bcl-2[*]
Tumour suppressor gene	Chronic granulocytic leukaemia (>20% *P53*, 15% *RB1*)	*P53*, *RB1*
	Acute myeloid leukaemia (?20%)	*WT1*

[*]Unique to leukaemia/lymphoma. Other genetic changes are observed in a wide range of tumours. PML, promyelocytic leukaemia; TCR, T cell receptor; Ig, immunoglobulin; *WT1* Wilms tumour gene; / defines a fusion product resulting from a chromosomal translocation.

Box 6.5   How molecular characterisation of a chromosomal breakpoint in an acute leukaemia explains an unusual response to treatment

All-*trans*-retinoic acid is a naturally occurring metabolite of vitamin A. Retinoic acid compounds play key roles in physiological pathways such as growth and differentiation. In terms of the myeloid cell, retinoic acid induces terminal differentiation of precursors. To function, the retinoic acid compounds bind to their nuclear receptors, of which there are many. One receptor is RARα, the gene for which is on chromosome 17q. In the late 1980s, a number of studies reported that an unusual form of leukaemia known as *acute promyelocytic leukaemia* responded dramatically to treatment with retinoic acid. This was surprising since the retinoic acid compounds are not cytotoxic drugs. The reason for the retinoic acid effect was not clear until 1992 when it was shown that the chromosomal translocation which occurred in acute promyelocytic leukaemia involved chromosomes 15 and 17. DNA studies have now confirmed that the translocation disrupts the *RARα* gene on chromosome 17. Interference with the function of the natural retinoic acid by the translocation meant that myeloid precursor cells were unable to proceed along the normal differentiation pathway. Providing an alternative source of retinoic acid overcame this and the leukaemic cells differentiated into mature neutrophils resulting in a *differentiation-induced* remission. There are a number of questions which still remain to be answered, e.g. what is the exact defect produced by the translocation and how does the retinoic acid overcome this? Why do the patients have a very dramatic response to retinoic acid but subsequently develop resistance? The answers will come as the molecular studies provide further information on retinoic acid and the translocation product involving the *RARα* gene.

---

its initiation. Apart from the proto-oncogenes, chromosomal rearrangements in acute leukaemia can produce unusual hybrid genes, e.g. a translocation between chromosomes 14 and 10 involving the T cell receptor and a homeobox gene *HOX-11*. The characterisation of DNA changes associated with a translocation in acute promyelocytic leukaemia has explained the mechanism for an alternative but successful treatment approach (Box 6.5).

Lymphoma

Two interesting gene fusions occur in lymphoma. Burkitt lymphoma results following a translocation between chromosome 8 (containing the proto-oncogene *myc*) and chromosome 14 (immunoglobulin heavy chain gene). Expression of *myc* is seen to be inappropriately high in this situation and not surprisingly the lymphoid cells with this translocation eventually become malignant. More recently, another translocation involving chromosome 14 has been described but this time it occurs with chromosome 18. The

result is juxtaposition of the immunoglobulin heavy chain genes and the *bcl-2* proto-oncogene. As mentioned in the section on apoptosis (p. 126), *bcl-2* inhibits this type of cell death. The lymphoid cell with the *bcl-2*/immunoglobulin hybrid expresses *bcl-2* at a higher than normal level and so the cell has a growth advantage (because apoptosis is inhibited) and eventually becomes the source of a malignant clonal population.

Viral-induced cancer

Hepatitis B virus – clinical features

The hepatitis B virus genome was cloned and sequenced in 1979. Two years later an association between chronic hepatitis B infection and cancer of the liver (called hepatocellular carcinoma or hepatoma) was described. Evidence for the above association came from epidemiological studies, naturally infected animal models such as the woodchuck and transgenic mice.

On a worldwide basis, 80% of all hepatocellular carcinomas occur in those who are infected with hepatitis B. Regions at particular risk are East Asia and sub-Saharan Africa. Hepatitis B virus carriers are estimated to have a 100-fold greater risk of developing hepatocellular carcinoma compared to uninfected individuals from the same region. Male carriers of the hepatitis B virus are particularly at risk of developing liver cancer with a latency period which ranges from 20–40 years. In over 90% of cases of hepatocellular carcinoma there is co-existing cirrhosis of the liver. The association between hepatitis B infection and liver cancer is even higher in woodchucks, with 100% dying of this cancer within 3 years of infection. Transgenic mice which express the hepatitis B virus X gene will develop liver cancer after a latency period of ~12 months.

Hepatitis B virus – DNA changes

The hepatitis B virus is a DNA virus. Therefore, unlike retroviruses, it does not have to integrate into the host genome to complete its replication cycle. However, hepatitis B viral integration does occur during persistent infection. Associated with this are a number of chromosomal changes including deletions and translocations. It is interesting that different patterns of DNA changes are seen with hepatitis B infection depending on the host. For example, in the woodchuck the proto-oncogenes *myc* and N-*myc* are activated following viral insertion in host DNA. In the human this is not seen, but the characteristic changes involve point mutations and deletions in the *P53* gene. Another gene which is suspected of playing some role in the oncogenic process is known as the X gene. It is thought to function by activating transcription of other genes.

The hepatitis B virus surface antigen (HBsAg) also acts to induce a chronic inflammatory response which is an important predisposing factor to abnormal mitosis and so to cancer. Approximately 50% of hepatocellular cancers have loss of heterozygosity at chromosomal loci 11p and

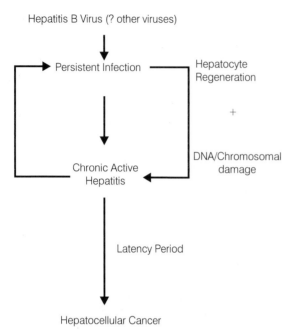

Hepatitis B Virus (? other viruses)

Persistent Infection ⟷ Hepatocyte Regeneration

+

Chronic Active Hepatitis ⟵ DNA/Chromosomal damage

Latency Period

Hepatocellular Cancer

Fig. 6.15   Genetic and environmental factors which could play a role in hepatocellular cancer.

13q which are the sites for the tumour suppressor genes *WT1*, *WT2* (Wilms tumour) and *RB1* (retinoblastoma). Allele loss in many other chromosomal locations has also been found (Table 6.3).

At present, two important components have been implicated in the hepatitis B virus-induced hepatocellular carcinoma: (1) the presence of active liver cirrhosis and (2) viral DNA effects on the host genome (Fig. 6.15). However, in the human, it is important to note that infection with the hepatitis B virus is not a necessary event for hepatocellular carcinoma and the hepatitis B virus itself is not sufficient to produce the cancer (see the following section on the environment and cancer).

Human papillomavirus

These viruses and their association with cervical cancer have been discussed in Chapter 5. High-risk or oncogenic viruses are characterised by their ability to integrate into the host genome compared to the more benign papillomaviruses which remain episomal. The high-risk viruses have early gene products known as E6 and E7 which can immortalise human keratinocytes and are considered to be responsible for initiating the development of cervical cancer. However, just as was described for colon cancer, the progression to the final cancer requires a number of genetic insults in addition to the effects of the E6 and E7 proteins. These proteins can bind to the gene products of *P53* and *RB1*, thereby promoting further the growth advantage of the infected cell by the mechanisms described earlier. Not surprisingly, mutations in the *ras* genes are also present in cervical cancer cells.

Environment and cancer

Experimental models

Chemical and physical agents which produce tumours also cause characteristic changes in DNA (Table 6.7). Animal models where cancer has been induced by a carcinogen confirm that the predominant mutations in the tumour cells mirror the type of chemical agent used. For example, breast cancers in rats which are induced by nitrosomethylurea contain G to A mutations in the H-*ras* proto-oncogene. In comparison, skin tumours induced by dimethylbenzanthracene in mice have A to T mutations in DNA.

Carcinogens

Consistent DNA changes in hepatocellular carcinoma from Southern Africa or East Asia are G (guanine) to T (thymine) transversions in codon 249 of the *P53* gene. This finding is suggestive of an *environmental agent* (for example, aflatoxin B, a food contaminant present in the above regions) as an alternative or contributing risk factor for hepatocellular cancer apart from the hepatitis B virus mentioned earlier. In experimental mutation assays, aflatoxin B1 binds preferentially to guanine residues in GC rich regions and induces the same G to T changes. G to T transversions may also be the result of oxidants which develop in association with the inflammatory process which follows the chronic hepatitis. As a comparison, *P53* mutations in colon cancer are predominantly transitions, i.e. they are more likely to be endogenous in nature rather than secondary to external mutagens.

Another example of environmentally related tumourigenesis is skin cancer. Ultraviolet light-induced mutations have characteristic in vitro changes which include frequent C to T mutations at dipyrimidines. Also diagnostic of ultraviolet damage are CC to TT dinucleotide

Table 6.7   DNA changes and carcinogens (from Jones et al 1991, Chang et al 1995).

Carcinogens	Changes in DNA
Some alkylating agents	G → A transitions[a]
Benzo[α]pyrene e.g. cigarette smoking	G → T transversions[b]
Melphelan	A → T transversions
Ionising radiation	Deletions
UV light	CC → TT dinucleotide changes

[a]Transition, change of a purine to a purine or a pyrimidine to a pyrimidine.
[b]Transversion, change of a purine to a pyrimidine or vice versa. A 'hot spot' for mutations in DNA are CpG dinucleotides since the cytosine can be methylated and then deamination of the 5-methylcytosine will lead to replacement of the cytosine (C) by thymine (T). In double-stranded DNA, the original C:G will now be T:A (G, guanine; A, adenine). This finding suggests a defect in DNA which has an *endogenous* basis. In contrast, exposure to *exogenous* mutagens is more likely to be associated with transversions and other changes listed above.

changes. These mutations are frequently found in the *P53* tumour suppressor gene from squamous cell carcinomas of the skin. Objective evidence is thus provided of the effect that sunlight has on the development of skin-related tumours. The association between cigarette smoking, cancer of the head and neck and the *P53* gene has also been clearly shown at the DNA level (Box 6.6). The DNA 'fingerprint' left behind by a carcinogen has both scientific and practical value. For example, the suspicion that an agent in the workplace was a potential carcinogen could be confirmed by finding the appropriate *P53* mutations in DNA from exposed workers.

Imprinting

Reference was made in Chapter 3 to parent of origin effects (imprinting) and the possible role these play in *genetic disorders*. It is now evident that similar effects are seen in some cancers. For example, the source of two unusual human tumours is parental-specific. Hydatidiform mole has the normal number of 46 chromosomes but all 46 are *paternal* in origin. Ovarian teratoma is the opposite in that all 46 chromosomes are *maternally* derived. Some cancers which form part of inherited syndromes (e.g. osteosarcoma, Wilms tumour, Beckwith–Wiedemann and hereditary paraganglioma) demonstrate features consistent with imprinting (Table 6.8). At the clinical level, it is important to consider the possibility of imprinting so that counselling given to families with inherited or familial cancers is accurate. The potential for imprinting to confuse the inheritance pattern may explain why the genetic component of some tumours has remained obscure (see Fig. 3.23).

Imprinting and carcinogenesis

The molecular basis for imprinting is unknown. It is intriguing that methylation differences are found in association with imprinting in transgenic mice, i.e. the functional allele is hypomethylated and the inactive allele is methylated. Whether methylation patterns are primary or secondary effects has yet to be determined. Another question which remains unanswered is whether differential methylation of alleles due to imprinting affects their susceptibility to mutation. Imprinting might also be a mechanism by which the 'second hit' occurs. If a locus is imprinted, only one of the two alleles is functional. In this circumstance, it would require a single 'hit' to inactivate the one functional allele.

Recently, *loss* of imprinting has been observed in some cancers, e.g. both maternal and paternal *IGF2* alleles are

---

Box 6.6 Molecular epidemiology and the effects of exogenous mutagens

Epidemiological studies have linked cigarette smoking and alcohol consumption to the development of squamous cell carcinoma of the head and neck. Statistical analyses can now be supplemented by data from laboratory-based experimentation. A study in which a cohort of patients with head and neck cancer were divided into three groups illustrates this. The groups were those who: (1) smoked and drank alcohol, (2) smoked but did not drink alcohol, (3) did not smoke or drink alcohol. The *P53* gene from tumour specimens was studied in the three groups. Two facts emerged: (i) mutations in *P53* were 3.5 times more common in those who smoked and consumed alcohol compared to the group which neither smoked or drank and (ii) the mutations observed in the non-smoking, non-drinking group were more consistent with *P53* defects that had occurred spontaneously rather than being induced by exogenous mutagens. Mutations which produce transversions are less common than transitions (see Table 6.7 for a description of these changes) since CpG sites are 'hot spots' for *naturally occurring mutations* in DNA. Finding a predominance of transversions would suggest an exogenous factor (see p. 136 for a discussion of *P53* and DNA transversions). Most mutations in the non-smoking, non-drinking group were at CpG sites although the numbers were insufficient for statistical validation. Another observation was that most *P53* mutations produced truncated proteins. Thus, the incomplete p53 protein would result in false-negative results if studies similar to the one described had utilised protein markers for *P53* rather than DNA (Brennan et al 1995).

Table 6.8 Evidence from inherited cancer syndromes that DNA imprinting occurs.
The parental origin of a gene can affect its subsequent expression.

Locus	Phenotype	Parent of origin effect
13q14 Retinoblastoma	Inherited form of retinoblastoma	Germline mutations occur more frequently during spermatogenesis than oogenesis*
	Sporadic form of retinoblastoma and somatic mutations	Both maternal and paternal chromosomal loci are affected
	Sporadic form of osteosarcoma and somatic mutations	Predominantly involves loss of maternal chromosomal loci
11p13 Wilms tumour	Sporadic form of Wilms tumour and somatic mutations	Predominantly involves loss of maternal chromosomal loci
11p15	Beckwith–Wiedemann syndrome	Predominantly involves loss of maternal chromosomal loci
11q23-qter	Hereditary paraganglioma (also called glomus tumour)	Tumour develops when paternally inherited; offspring of female patients are carriers but not clinically affected

*Not an imprinting effect but a higher susceptibility to mutation in sperm compared to ova.

expressed in Wilms tumours in which there is no loss of heterozygosity. In normal tissue, it is the paternal *IGF2* allele alone which is functional. *IGF2* (insulin growth factor 2) is a gene which has, as its name implies, growth stimulatory effects. Hence, the normal output from a single gene is increased when the maternal allele also expresses. A similar loss of imprinting occurs with *IGF2* and the Beckwith–Wiedemann syndrome. Two features of this disorder are noteworthy: (1) there is a predisposition to tumours including Wilms tumour and (2) there is prenatal overgrowth. Both these manifestations would be consistent with increased expression of the *IGF2* gene. The Wilms tumour and Beckwith–Wiedemann examples have been oversimplified since adjacent to *IGF2* is another gene with is imprinted in an opposite direction.

The gene is *H19*, the function of which is still unclear but it also loses its imprint in the tumours described. Oppositely imprinted but adjacent genes are also found with the Prader–Willi syndrome and Angelman syndrome (see Ch. 3). The functional significance of this observation is unknown.

The effect of 'relaxation' in imprinting is not clear but it may predispose to tumour formation since a gene which is not normally expressed is now functional. The story of imprinting and carcinogenesis is still in its early days. As well as explaining how tumours develop, the loss of imprinting opens the potential for a future line of treatment since re-establishing the imprint, if this were possible, would allow the additional gene which is expressing to be 'turned off'.

DIAGNOSTIC APPLICATIONS

Anatomical pathology

Microscopy

The standard approach in anatomical pathology is microscopic examination of a tissue section with one or more stains. A diagnosis is made on the basis of cell morphology and staining characteristics. A greater level of resolution is possible with electron microscopy. The most recent development is immunophenotyping which allows the identification of specific antigens by staining with monoclonal or polyclonal antibodies.

In situ hybridisation

To the above investigative approaches can now be added in situ hybridisation using DNA probes to detect specific sequences. An illustration of the value of in situ hybridisation is provided by the human papillomaviruses. A description of these viruses, their detection and association with genital tract lesions is found in Chapter 5. Diagnosis and typing of such viruses have become simplified *and* more accurate with tests such as Southern blotting and DNA amplification by PCR. In the research laboratory, the availability of in situ hybridisation has allowed the tissue and cellular localisation of mRNA transcripts from these potentially oncogenic viruses to be determined.

Flow cytometry

The procedures described above provide details on the type and geographical localisation of a tumour. The distribution of oncogenes or defects in DNA can also be determined. However, the studies described to date are often deficient when it comes to quantification, e.g. is an oncogene or its product amplified? Quantification is possible with flow cytometry. Components in this technique include: (1) a laser which permits measurements of large numbers of individual cells in a short time period and (2) monoclonal

antibodies which are labelled with fluorochromes. The immunological (antigenic) properties of the cells being assayed are enhanced further by measurements of additional parameters such as size and granularity. More sophisticated flow cytometers are able to sort out cell populations for further study. In this way, isolated nuclei, even from archival specimens, can be examined individually.

Just as the *P53* tumour suppressor gene is the most common DNA abnormality in tumour cells, so is *aneuploidy* (any chromosome number other than the normal 46) the usual cytogenetic finding. Determination of the DNA content of tumour cells is undertaken by flow cytometry. In this approach, nuclei prepared from tumour tissues are stained with DNA-binding fluorescent dyes such as propidium iodide. Flow cytometric analysis enables the various phases of the cell cycle (G_0, G_1, S, G_2 and M) to be defined (Box 6.1). Aneuploidy is assessed from the DNA content of the cell's G_0/G_1 phases relative to the DNA content of normal diploid cells. Aneuploid populations are seen as separate peaks from the normal diploid peak. Additional calculations are possible from the flow cytometry profile, e.g. the percentage of cells in S phase.

Cell cycle parameters described above are being used to attempt prediction of disease or treatment outcomes in a number of malignancies such as cancers of the breast and ovary, colorectal cancers and many others. Results can have both prognostic and therapeutic implications, e.g. patients with aneuploid ovarian carcinoma have a far worse prognosis than those with diploid tumours, irrespective of the extent of disease or histological appearance.

Fine needle aspiration and DNA amplification

A useful source of tissue for histopathological examination is obtainable through fine needle aspiration. This is a rapid, less traumatic procedure than conventional biopsy and carries negligible risk of tumour dissemination. The problem of whether sufficient material can be obtained for histological assessment has now been overcome to some extent by PCR. The utility of the above approach can be illustrated by reference to nasopharyngeal carcinoma. In this cancer, conventional biopsies can miss the primary lesion because the nasopharynx is difficult to visualise and the tumour infiltrates submucosally. In such circumstances, the Epstein–Barr viral genome can be sought in tumours by fine needle aspiration and PCR. Finding the virus enables the diagnosis of nasopharyngeal carcinoma to be made as well as distinguishing metastatic lesions from other head and neck tumours. Another advantage of PCR is that archival materials such as formalin-fixed, paraffin wax-embedded tissue blocks are suitable for DNA testing.

Solid malignancies

In the first edition of this book it was not necessary to include a section on DNA testing of solid malignancies because the information available at the time was still considered to be in the realm of research. However, this has rapidly changed in a few years with the cloning of many cancer-related genes and developments in DNA diagnostic methods. A number of solid malignancies will be used to illustrate DNA testing in cancer. Each will exemplify a particular feature, although the issues raised are common to all.

Familial adenomatous polyposis

At the risk of sounding repetitive, it should be noted that the *APC* gene (like most clinically important genes) is *large* and mutations are *multiple* with the majority being family-specific. By 1996, over 150 DNA changes had been described. Fortunately, exon 15 of the *APC* gene is a hot spot for mutations and approximately two-thirds are found within this region. Various strategies have now been developed for mutation testing in familial adenomatous polyposis. For example, direct DNA testing in genomic DNA will identify ~10–20% of mutations. Since most *APC* defects produce a truncated protein, the PTT test (protein truncation test) will then increase the detection rate to about 90%. Finally, a screening method such as SSCP and DNA sequencing can be used to look for the remaining mutations. This comprehensive DNA approach should allow most mutations in the *APC* gene to be detected (Figs 6.16, 6.17).

The laboratory *resources* required to undertake these tests are considerable. This may change with the more frequent use of DNA 'kits' although the *costs* involved may actually increase compared to the 'home-made kits' which experienced laboratories are often able to develop for their own purposes. In terms of familial adenomatous polyposis the resources and expenses would be justified since those who are at risk require long-term follow-up and annual bowel examination until mid-adult life. The likelihood that

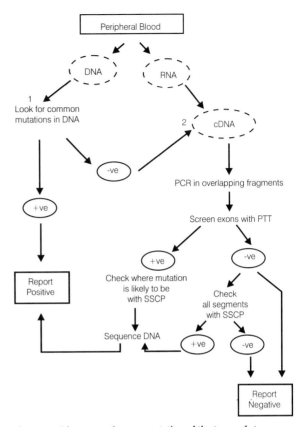

Fig. 6.16 Diagrammatic representation of the types of steps which might be used in designing a strategy for detection of multiple mutations which occur in large genes such as *APC*. DNA and RNA can be made from 10 ml of blood. **(1)** The mutations which tend to recur can be sought rapidly by testing DNA with PCR and ASO or restriction enzyme analysis (see Figs 2.18, 2.19 for explanations). If *positive*, a report is issued (assuming the appropriate laboratory safeguards and controls have been completed). If *negative*, it may become necessary to utilise RNA if the gene is too large to test the entire genomic region (i.e. exons and introns). **(2)** RNA is made into cDNA (exons only) and the cDNA is amplified in overlapping fragments. Individual exons or combination of exons are studied by PTT (see Fig. 2.20 for an explanation). If a truncated protein is found, the likely region for the mutation is identified with a screening test such as SSCP (see Fig. 2.17). DNA is then sequenced and the result given. A negative PTT test can proceed along two options. If the resources are limited and/or the gene is particularly difficult, a negative result (for the mutations tested) is issued. If resources etc. are not a problem, the entire cDNA can be tested one segment at a time by SSCP. Positive results from SSCP will then need to be confirmed by DNA sequencing. Since there is no prior evidence that a truncated protein has occurred, it will be necessary to be sure that the mutation which has been found at this time is in fact one likely to affect the protein's function.

a DNA mutation will be found is high and this will enable 50% of those who are at risk to be excluded from follow-up and a more selective, and hopefully more effective, surveillance to be continued in those who carry the mutant gene.

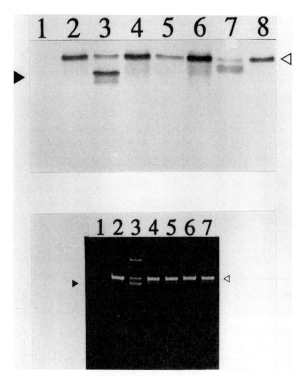

Fig. 6.17 Mutation detection in the *APC* gene.
The top gel is a PTT gel which has had ^{35}S incorporated into the protein synthesis, allowing the band patterns to be detected by autoradiography. Lanes 2, 4, 5, 6 and 8 show normal patterns of 68 kD in size (◁). Lanes 3 and 7 have additional (lower) bands which represent truncated proteins (▶). In lane 3 the protein is 63 kD and in lane 7 it is 65 kD. The segment of DNA in the *APC* gene which corresponds to the truncated proteins would next need to be characterised to identify the underlying mutation. The bottom gel is DNA stained with ethidium bromide following PCR for the codon 1061 mutation in the *APC* gene. This is a 5 bp deletion and so is readily detectable by PCR and electrophoresis. All tracks show the one normal band except for track 3 which has both the normal band and one below it. The latter identifies the 5 bp deletion. An additional (upper) band in track 3 is a heteroduplex which is an artefact but useful in detecting heterozygotes (see Fig. 2.17 for further explanation of heteroduplex).

Breast cancer

The *BRCA1* gene is comparable to *APC* in terms of size, mutations and the types of tests which are used to detect them. However, breast cancer has yet to reach the same stage as familial adenomatous polyposis in terms of routine DNA testing to detect at-risk individuals. This is because breast cancer is a more heterogeneous disorder than familial adenomatous polyposis even though a positive family history for the former is obtained. The genes and their relative contributions to breast cancer remain to be determined. However, in selected cases and, in the context of a *research environment*, DNA testing for breast cancer is an important development. Apart from the issues raised above with familial adenomatous polyposis, the finding of *BRCA1* and the subsequent media attention have identified other considerations in DNA testing for cancer.

Clinicians and the public have been exposed to *BRCA1* risk statistics in various forms, e.g. relative risk, absolute risk, lifetime risk, population risk, cumulative risk, attributable risk and so on. Not surprisingly they can be confusing. Moreover, in the clinic or the family physician's office, the patient simply wants to know how a mutation in *BRCA1* or another relevant gene will affect that person's life or her family. At present, health professionals are finding it difficult to understand what *BRCA1* is, how it works and the relevance of mutations in the gene. For example, the finding of a mutation in *BRCA1* does not mean that 100% will develop breast cancer since the data at present would suggest that the risk by age 70 years is about 85%. Therefore, the *counselling issues*, i.e. explaining the significance of DNA changes to a patient, are complex. From this has emerged the concept of a *multidisciplinary* cancer genetic clinic. The clinic involves a range of health professionals (geneticists, oncologists, nurses and counsellors) and, if the facilities are available and the clinical indications are present, the clinic can arrange for DNA mutation analysis of the breast cancer-related genes (see also Ch. 10 for further discussion about the 'molecular medicine' team approach).

Who should be referred to this type of clinic is another difficult question. Guidelines for health professionals are available (Box 6.7). However, it is essential that the person who does the referring (usually the family physician or surgeon) has some knowledge of DNA changes in cancer and the advantages/disadvantages of DNA testing so that he/she can continue to play an active role in the individual's ongoing care and follow-up, be it physical or psychological.

Retinoblastoma

Infants at risk require repeated ophthalmological examinations under anaesthesia to detect tumours at an early stage. DNA linkage analysis using retinoblastoma-specific DNA locus probes enables the wild-type to be distinguished from the mutant 13q14 loci in a pedigree analysis. Those at risk can be assessed and their risk modified depending on the DNA polymorphism patterns. An absolute exclusion or confirmation as susceptible is not possible unless a mutation can be detected, because if DNA polymorphisms located outside the retinoblastoma locus have been used this can lead to error through recombination. Although rare, retinoblastoma is curable and so optimal follow-up and early treatment are essential. The addition of DNA testing to the conventional medical management procedures can be a useful adjunct to identifying those who carry the mutant allele and those who have the normal one as well as identifying at-risk offspring of individuals who have the genetic form but have survived to adult life.

P53

The implications for DNA testing for germline *P53* mutations are more complex than with other genes which predispose to cancer, e.g. *RB1* (retinoblastoma), *APC* or

Table 6.9 Approaches used in the diagnosis and classification of the haematological malignancies.
Conventional tests remain the mainstay of diagnosis but are increasingly being supplemented by molecular technology.

Test	Information provided
Peripheral blood, bone marrow, lymph node	Morphological appearances of cells can be seen. Staining by conventional means such as May–Grunwald–Giemsa can be supplemented by specific stains, e.g. esterase detection distinguishes leukaemias which do not have a lymphoid origin
Cytogenetics	Analysis of the cell's chromosome constitution will detect translocations, e.g. the Philadelphia chromosome (chromosomes 9 and 22) and that seen in Burkitt lymphoma (chromosomes 8 and 14 usually)
Immunophenotyping	Monoclonal antibodies to antigens formed during differentiation are particularly useful to distinguish leukaemic cells as myeloid or lymphoid in origin and whether they are primitive or differentiated cells

HNPCC (colon cancer) because *P53* does not produce tumours of a certain type. The broad range of neoplasms found with *P53* makes clinical screening and follow-up difficult. Hence, the balance between resources, costs and clinical effectiveness needs to be considered even more carefully. Clearly, *P53* mutation analysis continues to be an essential research tool and increasingly has a role to play in epidemiological studies. In terms of individual disorders, the merits of each case require special consideration.

Haematological malignancies

Conventional diagnostic approaches

A similar strategy to that described in anatomical pathology is usually followed in the leukaemias or lymphomas (Table 6.9). A number of tissues can be studied and their appearances under the light or electron microscope are frequently diagnostic. Examination of the cells' chromosomal constitution will enable translocations to be detected and these can be specific for a number of disorders (Table 6.6). Immunophenotyping is now extensively used to characterise the malignant clones.

Oncogenes and tumour suppressor genes

Detection of mutated proto-oncogenes and tumour suppressor genes (see Table 6.6) provides important diagnostic and prognostic information in the leukaemias and lymphomas. Both DNA mapping (Southern blotting) and PCR can be used. Compared to conventional cytogenetic tests, the sensitivity of the molecular approaches (particularly PCR) is high, e.g. 1 cell in 100 000 with the Philadelphia chromosome can be identified. This is discussed further in the section on minimal residual disease (p. 143).

Rearranged immunoglobulin and T cell receptor genes

In the lymphoproliferative disorders (leukaemias involving lymphocytes or lymphomas) an additional molecular approach is available through examination of immunoglobulin and T cell receptor genes. The germline state for these genes will be found on non-lymphoid cells and can be detected by DNA mapping (Southern blotting) or PCR. Within the normal (*polyclonal*) population of lymphocytes in the peripheral blood, individual rearranged genes cannot be identified because the possible permutations are so many that single ones will not be distinguishable. On the other hand, a large number of cells which bear the identical rearranged pattern, i.e. a *monoclonal* population which would be present in the lymphoproliferative disorders, is detectable.

A rearranged immunoglobulin gene locus identifies a monoclonal population of B lymphocytes. The correspond-

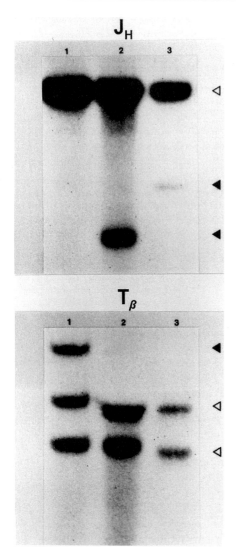

Fig. 6.18 Monoclonal population of lymphocytes indicative of a disease state.

The normal circulating lymphocytes will be polyclonal, reflecting the somatic rearrangements of their genes. Because of the multiplicity of gene rearrangements involved, the changes will not be detected at the DNA level. The presence of a monoclonal population will usually mean there is a haematological or immunological disorder involving these cells. DNA mapping patterns are able to detect monoclonal populations in B and T lymphocytes because the same gene rearrangement is now present in a large number of cells. Immunoglobulin (J_H) and T cell receptor β chain patterns are shown. Germline bands (non-rearranged genes) are indicated by ◁ and rearranged bands by ◀. Sample 1, T cell rearrangement; samples 2 and 3, different B cell rearrangements.

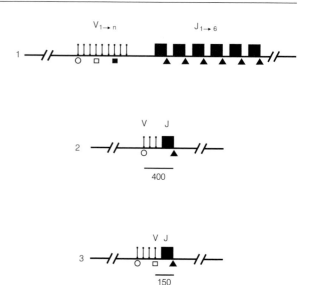

Fig. 6.19 A PCR-based strategy to identify immunoglobulin gene rearrangements occurring in lymphoproliferative disorders.

Primers to conserved sequences are mixed with primers from variant sequences. Amplification will occur from the conserved primer to one of the variant primers. (1) The immunoglobulin heavy chain locus showing an unlimited number of V genes and 6 J genes. DNA amplification primers designed from the 3′ ends of the J genes are constant (▲) whilst those from the V loci (O, □, ■) are different for each particular V gene. (2 and 3) Two rearrangements have occurred involving V and J genes. The 3′ constant primer will form an amplified product provided there is a nearby V-specific primer. The size of the amplified product will depend on the distance between the primers, e.g. 400 and 150 base pairs as illustrated here.

der. Rearrangements indicate a *clonal* population which does not necessarily mean it is *malignant*. The latter conclusion will come from a combination of genetic, haematological and clinical criteria.

The Southern blotting technique can distinguish ~1–5% of a cell population which is monoclonal. This relative insensitivity as well as the technically long and demanding steps in DNA mapping has meant that DNA amplification by PCR is a more attractive option. The exquisite sensitivity of PCR enables 1 target cell in 10^3 to 10^6 cells to be detected. Although better than Southern blotting, the sensitivity actually achieved depends on the underlying gene rearrangement because the amplification products of *coexistent polyclonal* lymphocyte populations and a number of other features associated with these rearranged genes (see below) can make it difficult to distinguish the *amplified monoclonal* product.

A disadvantage of the PCR technique to detect gene rearrangements is false-negative results. These do not reflect a deficiency in the technology but the underlying gene rearrangements which are being sought. For DNA amplification it is necessary to design primers which are specific in terms of DNA sequence for a particular region. Furthermore, the amplification primers cannot be too far

ing T lymphocyte-derived lymphoproliferative disorders demonstrate monoclonal patterns associated with rearrangements of the T cell receptor β, γ and δ genes (Fig. 6.18). The finding of gene rearrangements usually means there is a haematological or immunological disor-

apart (an approximate distance would be 0.5–1.0 kb) or amplification will not occur or be inefficient. Designing DNA primers for PCR involving immunoglobulin or T cell receptor genes is particularly difficult for a number of reasons: (1) the complexity of the germline structure, (2) the occurrence of DNA rearrangements, (3) additional somatic mutations once rearrangement has occurred, (4) the possibility of incomplete or aberrant gene rearrangements, and (5) clonal evolution in some leukaemias. To get around some of these problems, primers to sequences which tend to be conserved (e.g. regions 3′ to the J_H genes or 3′ to C_μ in the heavy chain genes) can be mixed with a number of primers derived from 5′ sequences that are likely to be altered (e.g. V_H). In this way, it is anticipated that amplification will occur from the 3′ conserved region to one of the many 5′ primers found in the mihture. Amplification will not be obtained from germline sequences since these are too far apart (Fig. 6.19). Using the above or variations of it, 75–90% of the gene rearrangements found in the lymphoproliferative disorders are detectable.

PCR is more suitable for the identification of translocation breakpoints which involve the immunoglobulin genes (e.g. *bcr-abl*, *bcl* rearrangements) since these have more specific, invariant sequences associated with them and so the appropriate amplification primers are easier to design. The sensitivity of PCR in this situation is high, e.g. a leukaemic cell mixed with 10^5 to 10^6 normal cells is distinguishable.

Minimal residual disease

High-dose chemotherapy and/or radiotherapy can be used to treat cancer and haematological malignancy provided the bone marrow can be rescued by autologous (from the same person) bone marrow transplantation (see Ch. 7). The infusion of autologous bone marrow provides sufficient haematopoietic stem cells for the marrow to regenerate. However, there is concern that even after intensive treatment, residual neoplastic cells can remain or the autologous marrow itself contains some malignant cells which then lead to recurrence of disease. For example, when acute leukaemia is first diagnosed, there are ~10^{12} leukaemic cells present. At the time of clinical remission as many as 10^{10} leukaemic cells can remain. Leukaemic cells which comprise <1% of the total marrow pool will not be detectable by conventional diagnostic approaches. This has led to the concept of 'minimal residual disease', i.e. disease which remains occult (hidden) within the patient but eventually leads to relapse.

To attempt removal of minimal residual disease in autologous marrow, 'purging' with cytotoxic drugs or by immunological approaches (e.g. monoclonal antibodies to B cells in the B-type lymphoproliferative disorders) has been attempted. Important issues, such as whether it is essential to remove all neoplastic cells and what the

effectiveness of purging is, cannot be determined unless a sensitive assay (such as PCR) is available. This has increasingly been applied to assess the state of occult neoplastic cells during treatment or in the situation of autologous marrow transplants (Box 6.8). Since PCR has the potential to detect 1 in 10^5 to 1 in 10^6 malignant cells, it is important to study a wide range of tumours, both haematological and solid, to answer the above questions.

A variation of the minimal residual disease theme involves the monitoring of progress following bone marrow transplantation from another donor (allogeneic transfer). This type of treatment is used in a number of leukaemias and bone marrow aplasias. Early identification of engraftment (presence of donor cells) or relapse (presence of host cells) is important to optimise post-transplantation treatment. However, the small number of cells usually present makes conventional diagnostic approaches difficult. PCR is ideally suited to this situation provided primers can be designed to detect DNA sequences which will distinguish host and donor cells. The extreme sensitivity of PCR means that only a few cells are necessary for assay and so the dilemma of engraftment versus relapse can be resolved early and rapidly.

Box 6.8 Assessment of minimal residual disease using PCR

In contrast to the immunoglobulin and T cell receptor rearrangements which arise from multiple sites within the gene loci, chromosomal translocations are clustered within fewer and better defined sites. Thus, DNA amplification by PCR can be used for detection since it is easier to construct amplification primers. Normal cells will not give rise to an amplified product because each primer comes from a different chromosome. The value of this approach can be illustrated by reference to acute promyelocytic leukaemia, which was mentioned in Box 6.5. Treatment of acute promyelocytic leukaemia with all-*trans*-retinoic acid produces a remission but PCR showed that patients in remission could be divided into: (1) those who remained positive for the *PML-RARα* hybrid transcript and rarely (2) those who lost this marker for the chromosomes 15 and 17 translocation. Relapses inevitably occurred (not surprising since most remained positive for the translocation product at the time of remission), but the duration of remission could be prolonged by treating with conventional chemotherapy after the retinoic acid. Duration of the disease-free interval and survival correlated well with absence of the *PML-RARα* marker. As well as identifying those patients who were at risk and so requiring additional treatment, the PCR test for this translocation marker would be useful in evaluating the presence of tumour cells in the bone marrow transplantation situation if this were being considered as part of the treatment for leukaemia.

FUTURE DIRECTIONS

Cancer pathogenesis

The genetic events which can predispose, initiate and then promote the development of tumours are many. To the two well-recognised mechanisms involving oncogenes and tumour suppressor genes can now be added the DNA repair genes. A fourth class of defects may involve epigenetic phenomena such as imprinting (Table 6.10). The next few years will lead to a much better understanding of the various molecular changes involved in the cancer pathway. An interesting recent observation has suggested that *P53* may play a role in preventing teratogenic effects during development. This focuses on the importance of understanding the normal physiological functions of these genes as well as the changes which occur following mutations in their DNA. There are many more cancer genes which need to be isolated and characterised. For example, research on prostate cancer, the second most common cause of cancer deaths in males, is gaining considerable momentum at the molecular level. The gene(s) for this tumour should not be too far away.

Screening for cancer

Familial adenomatous polyposis is a model which illustrates the preventative measures required for those at risk of cancer and the potential impact that DNA screening will have in this aspect of cancer management. However, the issues of who to test, when to test and how to resource the expensive laboratory and counselling services essential for optimal use of DNA information need careful consideration. The finding of the *BRCA1* gene has done a lot to promote informed but critical assessment of DNA testing in cancer. Universal screening programmes based on DNA analysis are a long way off but it would not be unreasonable to assume that the traditional ways, e.g. looking for blood in the stools, will seem old-fashioned when a kit enables mutations in the *APC* or related genes to be sought in the stools or, even better, the circulating blood. The dissatisfaction expressed with current cervical screening programmes could be overcome by DNA-based tests which screen for oncogenic papillomaviruses and tumour-related genes. Industry will also benefit since the mutagenic potential of chemicals, including food products, will be able to be determined more objectively by looking for characteristic DNA changes.

Disease prognosis

Prognosis for cancer can be highly variable. It is determined on the histological appearance and/or staging based on surgical or investigative findings. DNA ploidy, the percentage of cells in S phase and hormone receptor status provide additional criteria in determining outcome. In some cases, prognosis varies even within the categories defined by the above parameters. Additional indicators may be required to define disease subclasses. One approach which will assume increasing importance is an assessment of the underlying genetic defects. Already, DNA testing in neuroblastoma, one of the commonest tumours in infancy and early childhood, can predict response to treatment based on amplification of the N-*myc* proto-oncogene. Other genetic targets, e.g. assessment of *P53* and testing for drug-resistance genes, will enable prognosis and treatment regimens to be individualised rather than disease-based. Another way in which DNA testing will assist is

Table 6.10 Comparisons of oncogenes, tumour suppressor genes, repair genes and loss of imprinting as mechanisms for the development of cancer.
Perturbations in these genes or genetic mechanisms can be the forerunner to cancer.

Property	Oncogenes	Tumour suppressor genes	DNA repair genes	Loss of imprinting
Change to mutant allele	Gain of function	Loss of function	Loss of function	Gain of function
Mechanisms of action	Stimulatory, through signal transduction pathways	Inhibitory, through cell cycle and apoptosis	(i) Ensures fidelity of DNA replication (ii) Predisposes to mutations in the oncogenes, tumour suppressor genes	? Growth advantage
Mutations	Translocations, amplifications, point mutations (dominant effect)	Deletions, point mutations protein interference (recessive or dominant-negative effect)	Point mutations (recessive effect)	Not defined ? deletions imprinting control region (see Ch. 3)
Mutation origin	Somatic or germline*	Germline or somatic	Germline	Somatic

Ret is the only example of a proto-oncogene with a mutation in the germline which predisposes to multiple endocrine neoplasia.

illustrated by a study which compared traditional histopathological staging in squamous cell carcinomas of the head and neck with mutation analysis for the *P53* gene. The treatment of choice in this common cancer is surgical. Bad prognostic indicators include residual tumour tissue at the margin of resection and lymph node involvement. Following surgery, 13 of 25 cases that had been considered clear of tumour by microscopic examination were shown to have mutations in the *P53* gene in cells at the resection margin. Five of the 13 relapsed compared to no relapses in 12 patients who did not have *P53* mutations. In this situation, DNA testing provided a more accurate measure of incomplete tumour resection and so of prognosis compared to the traditional means of histopathological examination.

Therapy

The decision whether to utilise intensive treatment for cancer depends to some extent on the natural history of the underlying disorder. Some malignancies which are slow growing will not respond to treatment but follow an indolent course over many years and so need not be treated aggressively. Better understanding of the biology of cancer will be important in the future design of treatment regimens. Strategies to treat cancer based on gene therapy have rapidly expanded in the past few years. The future will determine their ultimate value. Preliminary results are impressive although it is likely that the methods for gene transfer will need considerable refinement.

FURTHER READING

Oncogenes and tumour suppressor genes

Bergh J, Norberg T, Sjogren S, Lindgren A, Holmberg L 1995 Complete sequencing of the *P53* gene provides prognostic information in breast cancer patients, particularly in relation to adjuvant systemic therapy and radiotherapy. Nature Medicine 1: 1029–1034

Bishop J M 1991 Molecular themes in oncogenesis. Cell 64: 235–248

Cavenee W K, White R L 1995 The genetic basis of cancer. Scientific American 272: 72–79

Chang F, Syrjanen S, Syrjanen K 1995 Implications of the p53 tumor-suppressor gene in clinical oncology. Journal of Clinical Oncology 13: 1009–1022

Harris C C, Hollstein M 1993 Clinical implications of the p53 tumor-suppressor gene. New England Journal of Medicine 329: 1318–1327

Hartwell L H, Kastan M B 1994 Cell cycle control and cancer. Science 266: 1821–1828

Hunter T, Pines J 1994 Cyclins and cancer II: cyclin D and CDK inhibitors come of age. Cell 79: 573–582

Knudson A G 1986 Genetics of human cancer. Annual Review Genetics 20: 231–251

Knudson A G 1993 Antioncogenes and human cancer. Proceedings of the National Academy of Sciences USA 90: 10914–10921

Krontiris T G 1995 Oncogenes. New England Journal of Medicine 333: 303–306

Lane D P 1994 p 53 and human cancers. British Medical Bulletin 50: 582–599

Lavin M, Watters D (eds) 1993 Programmed cell death: the cellular and molecular biology of apoptosis, Harwood Academic Publishers, Chur, Switzerland

Lefkowitz R J 1995 G proteins in medicine. New England Journal of Medicine 332: 186–187

Majno G, Joris I 1995 Apoptosis, oncosis, and necrosis. An overview of cell death. American Journal of Pathology 146: 3–15

Pines J 1994 The cell cycle kinases. Seminars in Cancer Biology 5: 305–313

DNA repair genes

Cleaver J E 1994 It was a very good year for DNA repair. Cell 76: 1–4

Jiricny J 1994 Colon cancer and DNA repair: have mismatches met their match? Trends in Genetics 10: 164–169

Passarge E 1995 A DNA helicase in full bloom. Nature Genetics 11: 356–357

Rustgi A K 1994 Hereditary gastrointestinal polyposis and nonpolyposis syndromes. New England Journal of Medicine 331: 1694–1702

Genetic models and cancer

Brennan J A, Boyle J O, Koch W M et al 1995 Association between cigarette smoking and mutation of the p53 gene in squamous-cell carcinoma of the head and neck. New England Journal of Medicine 332: 712–717

Cavenee W K, White R L 1995 The genetic basis of cancer. Scientific American 272: 72–79

Cline M J 1994 The molecular basis of leukemia. New England Journal of Medicine 330: 328–336

Easton D, Ford D, Peto J 1993 Inherited susceptibility to breast cancer. Cancer Surveys 18: 95–113

Ford D, Easton D F 1995 The genetics of breast and ovarian cancer. British Journal of Cancer 72: 805–812

Gayther S A, Warren W, Mazoyer S et al 1995 Germline mutations of the *BRCA1* gene in breast and ovarian cancer families provide evidence for a genotype-phenotype correlation. Nature Genetics 11: 428–433

Hoskins K F, Stopfer J E, Calzone K A et al 1995 Assessment and counseling for women with a family history of breast cancer. Journal of the American Medical Association 273: 577–585

Jones P A, Buckley, J D, Henderson B E, Ross R K, Pike M C 1991 From gene to carcinogen: A rapidly evolving field in molecular epidemiology. Cancer Research 51: 3617–3620

Perera F P 1996 Uncovering new clues to cancer risk. Scientific American 274: 54–62

Ponder B A J (ed) 1994 Genetics of malignant disease. British Medical Bulletin, Churchill Livingstone, Edinburgh, vol 50, p 517–745

Rainier S, Feinberg A P 1994 Genomic imprinting, DNA methylation and cancer. Journal of the National Cancer Institute 86: 753–759

Robinson W S 1994 Molecular events in the pathogenesis of hepadnavirus-associated hepatocellular carcinoma. Annual Review of Medicine 45: 297–323

Rustgi A K 1994 Hereditary gastrointestinal polyposis and nonpolyposis syndromes. New England Journal of Medicine 331: 1694–1702

Schwartz R S 1995 Jumping genes and the immunoglobulin V gene system. New England Journal of Medicine 333: 42–44

Swan D C, Vernon S D, Icenogle J P 1994 Cellular proteins involved in papillomavirus-induced transformation. Archives of Virology 138: 105–115

Tallman M S 1994 All-*trans*-retinoic acid in acute promyelocytic leukemia and its potential in other hematologic malignancies. Seminars in Hematology 31 (suppl 5): 38–48

Vogelstein B, Kinzler K W 1993 The multistep nature of cancer. Trends in Genetics 9: 138–141

Diagnostic applications

Campana D, Pui C-H 1995 Detection of minimal residual disease in acute leukemia: methodologic advances and clinical significance. Blood 85: 1416–1434

Evans D G R, Fentiman I S, McPherson K, Asbury D, Ponder B A J, Howell A 1994 Familial breast cancer. British Medical Journal 308: 183–187

Lo Y-M, D 1994 Detection of minority nucleic acid populations by PCR. Journal of Pathology 174: 1–6

Loda M 1994 Polymerase chain reaction-based methods for the detection of mutations in oncogenes and tumor suppressor genes. Human Pathology 25: 564–571

Naber S P 1994 Molecular pathology – detection of neoplasia. New England Journal of Medicine 331: 1508–1510

Ponder B A J 1994 Setting up and running a familial cancer clinic. In: Ponder B A J (ed) Genetics of malignant disease. British Medical Bulletin, Churchill Livingstone, Edinburgh, vol 50, p 732–745

Toribara N W, Sleisenger M H 1995 Screening for colorectal cancer. New England Journal of Medicine 332: 861–867

van der Luijt R, Khan P M, Vasen H et al 1994 Rapid detection of translation-terminating mutations at the adenomatous polyposis coli (APC) gene by direct protein truncation test. Genomics 20: 1–4

Future directions

Brennan J A, Mao L, Hruban R H 1995 Molecular assessment of histopathological staging in squamous-cell carcinoma of the head and neck. New England Journal of Medicine 332: 429–435

Cannon-Albright L, Eeles R 1995 Progress in prostate cancer. Nature Genetics 9: 336–338

Kuss B J, Deeley R G, Cole S P C et al 1994 Deletion of gene for multidrug resistance in acute myeloid leukaemia with inversion in chromosome 16: prognostic implications. Lancet 343: 1531–1534

Nicol C J, Harrison M L, Laposa R R, Gimelshtein I L, Wells P G 1995 A teratologic suppressor role for *P53* in benzo[α]pyrene- treated transgenic *P53*-deficient mice. Nature Genetics 10: 181–187

THERAPEUTICS

RECOMBINANT DNA (rDNA)-DERIVED DRUGS

Blood coagulation factors

A review of the treatment options in haemophilia demonstrates the potential for molecular technology in the area of therapeutics. Haemophilia is an X-linked genetic disorder characterised by spontaneous bleeding into muscles, joints and mucous membranes. Deficiencies in two of the blood coagulation factors (factor VIII in haemophilia A; factor IX in haemophilia B or Christmas disease) are the underlying defects. Severity of haemophilia depends on the level of coagulation factor with a reduction to <1% of normal activity producing a potentially life-threatening disorder. In males, the frequency of haemophilia A is ~1 in 10 000 and for haemophilia B it is ~1 in 50 000 (see Ch. 3 for further discussion of haemophilia).

There are over 12 coagulation factors in the pathway involved with haemostasis (control of bleeding). The coagulation factors act as a cascade with one activating a second which then activates a third and so on. Factor VIII's role in the coagulation pathway is to accelerate the activation of factor X by activated factor IX in the presence of calcium and phospholipid. Factor VIII in plasma is bound to another protein called von Willebrand factor. The latter is essential for the platelet's role in haemostasis and stabilises the factor VIII molecule. Knowledge of the factor VIII's molecular structure was greatly expanded in 1984 when the gene was cloned. From the gene's structure it was possible to predict potential protein domains and functions. The in vivo behaviour of factor VIII was further understood following expression of its cDNA.

Structure and function of the factor VIII molecule

The factor VIII protein is encoded by a gene which is 186 kb in size and contains 26 exons. mRNA specific for factor VIII is 9 kb in length. The protein comprises 2332 amino acids and is synthesised as a single chain which is processed

as it moves from the rough endoplasmic reticulum to the Golgi apparatus. As this occurs, the mid-portion of the molecule is excised. The two heterodimers formed are held together by metal ions. There are a number of domains associated with the 90–200 kDa amino-terminal *heavy chain* and the ~80 kDa carboxy-terminal *light chain* (Fig. 7.1).

Treatment of haemophilia with plasma-derived products

A summary of historical developments in the treatment of haemophilia is given in Table 7.1. Landmarks include the isolation in 1964 of a specific factor VIII-enriched product known as *cryoprecipitate*. From this came antihaemophiliac factors of greater concentrations and stability, so that during the 1970s more effective programmes utilising home therapy became established. In this way affected individuals were able to treat themselves the instant a bleeding episode was noted.

Complications occurring with the plasma-derived antihaemophiliac factors remained significant (Table 7.2). The initial problems associated with hepatitis B infection from blood products were thought to have been resolved once blood banks introduced screening programmes and viral-inactivating steps were incorporated into commercial production. Nevertheless, the subsequent recognition of other viruses, e.g. HIV, hepatitis C and parvovirus, focused again on the problems of human-derived products. For example, hepatitis C as the cause of non-A, non-B hepatitis was initially considered to be a complication which was more likely to occur if plasma had been obtained from paid donors. Subsequently, with the availability of more sophisticated tests for hepatitis C detection, it became apparent that both paid and voluntary donors were sources of this virus. The increasing number of haemophiliacs infected with HIV from the late 1970s illustrated further limitations of plasma products, even those that had undergone 'viral

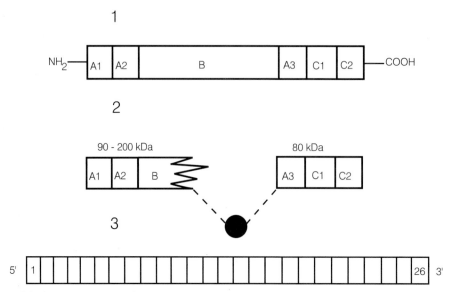

Fig. 7.1   Structure of the factor VIII protein and its gene.
(1) Cloning of the factor VIII gene showed that its 2332 amino acids are organised in domains identified as A1, A2, B, A3, C1, C2. The A domains have a similar structure to ceruloplasmin, a copper-binding plasma protein. The B domain is not essential for factor VIII procoagulant or cofactor activity. The C domains have phospholipid-binding function. (2) In plasma, factor VIII is predominantly found as a heterodimer comprising a heavy chain and a light chain. The heavy chain has some variability depending on the amount of B domain which is present. (3) The 26 exons which make up the factor VIII gene.

inactivation steps'. These processes included various combinations of heating and/or organic solvent exposure. This increased the production costs but gave no guarantee that all viruses (both known and *unknown*) would be neutralised. For example, parvovirus can withstand temperatures of up to 120°C.

Liver disease as a cause of morbidity and mortality in haemophilia is well documented. An association between hepatitis B virus, hepatitis C virus and liver disease is established. Immunosuppression involving both B and T cell lymphocyte function and independent of infection with HIV was also found in haemophiliacs. A mechanism proposed for immunosuppression implicated the many contaminating plasma proteins present in the factor VIII or factor IX products.

An additional and serious complication in haemophilia occurs in ~10–15% of patients who develop factor VIII or factor IX inhibitors (antibodies) as a consequence of exposure to new antigens (neoantigens) to which they have not developed immunological tolerance. These individuals are placed at a much greater risk of dying from an uncontrollable bleeding episode since conventional factor replacement becomes ineffective.

Finally, a critical consideration with any human-derived product is its availability, which will always be dependent on obtaining a regular supply of pooled human plasma. This can never be guaranteed.

Modern therapeutic developments in haemophilia

Monoclonal antibodies to factor VIII were included in production protocols during the 1980s. The substance

resulting had both a higher potency and greater purity. The reduction in contaminating proteins was considered to be an added bonus if this effect was significant in the immunosuppression observed in haemophiliacs.

In 1987, the first patient was treated with an rDNA-derived human factor VIII. The steps involved in preparing this product are summarised in Figure 7.2. The use of mammalian cell lines such as CHO (Chinese *hamster* ovary) enabled complex post-translational steps, e.g. glycosylation, to be undertaken. Recombinant factor VIII produced in this way was shown to have fewer protein contaminants although some were invariably present. The long-term effects of chronic exposure to these substances remain to be determined. The activity of the recombinant product was equivalent to monoclonal antibody-purified factor VIII (Table 7.3).

Expression studies of the rDNA-derived factor VIII product have also shown that optimal stability requires association with the von Willebrand factor. Therefore, one commercial product involves co-transfection of cDNAs for both factor VIII and the von Willebrand factor. Another observation that removal of the B-domain facilitates expression of recombinant factor VIII would have potential commercial application if manipulation of the gene in this way enabled more efficient in vitro production.

The first multicentred human trials of recombinant human factor VIII were started in 1988 and reports issued 5 years later indicated very promising laboratory and clinical responses. It is evident from these studies that recombinant and plasma-derived factor VIIIs are biologically identical. The potential risk from immunisation by

Table 7.1 Historical developments in the treatment of haemophilia A.

1840	Bleeding episode treated with normal fresh blood
1920s	Plasma rather than whole blood shown to be effective
1930s–1950s	Fractionation of plasma producing various components with antihaemophiliac activity. 1937: factor VIII implicated in haemophilia A
1964	Cryoprecipitate isolated. Cryoprecipitate is produced by allowing frozen plasma to thaw. When thawed a cold insoluble precipitate (cryoprecipitate) remains. This contains a high concentration of factor VIII
1970s	High potency freeze-dried factor VIII concentrates available (home therapy now feasible)
1984	Biotechnology companies Genentech (San Francisco) and Genetics Institute (Boston) clone and express the factor VIII gene
1980s	More effective viral inactivation steps incorporated in manufacture of factor VIII products
	Monoclonal antibody-purified factor VIII becomes available
	Alternative haemostatic pathways used to bypass the effect of inhibitors
1988	Clinical trials using recombinant human factor VIII start
1990s	First reports of multicentred trials with recombinant factor VIII. Clinical and laboratory responses very satisfactory. Risk of inhibitor development still unclear – suggestion that it may be increased with recombinant DNA product. Potential effects of contaminants (DNA, protein) from the recombinant DNA process remain to be defined
	Recombinant factor VII used to bypass factor VIII inhibitors
	Novel means to produce recombinant DNA-derived products are reported, e.g. transgenic expression of factor IX in sheep milk; autologous transplantation with factor IX gene in fibroblasts

Table 7.2 Some problems associated with plasma-derived haemophilia treatment products.

Infection	Hepatitis B virus Hepatitis C virus (60–80% seropositive) HIV-1 (34% seropositive) Parvovirus (resistant to 120°C – conventional heat inactivation inefficient) Unknown infectious agents
Liver disease	Up to 20% of haemophiliacs have progressive and potentially fatal liver disease
Immunosuppression	Contaminating proteins in factor VIII concentrates (including relatively pure ones) are implicated as the cause for immunosuppression which is independent of infection with HIV; both B and T cell lymphocyte functions are impaired
Inhibitor development	Exposure to neoantigens leads to the risk of antibodies developing against the coagulation factor(s). This complication is not excluded in the monoclonal antibody-purified (or even recombinant) products
Cost	Purification; viral inactivation
Availability	Plasma-derived factors limited since they rely on availability of pooled human plasma

these were expensive, had all the potential complications associated with factor VIII and, in addition, were reported to produce thrombotic complications. Activated factor VII on its own appeared to be a useful way to bypass the factor VIII block but isolation of this substance from plasma was commercially impractical. This problem was resolved with the development of an rDNA-derived product. Recombinant human factor VII has now been successfully used in a number of patients with factor VIII antibodies including one report involving a haemophiliac with antibodies to both human and porcine factor VIII who was also a Jehovah's Witness. This individual developed a life-threatening cerebral haemorrhage but refused treatment with human blood products. However, rDNA-derived substances were acceptable and recombinant factor VII was eventually used with resolution of his bleeding problem.

Future directions
The feasibility of rDNA-derived factor VIII and factor VII coagulation products is now being assessed. The availability of a regular and controllable supply will allow better planning, more efficient and early treatment of bleeding problems. The potential long-term immunological consequences of rDNA-derived products remain to be determined. However, from experience already gained with rDNA-derived insulin, growth hormone and the hepatitis B vaccine, these are unlikely to be major problems. Costs for the recombinant product are high but as was seen with the hepatitis B vaccine they will gradually

exposure to neoantigens produced by the rDNA process remained theoretical but did not appear to be a major problem in the trials reported to date. Whether recombinant products are more or less likely to induce the development of inhibitors is still uncertain and will take some years to assess.

Factor VIII inhibitor treatment
Plasma-derived activated coagulation factors (mixtures of activated factors X and VII and tissue factor) were produced to overcome the block on factor VIII activation resulting from the development of antibodies. However,

Table 7.3 Purity of various factor VIII preparations.

Source	Activity (units/mg protein)
Plasma	0.01
Cryoprecipitate	0.1
Concentrates 1970s	~0.5
Concentrates 1980s	~1.5
Immunopurified	~2000
Recombinant	~2000

fall. It would not be unreasonable to predict that most plasma-derived substances for the treatment of coagulation disorders will ultimately be replaced by rDNA-derived products.

Hormones and growth factors

Erythropoietin

Anaemia is an important complication of chronic renal failure and has a significant effect on morbidity. Components of this anaemia include decreased production of the hormone erythropoietin (usually abbreviated to EPO), iron deficiency and chronic infection or illness, which can impair bone marrow function. EPO is the primary regulator of red blood cell production. The majority of the hormone is produced in the kidney and its level is regulated by the tissue oxygen tension present in the kidney. EPO feeds back to the bone marrow where it acts on the committed erythroid progenitors and precursors. In the presence of hypoxia, the level of EPO increases and the red blood cell mass expands. Once hypoxia is corrected, EPO production in the kidney is suppressed.

EPO was discovered at the beginning of the twentieth century. It had also been suspected for some time that reduced production of this hormone played a major role in the anaemia of chronic renal failure. However, it was not possible to take this any further in the investigative or therapeutic sense since the amount of EPO which could be isolated from kidneys or the urine was insignificant. Thus, EPO did not become a therapeutic option until the gene was cloned in 1985. Recombinant EPO is now produced by expression of its cDNA in a mammalian cell line system similar to that described for haemophilia. The necessity for a sophisticated expression system reflects the requirement for glycosylation which would not be possible with the more rudimentary bacterial expression system.

Clinical trials have now confirmed that recombinant EPO given to anaemic dialysis-dependent patients with end stage renal failure is capable of raising the haemoglobin level and so reduce or remove the necessity for blood transfusions. Thus, the potential complications associated

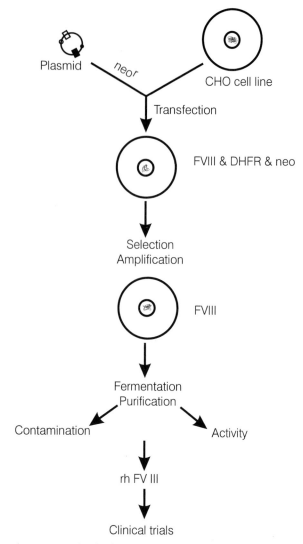

Fig. 7.2 Steps involved in producing recombinant human factor VIII.

The recombinant plasmid has both cDNA for factor VIII (■) and a dihydrofolic acid reductase gene (DHFR = □) as well as their respective promoters (● ,○). This plasmid is transferred into a mammalian cell line such as CHO to enable post-translational changes to occur. Another plasmid containing a gene for resistance to the antibiotic neomycin (neo^r) is also co-transfected. The end result is the CHO genome which has integrated into it factor VIII, DHFR and neo^r. Specific selection for CHO cells which have the neo^r gene is possible by culturing in a medium containing G418, a neomycin analogue which is lethal to cells unless they carry a gene for neomycin resistance. Amplification for the factor VIII-containing CHO cells is then obtained by culturing the CHO cells in medium with increasing concentrations of methotrexate. This will select CHO cells which have the DHFR gene (and so also the factor VIII gene) since DHFR-deficient CHO cells will not grow in the presence of methotrexate. The factor VIII-expressing CHO cells are then fermented in large, commercial volumes. Protein isolated from these cells is next purified by immunoaffinity chromatography with monoclonal antibodies to factor VIII. The end product is checked for contaminants (mouse and CHO proteins, DNA) and its functional activity assessed. Recombinant human factor VIII is then available for clinical use.

with blood transfusions (infection, immunosensitisation to HLA antigens, iron and circulatory overload) can be avoided with the recombinant hormone which is given subcutaneously or intravenously. The improvement in the quality of life for this chronically ill population has justified the costs of recombinant EPO.

Other indications for recombinant EPO are presently being assessed. These include the specific *normocytic normochromic anaemia* found in cancer, severe infection, or chronic inflammatory disorders such as rheumatoid arthritis. Although this form of anaemia had been identified for many years, it was difficult to determine what role EPO played since an accurate plasma assay was not available. However, once the gene was cloned it became possible to measure EPO very accurately. In cancer, there is evidence that the level of EPO is reduced and this accounts to some extent for the co-existent anaemia. A number of trials have now shown that the anaemia can be corrected or blood transfusion requirements reduced in 32–85% of cancer patients who are treated with recombinant EPO. What remains to be determined is whether there is an improvement in the quality of life and if the costs incurred will be justified.

Side-effects reported with recombinant human EPO include local reactions to the injection as well as more serious complications such as hypertension, seizures and venous thromboses. These are considered to reflect an inappropriate increase in the red blood cell mass which occurs secondary to the EPO effect.

Growth hormone

Human growth hormone, a protein of 191 amino acids, is essential for growth. Because this hormone is species-specific its only biological source is human. Following the successful treatment of a pituitary dwarf in 1958 with human growth hormone, programmes were established to isolate this substance from pituitaries obtained from cadavers. However, the programmes were ceased in the mid-1980s when a number of recipients died from Creutzfeldt–Jakob disease, a fatal slow virus infection of the central nervous system (see Box 10.2). A direct association between the pituitary extract and the slow virus infection was soon established, leading to withdrawal of the human-derived growth hormone.

Growth hormone has now been replaced with a re-combinant product following the cloning and expression of the gene in 1979. Because the protein does not require sophisticated post-translational modifications to its structure, recombinant growth hormone is prepared using a relatively simple bacterial expression system. However, there are two potential disadvantages from this expression system. They are the necessity for extensive purification to remove impurities of bacterial origin, particularly endotoxins, and the presence of an additional methionine amino acid at the start of the protein. The latter occurs because the eukaryotic start codon (ATG) is translated in the prokaryotic system into a methionine (see Ch. 2).

Clinical trials during the mid-1980s have demonstrated the efficacy of the recombinant growth hormone and it has remained in continuous use since. No significant side-effects have become apparent. The additional methionine does not, as originally feared, lead to a major increase in antigenicity of the product.

Haematopoietic growth factors

Bone marrow haematopoietic (blood-forming) cells are derived from the proliferation and differentiation of pro-genitor cells that form specific lineages following their interaction with cytokines. The pluripotential stem cell is the ultimate source of the lymphoid and myeloid precursor cells. The latter differentiates into the platelet, erythroid, neutrophil and macrophage lineages. As the progenitors mature into committed stem cells, they become responsive to a more limited range of cytokines, although two or more are often required for optimal effect. The major *cytokine* groups are the colony stimulating factors (CSFs) and the interleukins (ILs). These substances are involved in the regulation, growth and differentiation of a variety of cellular components. Haematopoietic *growth factors* include the CSFs, the ILs and EPO.

Knowledge that haematopoietic growth factors existed has been available since the 1960s. However, the minute amounts able to be isolated *and* the numbers involved (e.g. there are four different CSFs) made working with these substances extremely difficult. In the 1970s, traditional biochemical methods enabled small quantities of M-CSF and GM-CSF to be produced, followed by G-CSF and multi-CSF in the early 1980s. However, nothing further could be done because of the amounts available. The in vivo significance of these products remained unclear until the relevant genes were cloned. The first to be cloned (by AMGEN, a US biotechnology company) was G-CSF in 1986. The others soon followed. Expression of the cloned genes enabled in vivo animal studies followed by primate and then human trials.

The CSFs are glycoproteins which are capable of stimulating the formation and activity of white blood cells including granulocytes and macrophages. The prefix indicates the target cell, i.e. G-CSF (granulocyte) and M-CSF (macrophage or monocyte). The remaining two CSFs (GM-CSF and multi-CSF or interleukin 3) are less lineage-restricted since they affect both granulocytes and macrophages. GM-CSF also stimulates the precursor cells in the erythroid (red blood cell) and megakaryocyte (platelet) lineages. Many cell types, including fibroblasts and endothelial cells, produce one or more of the CSFs. Base-line production is low but rises rapidly following exogenous stimuli such as bacterial endotoxins. In turn, the CSFs provide the proliferative signals for cellular differentiation. Interleukins (e.g. IL-1, IL-4, IL-5, IL-6) are also produced by a variety of cells and have some

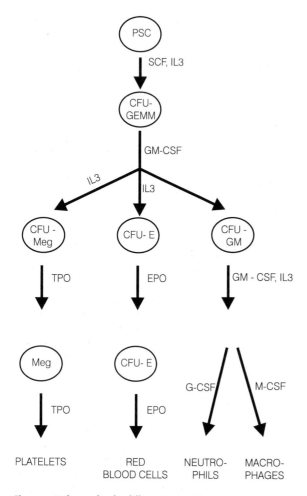

Fig. 7.3 Pathways for the differentiation of haematopoietic cells and the associated cytokines and growth factors.
The ultimate source of all haematopoietic cells is the pluripotential stem cell (PSC). Further along the pathway is another more differentiated myeloid progenitor stem cell known as CFU-GEMM (CFU, colony forming unit; GEMM, the four types of cells which will result, i.e. granulocyte G, erythroid E, megakaryocyte Meg and macrophage M). Differentiation into three pathways occurs in response to cytokines with cells responding to more specific ones as they become committed to a particular cell lineage. SCF, stem cell factor; IL, interleukin; CSF, colony stimulating factor; TPO, thrombopoietin; EPO, erythropoietin. IL-3 is also known as multi-CSF. Not shown in this diagram is the lymphoid lineage which derives directly from the PSC and the eosinophil or basophil lineages which come from the CFU-GEMM compartment.

haematopoietic growth factor activities as well as making important contributions to the immune and inflammatory responses (Fig. 7.3).

The production of recombinant-derived human growth factors, e.g. rh G-CSF, rh GM-CSF, rh IL-3 (rh = recombinant human), is providing important information on their physiological functions. The availability of commercial quantities has enabled potential therapeutic roles to be assessed in clinical situations associated with low white

blood cells (leukopenia), e.g. following bone marrow transplantation and after chemotherapy for cancer or leukaemia. G-CSF has shown therapeutic benefits by reducing the duration of leukopenia. Thus, the risk of infection-related complications is reduced, which in turn will lead to a decrease in morbidity, mortality and overall in-patient hospital costs since in-patients can be discharged earlier. Alternatively, more aggressive chemotherapy will become possible thereby increasing the chance that residual cancer cells will be killed. Zidovudine, used in the treatment of AIDS, has bone marrow suppression as one of its side-effects. An additional use of the recombinant growth factors would be in situations such as AIDS where pre-existing anaemia and immunodeficiency are exacerbated by treatment with a drug such as zidovudine. In this circumstance, the ability to reduce the degree or duration of marrow depression would have a beneficial effect on morbidity and mortality.

An unexpected benefit to come from rh G-CSF has been the ability to induce progenitor cells to move out from the bone marrow compartment. This has allowed the use of peripheral blood rather than bone marrow for autologous transplantation following intensive chemotherapy for cancer. Apart from avoiding an anaesthetic and the pain associated with multiple bone marrow aspirations, the use of G-CSF-stimulated peripheral blood leads to a more rapid haematological recovery. Another potential advantage involves studies which are underway to assess the efficacy of *cord* blood for transplantation. Cord blood cells avoid the morbidity associated with repeated bone marrow sampling particularly if the donor is an infant or child. The risk for graft versus host disease may also be less (see Chs 4 and 10 for further discussion). Potential immunomodulatory and antileukaemic effects of the growth factors are also under investigation.

Despite all the apparent advantages of the recombinant growth factors, a number of questions remain unanswered. What are the optimal drugs to use, e.g. single factors or combinations and if so which combinations? The routes of administration and regimens require identification. When are growth factors given in respect to bone marrow transplantation and so on. Potential long-term side-effects of the recombinant products remain to be assessed. It will also be necessary to determine the cost-effectiveness and cost-benefit analyses for these products. There are many other cytokines which are or could be produced by rDNA means (e.g. M-CSF, IL-3, SCF, IL-1, IL-6, IL-11). They act at an earlier stage of haematopoietic cell differentiation and as yet have not been approved for clinical trials (Box 7.1).

The past few years have seen dramatic developments, with the cloning of the many haematopoietic growth factor genes. In 1994, the gene for another factor, thrombopoietin, was isolated. It has a specific effect on platelets and interestingly the N-terminal portion of this gene has considerable homology with EPO. The evolutionary and functional relationships between EPO and thrombopoietin as well as the therapeutic significance of the latter will no doubt soon follow.

Biotechnology developments

Blood substitutes

A goal as yet unattained in the biotechnology industry is the development of a blood substitute. This substance will need to possess two important properties: (1) carry and release oxygen, (2) function as a plasma expander. The urgency to obtain such a product comes from the increasing concern about contaminants in human blood *and* an anticipated diminishing supply as there is more selective donor screening and donors themselves are becoming less willing to donate. An artificial blood substitute would have a multi-billion dollar market because of its widespread applications, e.g. trauma, surgery and medical conditions associated with anaemia.

A plasma-derived synthetic haemoglobin has been manufactured. Haemoglobin is extracted from out-dated human blood donations or from bovine blood. It next has to be modified chemically to overcome two problems which arise once haemoglobin is separated from its red blood cell environment. These are an increase in oxygen affinity (i.e. oxygen is more difficult to release from the haemoglobin) and an instability of the $\alpha_2\beta_2$ tetramer (see Ch. 1). These synthetic haemoglobins are known as 'haemoglobin-based oxygen carriers'. Modifications which will overcome the oxygen affinity and instability problems

include conjugation of haemoglobin to substances such as dextran or polyethylene glycol and polymerisation of these compounds. Drawbacks to the approaches described include the requirement for blood donor-derived products (demand will thus be dependent on donations) and the potential for viral or other infections to be transmitted. The immunogenicity of bovine-derived products remains to be determined.

The first report of a recombinant human haemoglobin product came in 1990. Subsequently, recombinant human haemoglobins have been produced using *E. coli* and yeast expression vectors. In both cases, the products have similar structure and function to native haemoglobin. However, limitations such as low yield, high oxygen affinity and instability have made these recombinant substances commercially unattractive. An attempt to overcome the problem of low yield involved the injection of human haemoglobin genes into fertilised ova from pigs. Pigs were chosen because of the similarities between porcine and human haemoglobins. Transgenic animals formed were demonstrated to have circulating human haemoglobin in the blood but unfortunately the expected high yield was not achieved and additional purification/stability steps which would have been required made the pig model an unattractive commercial option.

The ability to alter in vitro the α or β globin genes prior to their expression has also been utilised in an attempt to overcome the high oxygen affinity and stability problems mentioned. In one product, a genetically engineered recombinant haemoglobin was shown to have both higher stability and lower oxygen affinity. The stability was improved since the two α globin subunits were fused by expressing the genes as a tandem duplication. However, the preparation still had a very short half-life compared to the survival time for a red blood cell. The potential immunogenicity of the α globin dimer was also a concern. The reduced oxygen affinity was obtained by incorporating an asparagine to lysine mutation at codon 108 in the β globin gene. This produces a variant haemoglobin (equivalent to the naturally occurring mutant called Hb Presbyterian) which has reduced oxygen affinity. Thus, oxygen could be released more easily to surrounding tissues.

There is some way to go before a synthetic haemoglobin is shown to have adequate physiological function and to be free of side-effects. Nevertheless, the difficulties described are more likely to be overcome by molecular technology than conventional approaches. In the future, the successful commercial production of a range of recombinant therapeutic agents can be expected to continue (Table 7.4).

Novel expression systems

The initial high costs in preparing genes, vectors and evaluating both *efficacy* and *safety* through a range of in vitro and in vivo parameters will be offset by long-term

Table 7.4 Some recombinant DNA-derived human therapeutic agents.

Class	Recombinant product
Drugs	Erythropoietin
	Insulin
	Growth hormone
	Coagulation factors (VIII, VII)
	Plasminogen activator
	DNase
Vaccines	Hepatitis B
Cytokines	GM-CSF
	G-CSF
	Interleukins
	Interferons

and reliable sources of therapeutic products. More efficient expression systems would reduce costs even further. Large transgenic animals, e.g. sheep, goats and pigs, are being investigated with this in mind. Recombinant drugs which are expressed in the milk or blood of these animals might provide a high-yield source. Examples being evaluated at present include coagulation factor IX, α_1 anti-antitrypsin (sheep milk) and tissue plasminogen activator (goat milk). The long-term developments in the transgenic 'pharmyard' as it has been called, are promising.

GENE THERAPY

Gene therapy can be defined as the transfer of genetic material (DNA/RNA) into the cells of an organism to treat disease or for research purposes, e.g. marking studies. Gene therapy in the human refers to *somatic cell gene therapy*, which means the target cell is not part of the germline so that transmission to future generations cannot occur. Germline gene therapy (an example of which would be the transgenic animals described in Ch. 2) has been proscribed for various reasons which are discussed further in Chapters 9 and 10. When first proposed as a therapeutic option, gene therapy was considered only in the context of genetic disorders. Today, gene therapy has a broader role to play in treatment, e.g. in cancer and the infectious diseases. Disorders for which gene therapy has been tried or considered include:

Genetic diseases

- Immunodeficiencies, e.g. adenosine deaminase deficiency
- Cystic fibrosis
- Familial hypercholesterolaemia
- Storage disorders, e.g. Gaucher disease
- Coagulopathies, e.g. haemophilias A, B
- Haemoglobinopathies, e.g. β thalassaemia, sickle-cell disease.

Acquired diseases

- Cancer, e.g. melanoma, brain and renal tumours
- AIDS
- Vascular disease
- Neurological disorders, e.g. Parkinson disease, Alzheimer disease.

A number of criteria have been proposed to identify the types of genetic disorders for which gene therapy would be indicated. These include: (1) a life-threatening condition for which there is no effective treatment, (2) the cause of the defect is a single gene and the involved gene has been cloned, (3) regulation of the gene need not be precise, and (4) the technical problems associated with delivery and expression of the gene have been resolved (see below). Similar considerations would hold for the acquired disorders although in these circumstances a 'cure' might not be the prime treatment consideration and so the same stringent criteria would not necessarily apply.

Monitoring of potential gene therapy protocols by various government and institutional biosafety committees has been intense. It was not until September 1989 that the US National Institutes of Health (NIH) approved the first *marker* study involving transfer of DNA into patients with melanoma, a malignant skin cancer (details on p. 156). In September 1990, the first *therapeutic* transfer of a genetically engineered cell was undertaken in a 4-year-old child with the potentially fatal genetic disorder, adenosine deaminase deficiency (details on p. 157). By 1995, over 100 gene therapy trials had been approved by the NIH's advisory body known as the RAC (Recombinant DNA Advisory Committee).

Gene delivery

Strategies for gene delivery

There are two ways to transfer DNA (RNA) into cells (Fig. 7.4). A prerequisite for ex vivo transfer is the necessity to culture cells in vitro. Therefore, not all cells will be useful targets for this type of gene therapy. Another requirement is the ability to return the genetically altered cells to the patient, i.e. the cells need to be transplantable. The above considerations have meant that considerable work involving ex vivo transfer has focused on haematopoietic cells. Apart from the fact that ex vivo transfer may be the only suitable approach available in

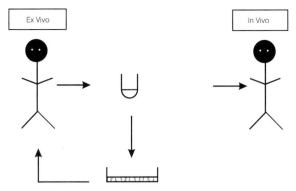

Fig. 7.4   Transferring DNA into cells.
The ex vivo approach involves the removal of cells from the patient. DNA (or RNA) is next introduced into the cells which are then cultured to obtain adequate numbers. The genetically altered cells (and perhaps physically or antigenically altered following the ex vivo manoeuvres) are finally returned to the patient. In some circumstances, ex vivo transfer is the only feasible option, e.g. haematopoietic cells. In terms of safety, there is more confidence with ex vivo transfer since only the appropriate cells will take up the DNA/RNA. A more physiological approach and challenge for the future is in vivo transfer which involves direct entry of DNA (or RNA) into the patient. Targeting is vital in this form of transfer.

many cases, it has another advantage in terms of safety, i.e. there is more control over which cells will take up the foreign DNA. However, in vivo transfer is considered to be more physiological and may be the only option in some circumstances, e.g. disseminated cancer. In vivo transfer remains a future priority awaiting developments which ensure that the right cells express the transferred DNA and they do so in adequate numbers. The concept of 'targeting' becomes a real issue when in vivo transfer is considered (discussed further below).

The ultimate aim in gene transfer is to get DNA into specific tissues. Again there are two major approaches – *physical* and *viral* (biological) means. The cell and nuclear membranes can be made more permeable to DNA following co-precipitation of DNA with calcium phosphate or an electric shock (called electroporation) (Table 7.5). Using micropipettes it is possible to inject DNA into the nucleus of a cell. More novel approaches to facilitate movement of DNA into a cell include: the injection of DNA directly into muscle cells; insertion of DNA via cationic liposomes in the process known as lipofection, i.e. synthetic spherical vesicles which have lipid bilayers and so are able to cross the cell membrane; the coating of DNA with proteins; and the 'gene gun' – DNA-coated microprojectiles. Physical methods can be relatively inefficient when it comes to cells taking up DNA. More importantly, DNA inserted into the host genome in this way is usually present as multiple copies, i.e. there is no control over the sites of insertion and so the function of normal genes could be affected. Finally, the expression of the introduced gene is only transient.

The current preferred method for gene transfer involves

Table 7.5   Delivery systems for gene transfer.
Approaches to get DNA into cells to allow genes to be expressed. The large number of approaches available would suggest that no single method is ideal. Even within the viral vectors there are useful and less useful attributes which can be found with each virus.

Type of approach	Delivery method
Physical	*Improving membrane permeability to DNA*: calcium phosphate co-precipitation, electroporation (electric shock)
	Microinjection: into the cell nucleus, into muscles
	Other methods: insertion via liposomes, coating DNA with proteins, gene gun, microencapsulation
Viral	*Integrated into host's genome*: retrovirus, adeno-associated virus
	Not integrated into host's genome: adenovirus, herpes simplex 1 virus[*], smallpox virus

*Although the herpes simplex 1 virus does not integrate (and so a transient effect would be expected) its wild-type form displays latency. Although the molecular basis for latency remains to be defined, it means that this virus could potentially be made to express its inserted gene over a longer time frame without the necessity for integration to occur.

the use of viruses particularly the retroviruses (Box 7.2, Fig. 6.2). Conventional RNA and DNA viruses are less suitable as vectors because nucleic acid is rapidly degraded if it cannot integrate into the DNA of the host genome. A further advantage of the retroviral vectors relates to their high efficiency of gene transfer which in theory can approach 100% although in practice is considerably less than this. Following retroviral-induced insertion, the gene of interest is present as *one copy* in a single, *random* site in the host's genome. Thus, progeny of the infected cells are more likely to retain the gene, i.e. expression is less likely to be transient.

There are a number of potential problems with retroviruses as vectors for gene transfer. These include the concern that the retroviruses will revert to replication-competent organisms which would give them the theoretical risk of inducing cancer. DNA insert size is limited, which can be a problem if a large gene is involved. Since retroviral vectors are produced from living cells there is the worry that contaminants derived from these cells will be present. Because of these problems a number of other viruses have been developed for use in gene therapy. Two of the viruses (retrovirus, adeno-associated virus) will integrate into host DNA and so have the potential for long-term expression of the transduced gene, i.e. a cure is possible. This advantage must be balanced by long-term side-effects if the integration has interfered with the function of a normal gene. The other viral vectors (as well as the physical means) do not lead to integration and so the associated genes are only expressed for a short term, i.e. treatment will need to be repeated.

Target cells for gene transfer
Another consideration in gene therapy is the *target* cell. If

Box 7.2 Use of retroviruses to transfer genes

Wild-type retroviruses can convert their RNA into double-stranded DNA which can then integrate into the host's genome (see also Fig. 5.2, Box 7.7). Viral proteins encoded by the *gag, pol* and *env* genes make up ~80% of the retroviral genome. These RNA segments can be deleted and replaced by a 'foreign' gene, e.g. human adenosine deaminase (ADA). Now the recombinant retrovirus is no longer infectious because it cannot make its own structural proteins. This is a good thing for gene therapy. Persistent infection by the genetically engineered retrovirus would not be permissible since it might lead to neoplastic change, the wrong cells expressing the gene or the germ cells becoming infected and so passing the gene on to future generations. To become a useful vector for DNA transfer the retrovirus must infect in a *controlled* way. This can be done with 'packaging cells'. These contain a 'helper' retrovirus which has also been genetically manipulated to produce empty virions, i.e. structural proteins are present but a complete infectious virion cannot be made. However, the retroviral vector with its inserted ADA gene can utilise the structural proteins produced by the helper virus in the packaging cells to form a complete (infectious) virion which can undergo *one* round of infection. This would be enough to get the genetically engineered retroviral RNA into the target cells and thence the latter's DNA. Advantages of the retroviral vector for DNA transfer include: (1) a single virus infects one cell, (2) transduction efficiency is adequate although needs to be improved, (3) integration into the host genome means there is long-term expression of the inserted gene, and (4) the virus is non-immunogenic. Disadvantages are: (1) the target cell must be dividing before the retrovirus can integrate into the cell's genome and (2) integration is random.

a retroviral vector is used for transduction, an important prerequisite for the target cell is for it to be dividing so that the retrovirus can integrate into the host genome. The target cell should also be appropriate to the type of expression required. For example, a neurological disorder may derive no benefit from the transfer of genes into haematopoietic cells. Finally, the target cell needs to be long-lived to prolong the effects of gene therapy. For the latter condition, the use of pluripotential stem cells would be ideal since integration of a gene into this type of stem cell should produce a cure or at the very least a long-term effect. Because of the potential availability of stem cells and the relatively advanced state of bone marrow transplantation, considerable work to date has focused on the haematopoietic stem cells as targets for gene transfer.

The bone marrow pluripotential stem cell is a rare cell which to date has only been satisfactorily isolated and characterised in the mouse. Gene transfer into human bone marrow-derived stem cells has been possible because

of the infectious capability of the retroviruses. Nevertheless, expression observed in these instances has been low and of short duration. Thus, gene therapy would not be appropriate in disorders such as the β thalassaemias for which significant gene expression would be required to produce an adequate supply of protein. This may be overcome in the near future with recent developments in molecular technology. These include: (1) the potential to stimulate division of the pluripotential stem cells with the recombinant human growth factors, thereby making these cells move out of the G_0 phase of the cell cycle and so become more accessible to infection with a retrovirus (Box 6.1); (2) the use of monoclonal antibodies to identify surface antigens found on primitive cells, e.g. CD34-positive cells; and (3) the availability of DNA sequences which can significantly up-regulate (i.e. increase) gene expression. Enhancer elements recently located in the β globin gene locus are able to increase expression of transfected genes. The incorporation of these enhancer elements in retroviral constructs is now being attempted (see Ch. 1 for a further description of the enhancer elements).

Other cells with potential for gene therapy are summarised in Table 7.6. These cells display a number of features which make them useful targets for the transfer of genetic information. However, none promises a cure even if effects are long-lived because they are not stem cells, i.e. life span is finite as the cells do not have an unlimited potential for self-renewal and they do not differentiate into various types of secondary cells.

Gene marking

Gene therapy protocols are of two types: (1) those which are expected to lead to a therapeutic result and (2) those which are essentially of research interest. The latter usually involve marking studies in which the fate of a particular cell is followed by tagging that cell with a foreign gene.

Tumour infiltrating lymphocytes

Tumour infiltrating lymphocytes are lymphoid cells which invade solid tumours. They can be grown in culture and have tumour killing potential. For example, removal of these cells from a melanoma and growth in the presence of an interleukin (IL-2) enables the tumour infiltrating lymphocytes to be cultured and then reinfused back into patients. The lymphocytes target to the melanoma where they can induce regression of the tumour. The first human gene transfer involved a marking study with the addition of a neomycin gene via a retroviral vector into tumour infiltrating lymphocytes obtained from patients with advanced melanoma. The transduced lymphocytes were then reinfused into the patients and their survival duration and potential toxic effects noted. The results were satisfactory in that the neomycin-containing tumour infiltrating lymphocytes persisted in the circulation; they were able to

Table 7.6   Target cells for gene transfer.
Target cells need to be long-lived and of a suitable type for the expression required.

Cell	Utility
Haematopoietic stem cells	Sources are bone marrow and umbilical cord blood. The possibility for gene transfer into the pluripotential stem cell means a cure is feasible. Haematological and immunological defects are the types of disorders which could be corrected. Particularly suited to ex vivo transfer.
Lymphocytes	The major advantages of the lymphocyte are its role in immunity, its relatively long life and its ease of access in the blood. This cell has been the target for gene transfer in melanoma and adenosine deaminase deficiency. Also suited to ex vivo transfer.
Respiratory epithelium	Not suited to ex vivo transfer. Slowly dividing respiratory epithelial cells less effective targets for in vivo transfer with retroviruses. At present good transfer possible with adenoviruses and lipofection.
Hepatocytes	Ex vivo: cultures can be obtained from liver tissue. Cells transduced with retrovirus can be reimplanted in the liver via the portal circulation. In vivo: novel methods to get DNA into cells involve coating DNA with a protein receptor which is recognised by hepatocytes.
Fibroblasts, keratinocytes (skin)	Easy to access and grow in culture. Can produce biologically active compounds, e.g. coagulation factor IX. Main problem is short-term effect which may be due to graft rejection. Suitable for ex vivo or in vivo transfer.
Skeletal muscle	A problem with skeletal muscle is that it is post-mitotic (and so retroviral vectors less effective) and multinucleate. In vivo: injection of DNA in plasmid form into muscle cells enables expression of the DNA without it incorporating into host genome. In vitro: adenoviral vectors have been used to insert a mini dystrophin gene into cells. Approaches require further assessment.

target to the melanoma cells and there were no apparent side-effects directly attributable to the retroviral vector.

The success of this limited form of gene transfer enabled the next step to proceed. In this case, the gene for tumour necrosis factor (TNF) was inserted into the tumour infiltrating lymphocytes using a retroviral vector. Tumour necrosis factor is a potent agent which can cause regression of tumours but has limited use because of its toxic side-effects. Thus, the aim was to deliver tumour necrosis factor directly to melanoma cells and so reduce systemic toxicity. Since January 1991, a number of patients have received tumour necrosis factor. The therapeutic benefit of tumour necrosis factor delivered by genetically engineered lym-

Box 7.3   What causes relapse in leukaemia following intensive chemotherapy and autologous bone marrow transplantation?

Autologous bone marrow transplantation remains an option for treating some leukaemias with a poor prognosis. In this situation, marrow is harvested from the patient prior to intensive chemotherapy which will inevitably be complicated by marrow aplasia. The patient is 'rescued' from the otherwise fatal aplasia by reinfusing his/her own marrow which has either been purged of leukaemic cells by various in vitro manoeuvres or is a remission marrow, i.e. leukaemic cells are not present to any extent by conventional assessment. The combination of intensive chemotherapy and autologous marrow transplantation will cure leukaemia in some individuals but others will eventually relapse. An important but unanswered question relates to the source of this relapse. Relapse could have occurred in *transplanted cells*, i.e. purging of leukaemic cells from the marrow had been inadequate or the remission marrow contained a signficant number of leukaemic cells. On the other hand, the source of relapse could have been *leukaemic cells* which had persisted in the patient's bone marrow even after intensive chemotherapy. The two possibilities could not be investigated by conventional means since: (1) cells used for transplantation were identical to those in the patient (i.e. autologous) and so could not be distinguished and (2) the population of cells to be studied was extremely small. Both problems have now been overcome by molecular technology. First, PCR enables a very few cells to be studied. Second, autologous cells can be marked with a gene, e.g. the neomycin gene, prior to re-infusion into the patient. PCR could then be used to identify whether relapsed cells had the neomycin gene. Results from such marking studies have confirmed that relapse can involve the transplanted cells and so purging protocols have been inadequate and/or remission marrows contain sufficient leukaemic cells to produce a relapse.

phocytes has been disappointing. Nevertheless, the results of both studies demonstrated that retroviral-mediated gene transfer was feasible. Another research focus for gene marking studies involves the tagging of transplanted cells to determine if relapses following treatment for cancer occur in the host or donor cells. This use of gene transfer will provide answers to a number of interesting biological questions (Box 7.3).

Introducing new genes

Adenosine deaminase (ADA) deficiency

In 1990, a 4-year-old child with the potentially fatal autosomal recessive disorder, adenosine deaminase deficiency, received an infusion of her own lymphocytes which had been genetically altered by a retrovirus containing a normal ADA gene (Fig. 7.5). ADA deficiency was chosen

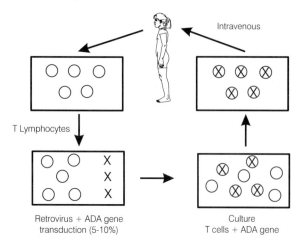

Fig. 7.5 Protocol for adenosine deaminase (ADA) gene therapy. T lymphocytes (O) are removed from the patient's circulation and infected ('transduced') with a retrovirus containing the wild-type ADA gene (x). Transduction efficiency at 5–10% is not high and so cells will need to be cultured. The patient is re-infused with her own lymphocytes after sufficient numbers of the transduced cells are grown in culture. Since pluripotential stem cells have not been used a cure is unlikely and so the procedure must be repeated depending on the life span of the transduced lymphocytes.

Labels in figure: Intravenous; T Lymphocytes; Retrovirus + ADA gene transduction (5-10%); Culture T cells + ADA gene

for a number of *clinical* reasons. It constituted an important cause of the severe combined immunodeficiency syndromes in children. Death within the first 1–2 years of life was common. Medical treatment was suboptimal. This included a recently released drug called PEG-ADA which comprised the natural product (ADA) coupled to polyethylene glycol (PEG) to increase its half-life. PEG-ADA was very expensive and follow-up time had not been sufficient to confirm its long-term efficacy. Another therapeutic option for ADA deficiency was bone marrow transplantation. This usually produced cure if the transplant was successful. However, less than one-third of patients had an appropriately matched sibling for transplantation. The above 4-year-old girl had been treated with PEG-ADA but had responded inadequately to this form of therapy. In these circumstances, approval was given to attempt gene therapy.

Features at the *DNA level* which made ADA deficiency a good candidate for gene therapy included: (1) the target or affected cells were lymphocytes and so accessible through the blood, (2) T lymphocytes have a relatively long life span, (3) the gene had been cloned and was relatively small at 3.2 kb in size, and (4) it was expected that a moderate level of gene expression would be sufficient to reduce mortality in this condition.

At first, stem cells isolated from the bone marrow of the 4-year-old girl were proposed as the targets for gene transfer. Subsequently it was found that mature T lymphocytes isolated from peripheral blood were more practical alternatives to the elusive stem cells (estimated to comprise 1 in every 10 000–100 000 cells). Multiple infusions of genetically engineered autologous T lymphocytes were given at 1–2 monthly intervals. Since the target cell in this case was no longer the pluripotential stem cell, a cure became unlikely. The child initially required maintenance infusions at 3–4 monthly intervals. It remains to be determined how frequently lymphocyte transfusions will be necessary. Long-term follow-up will be necessary to assess the advantages as well as potential side-effects.

By 1995, 10 patients with ADA deficiency had undergone various forms of gene therapy. In some cases it has been possible to identify transduced cells in the patients' peripheral blood or bone marrow for 6 months or more following the last treatment course. Cellular and humoral immune responses in the patients have improved and this has been accompanied by a better quality of life for these children. For ethical reasons, none has had his/her PEG-ADA treatment stopped, although in some cases the dose has been reduced without ill effects.

Cystic fibrosis
An abnormality in the chloride transport channel in cystic fibrosis leads to failure of chloride to exit from the cell and so water is forced back into the cell. This produces an accumulation of thick, dry mucus in the lung, pancreas and other organs which is the hallmark of this disorder. A major cause of morbidity and mortality in cystic fibrosis occurs secondary to the chronic respiratory infections which ultimately lead to respiratory failure. Thus, local 'gene therapy' at the level of the respiratory epithelium would be beneficial. This would be a viable option since it had been shown that it was possible to correct the chloride channel defect by transfer of the cystic fibrosis transmembrane regulator (*CFTR*) gene into *CFTR*-deficient cell lines as well as in the airways epithelium of rats. Primate studies showed no toxic effects following infection with a recombinant virus which contained the *CFTR* gene.

The conventional gene therapy approach which has been illustrated in Figures 7.4 and 7.5, i.e. take cells, genetically manipulate ex vivo and then return them to the appropriate environment, is more difficult to achieve in the respiratory epithelium. Moreover, the epithelium divides slowly and so a retroviral vector would be less satisfactory. In this circumstance, an alternative vector is required. One virus which would have appeal in the context of respiratory epithelium is the adenovirus. This virus is ubiquitous in humans and will target to respiratory epithelium. Adenoviruses used are non-pathogenic integrating DNA viruses that are replication-deficient. The adenoviral genome will accept a relatively large DNA insert such as the ~6.5 kb *CFTR* mRNA and it produces a stable recombinant. In contrast to the retrovirus, host cell division is not a requirement for integration to occur.

Despite its obvious advantages, a number of questions related to the adenovirus as a vector for gene transfer remain unresolved. It is not clear how (and if) integration

Box 7.4 Liposomal mediated gene transfer for cystic fibrosis

A risk with viral vectors which integrate into the host genome is the potential for *insertional mutagenesis*. One estimate places this at 1 in 100 000 for each transduced cell. However, multiple events are usually required before a cancer develops and so the risk should be considerably less. Retroviruses have also been used since 1988 in human studies and for a longer time in animals. No instances of cancer related to gene therapy have been reported in humans. Some examples have occurred in animals but these are considered to have arisen from contamination of retroviruses with helper viruses. Nevertheless, the concern about potential side-effects resulting from the use of retroviral vectors has been an important motivation to find alternative DNA delivery systems. Other viral vectors also have their problems. For example, in the case of cystic fibrosis, the advantages of the adenovirus in terms of its tropism for respiratory epithelium and insertion into non-dividing cells is balanced by transient cell expression and stimulation of inflammatory as well as host immune responses which are likely to reduce the long-term efffectiveness of the adenovirus. There are now a number of clinical trials in progress which will assess the value of the adenoviral–*CFTR* construct to treat human cystic fibrosis. In parallel with these are studies which utilise liposomal mediated gene transfer as an alternative gene delivery system. Liposomes are synthetic spherical vesicles with a lipid bilayer. A gene or DNA segment is added to a cationic lipid suspension and then mixed with the cell of interest in a process known as *lipofection*. Liposomes then transfer genes to the target cells by fusing with the plasma membrane. Liposomes do not contain protein and so will not induce an immune response. However, once in the cell there is no specific mechanism to transfer DNA to the nucleus and so liposomes are relatively inefficient gene delivery systems. Preliminary results from the gene therapy studies underway have not identified which is the superior way to express DNA in transduced cells. Long-term follow-up may resolve this dilemma.

Box 7.5 Gene therapy for familial hypercholesterolaemia

Familial hypercholesterolaemia (FH) is an autosomal dominant disorder associated with mutations in the gene for the LDL (low density lipoprotein) receptor. Hypercholesterolaemia leads to affected homozygotes dying prematurely from coronary artery disease in childhood. Heterozygotes are also at increased risk from coronary artery disease. Drug options for FH are limited. Liver transplantation, an extreme form of treatment, can lead to a cure. Although the LDL receptor gene is expressed in many cells its most important site is in the liver. A rabbit animal model for FH has been successfully used to develop a gene therapy strategy which reduced the animal's serum cholesterol over a period of months. One ex vivo approach to human gene therapy involved resection of ~250 g (~1/6) of the liver followed by transduction of hepatocytes with a retrovirus containing the wild-type LDL receptor gene. After culture, the transduced hepatocytes were returned to the patient via the portal circulation where they targeted to the liver. Results from the first five patients showed that the surgical steps were feasible but the expression of the introduced LDL receptor gene was inconsistent and not enough to make a significant difference to the serum cholesterol level. More novel protocols, including attempts at in vivo gene therapy, are presently being designed. For example, the LDL receptor gene is ligated into a plasmid coated with a protein which has a receptor recognised by liver cells. After systemic administration, the LDL receptor gene vector attaches to hepatocytes and a proportion of vector DNA reaches the nuclei of liver cells.

in a number of centres. An alternative to the viral vector approach for gene transfer utilises liposomes and a specific trial to compare this form of gene therapy is also underway (Box 7.4). Another genetic disorder which has been treated by gene therapy is familial hypercholesterolaemia (Box 7.5).

Cancer

Gene therapy for cancer can be considered in a number of categories. Protocols are designed to:

- Alter natural immunity to cancer cells
- Kill or interfere with growth of cancer cells
- Insert a wild-type tumour suppressor gene, e.g. *P53*
- Interfere with oncogene expression, e.g. K-*ras*
- Enhance tolerance to high doses of chemotherapy or delay resistance.

Important cells in the host's immune response to tumour cells include lymphocytes, macrophages and neutrophils. As indicated in Figure 7.3, the major group of cytokines involved in the haematopoietic pathways producing these cells are G-CSF, GM-CSF and the interleukins. Other

between virus and host occurs and so beneficial effects may not be permanent. Nevertheless, this may not be a major problem since it would be possible to undergo repeated exposures, e.g. a nebulising spray or bronchoscopic insertion. The long-term safety of an attenuated adenoviral vector and the immune response to the cystic fibrosis protein produced remains to be seen. Another potential source for a host immune response comes from low level expression of some viral (i.e. foreign) endogenous genes. To counter the latter, molecular biologists are constructing more sophisticated adenoviral vectors which do not co-express their own genes. Overall, the approach involving introduction of the cystic fibrosis gene product to a specific anatomical region is highly attractive and is being pursued

Box 7.6 Gene therapy for cancer using a prodrug 'suicide gene' strategy

To overcome the problem of non-selectivity with cytotoxic treatment, a gene therapy approach which utilises the conversion of an inactive compound (prodrug) to an active metabolite has been devised. For example, thymidine kinase (tk), an enzyme from the herpes simplex 1 virus (HSV-1), is harmless to mammalian cells. However, tk converts the antiviral agent ganciclovir to a substance which inhibits DNA synthesis. Similarly, 5-fluorocytosine (the prodrug) can be changed to 5-fluorouracil (a cytotoxic agent) by reaction with cytosine deaminase. Following gene transfer, cells which express the 5-fluorocytosine will be destroyed when exposed to cytosine deaminase while the remaining cells willl be safe. Hence, this form of gene therapy has been described as involving a 'suicide gene'. An example of how a 'suicide gene' can be used in the treatment of cancer follows.

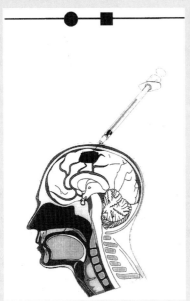

Glioblastoma multiforme is a malignant brain tumour which responds poorly to conventional treatment. Using imaging techniques, it is possible to inject the tumour with a herpes simplex virus (HSV) vector containing its promotor sequence (●) and the tk gene (■). The patient is next given ganciclovir and only those cells which have tk will be able to convert the ganciclovir to its toxic metabolite. As well as the tumour cells which express HSV-tk it has become apparent that neighbouring cells are also susceptible to damage. This 'bystander effect' contributes to the tumour killing potential. Another benefit may come from use of HSV-1 rather than a retrovirus since the herpes virus in its natural infection can establish latency within the central nervous system (as shown by recurrence of 'cold sores' and herpes zoster). If the genetically engineered HSV-tk can be induced to behave in this manner, a longer course of controlled treatment would become possible.

substances which can be used to stimulate immune responses are γ-interferon and HLA antigens. Thus, gene therapy protocols in cancer treatment utilise various permutations of the above genes. The genes are inserted into tumour cells or autologous fibroblasts which are then injected back into the patient. In this way it is hoped to stimulate an immune response specifically to the tumour cells. Another approach which involved a more direct local destruction of cancer cells was initially attempted by inserting the gene for tumour necrosis factor, a potent cytolytic agent, into the tumour infiltrating lymphocytes mentioned earlier. Sophisticated vectors have been designed to provide concentrated but localised doses of cytotoxic drugs (Box 7.6). In vivo transfer protocols to replace *P53* when it is deficient in lung cancer or interfere with the function of *K-ras* when it is overexpressed in the same tumour are also in the trial phase.

Two key problems associated with chemotherapy in cancer are bone marrow toxicity and the development of drug resistance. To protect the bone marrow, chemoprotection gene therapy protocols are specifically designed to target stem cells and introduce into them genes such as *MDR1* (multidrug resistance 1). This codes for P-glycoprotein and provides cells with resistance to a wide range of cytotoxic drugs. *MDR1* is discussed further in the last section of this chapter as well as in Chapter 6. Another potential chemoprotective gene is that for ATase (alkyl transferase). This gene is involved in DNA repair and will protect the stem cells from a range of alkylating agents. If bone marrow stem cells can be protected with the genes just described, more aggressive chemotherapeutic regimens will become possible. Limitations to the success of this approach include the isolation of stem cells and the relatively inefficient transfer of genes into cells, even with the retroviral vectors.

Modulating the expression of genes

Fetal haemoglobin (HbF)

As described in Chapter 1, there is progressive switching of globin genes during development. The fetal genes (Gγ and Aγ) are replaced by adult β and δ globin genes at ~6 months of age. A delay in switching from fetal to adult globin genes has been seen in infants of diabetic mothers and in association with the sudden infant death syndrome. The underlying mechanisms are unclear although an elevated plasma α-amino-*n*-butyric acid has been proposed as potentially significant in the diabetic situation. A second group of disorders leading to incomplete HbF switch are genetic in origin, e.g hereditary persistence of fetal haemoglobin (HPFH). This has been described in Chapter 3 and the molecular defects reviewed. Thus, the HbF (fetal) to HbA (adult) switch can be altered. This has clinical relevance to the β thalassaemias and sickle-cell disease.

Patients with the severe homozygous form of β thalassaemia do not manifest clinical problems until after 6 months of age when their normal fetal genes are replaced by the non-functioning adult β gene. This leads to an imbalance in the normal adult α/β globin protein ratio of 1 and the red cell's life span is correspondingly reduced. Since HbF is physiologically normal, the potential to postpone or induce an incomplete switch of the fetal genes is appealing since this would correct the clinical problems and complications of β thalassaemia. This mechanism would involve replacement of β globin protein by γ globin and so the α/β+γ ratio would return towards 1. Similarly, the complications associated with HbS (sickle-cell haemoglobin) can be reduced in the presence of HbF which interferes with the polymerisation of HbS.

A number of chemotherapeutic agents can increase the level of HbF. In humans, 5-azacytadine and hydroxyurea produce a consistent elevation in HbF although there are clear genetic effects also operating since the changes differ considerably from individual to individual. The obvious drawback to treatment with these agents is their potential side-effects, particularly marrow toxicity, and the long-term concern that leukaemia or another form of cancer will develop. The mode of action of these drugs remains unclear. The initial observation that 5-azacytadine demethylated DNA and so potentially exerted its effect through gene reactivation is no longer considered to be its mechanism of action. Since fetal haemoglobin in adults is derived from a population of cells called F cells, it is proposed that the cytotoxics exert their effect through changes in the cell cycle brought about by pulsed doses of cycle-specific drugs. Thus, rapid erythroid regeneration after each drug pulse encourages predominantly the formation of F cells.

Clinical trials are presently underway to monitor the effects of hydroxyurea in patients with severe forms of sickle-cell disease. Hydroxyurea has been preferred because of its ease of administration and lesser potential toxicity. Preliminary data show useful responses with a reduction in the number of painful crises. Hydroxyurea has now been proposed as a form of preventative therapy in some cases of HbS disease. More specific agents which can interfere with the HbF to HbA switch or reactivate HbF once it has switched will provide an important therapeutic benefit for the haemoglobinopathies which are found in many parts of the world.

Antisense technology

Perturbation of DNA function is possible with *antisense RNA* (Fig. 7.6). Three mechanisms have been proposed to explain how antisense RNA works: (1) steric interference with ribosomal activity, (2) activation of substances which digest mRNA, and (3) binding to mRNA and so preventing its translation. The potential for antisense technology has already been demonstrated in agriculture, e.g. genetically engineered tomatoes that are mush-resistant because the gene producing polygalacturonase, which breaks down the

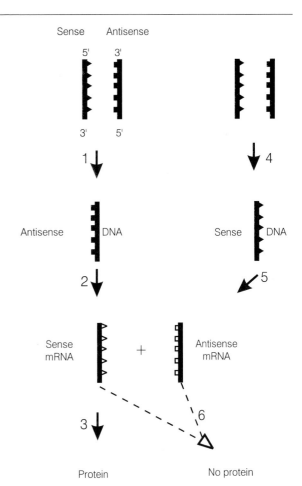

Fig. 7.6 Antisense technology as a form of gene therapy.
Antisense RNA can inhibit the expression of a gene and would be effective therefore in diseases where unwanted gene expression occurs. The underlying mechanism(s) which produce the antisense effect are not fully known. One explanation involves a binding of the antisense RNA (or an antisense oligonucleotide) to sense mRNA and so inhibition of the latter during translation (protein synthesis). The normal transcription → translation pathway is illustrated on the left of the diagram (**1–3**) with the antisense DNA strand providing the template for transcription by RNA polymerase to make (sense) mRNA. On the right (**4–6**), the DNA sense strand has been copied. This could be a normal response by an organism to foreign DNA or a genetically engineered gene which has been 'flipped' around so that the sense sequence becomes the template for mRNA synthesis. Alternatively, oligonucleotides with the antisense sequence are introduced into the cell. The end result is the same with the antisense mRNA/oligonucleotide binding to the sense mRNA and inhibiting its activity.

cell wall, has been inhibited by antisense RNA. A number of plants have now been genetically engineered in similar fashion to give them resistance to particular viruses. In the human, antisense strategies would be useful in diseases for which there is *inappropriate expression* of a gene, e.g. oncogenes which lead to a cancer, or in diseases associated with the presence of *foreign genes*, e.g. an invading pathogen. Although multiple events are involved in the evolution of a malignant cancer cell, the potential for a gene therapy approach to interfere with progression of one key step in this pathway may be sufficient to inhibit or retard the development of a cancer. The identification of mRNAs which are pathogen-specific, e.g. the common 35 nucleotide mRNA leader sequence in all trypanosomes or the prokaryote-specific rRNAs of bacteria, would be useful targets for therapy via antisense technology. In contrast, an antisense approach would not be effective in genetic defects associated with a *lack* of expression.

The synthesis of *antisense oligonucleotides*, rather than the use of a gene which makes antisense RNA, provides an alternative strategy with which to manipulate gene function. These oligonucleotides have a DNA sequence complementary to target DNA or RNA. Antisense oligonucleotides bind to their targets and inhibit transcription or translation. An oligonucleotide 20–30 base pairs in size will usually ensure attachment occurs to a *unique* sequence within the genome. In vitro, antisense oligonucleotides are able to decrease the tumourigenicity of cell lines. Antisense oligonucleotides have been synthesised to a number of potential oncogenic sequences and have specifically inhibited their expression in cell lines, e.g. the *bcr-abl* hybrid transcript which is unique to the leukaemic cells in chronic granulocytic leukaemia.

An extension of antisense technology is DNA *triple helix formation*. This relies on the ability to accommodate a third strand of DNA within part of the DNA duplex to form a triple helix. Binding of the third base occurs to the already formed A/T or G/C pairs. Triple helix formation was first demonstrated in the 1950s. Confirmation of its existence in double-stranded DNA came in the late 1980s. Subsequently, in vitro studies were able to show repression of transcription by the proto-oncogene *myc*. Triple helix formation can inhibit DNA replication or DNA/protein interactions, thereby affecting transcription. Potential targets for triple helix formation are similar to those described for antisense approaches.

Antisense technology has in part been driven by developments in automated DNA synthesisers. These have enabled the synthesis of large amounts of relatively cheap and good quality oligonucleotides. Successful chemical modification of the oligonucleotides reduces their potential for endogenous breakdown by cellular and nuclear nucleases. Considerable technical problems still remain to be resolved. For example, antisense molecules are not catalytic. Binding to target, which is reversible, could turn out to be ineffective. Delivery of antisense compounds also needs to become more efficient. Membrane-based vehicles, e.g. liposomes, similar to those described in gene therapy, are able to internalise antisense oligonucleotides into cells. Once in the cell the oligonucleotide can act within the cytoplasm or gain access to the nucleus if DNA is the target. The pharmacokinetic properties and potential toxicity of oligonucleotides are still to be fully determined. Studies to date have utilised non-physiological conditions

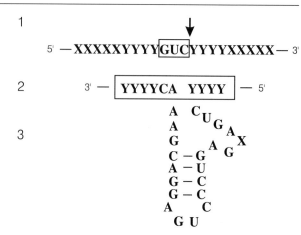

Fig. 7.7   The structure of a ribozyme.
(**1**) The RNA sequence to which the ribozyme will target is shown as a series of Ys. The GUC sequence lies adjacent to the site of ribozyme cleavage (↓). (**2**) The complementary sequences (Ys) on the ribozyme which ensure its specificity for a target. (**3**) The ribozyme's catalytic domain.

since saturating quantities of oligonucleotides have been required to achieve adequate intracellular and nuclear concentrations. Nevertheless, the novelty of this technology and the successful in vitro studies observed to date make it likely that in vivo testing will produce interesting results.

Ribozymes

Another form of gene therapy involves the use of ribozymes. Ribozymes are naturally occurring RNA species that cleave RNA ('endoribonucleases') at specific sequences (see Ch. 1). Ribozymes would have similar applications to those described for antisense oligonucleotides, i.e. DNA or RNA species, whether they are from tumours or infectious agents, could be specifically inhibited. A potential advantage of ribozymes over antisense molecules lies in the former's catalytic activity so that following binding there is cleavage of target RNA. Specificity of the ribozyme rests with the hybridising (antisense) arms located on either side of the molecule's catalytic activity domain (Fig. 7.7). In vitro studies have demonstrated that expression of the p24 and *Gag* proteins of HIV can be suppressed by ribozymes. Clinical trials are now underway.

Technological constraints for ribozymes include their design which makes construction of oligonucleotides more difficult. They are also susceptible to degradation by RNases. Replacement of some RNA components of the ribozyme with DNA sequences will reduce the latter problem. More efficient methods for delivery of ribozymes into cells will also need to be developed. A combination of ribozyme and antisense technology is a promising future development. In this strategy, ribozymes are incorporated into antisense oligonucleotides, thereby providing the latter with catalytic activity. These catalytic antisense molecules have twice the efficiency of the conventional substances.

Box 7.7 Gene therapy for AIDS

HIV enters the T lymphocyte after attaching to CD4 receptors. In the cell, viral RNA (vRNA) is made into DNA (vDNA) by reverse transcriptase. The vDNA integrates into the host genome. The various HIV proteins and viral RNA form mature virions which then leave the T lymphocyte to infect other cells. There are a number of steps in this pathway which can be attacked by a gene therapy approach.

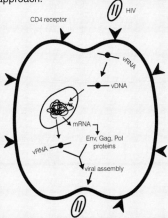

Features of HIV-1 infection which make treatment of AIDS difficult include: (1) the viral replication cycle is associated with the normal activation and function of the CD4+ class of lymphocytes (T helper lymphocytes); (2) despite an early quiescent clinical phase, HIV is actively replicating; (3) multiple tissues are affected and the focus of the viral infection, the immune system, bears the brunt of the viral-induced damage so that eventually the infected host is no longer able to mount adequate responses to a range of infections; and (4) as if the above is not enough, mutations in the HIV genome occur frequently and drug-resistant variants emerge rapidly. Estimates from the World Health Organization for numbers infected with HIV-1 are 20 million at present with 40 million adults and 10 million children by the year 2000. Not surprisingly these data have given AIDS a high priority in terms of gene therapy strategies. Possible ways to attack HIV-1 by gene transfer include: (1) enhance the host's immunological system, e.g. nucleic acid-based vaccines to various viral proteins or the injection into the patient of lymphocytes which have been genetically engineered to function more appropriately or effectively, and (2) interfere with the HIV's function, e.g. at the *RNA level* this would involve ribozymes and/or antisense technology. Another mechanism is known as RNA 'decoys'. Here transduced cells produce RNAs which compete intracellularly with the wild-type viral RNAs for protein products produced by the regulatory genes *rev* and *tat*. Interference with these genes affects viral transcription. At the *protein level* a gene therapy approach has been tried which involves the production of mutant proteins ('transdominant proteins') which will compete with the viral wild-type proteins and so interfere with viral function.

However, considerable technological developments are still required before the novel therapeutic approaches just described will have successful clinical outcomes. Infection with HIV is a good model to summarise ways in which gene therapy can interfere with a gene's function (Box 7.7).

Replacing genes

The adenosine deaminase deficiency example given earlier illustrates gene therapy in which a normal gene is added to the patient's deficient cells. Hence, it represents *addition* rather than *replacement* therapy. One potential problem with addition therapy is that the inserted genes are randomly integrated into the genome. This can lead to inefficient expression, inappropriate expression or interference with the function of nearby normal genes. Apart from the ethical issues, the inability to direct where DNA will be transferred into the genome is a significant factor in the prohibition of germline gene therapy. Thus, a more satisfactory approach to gene therapy would involve the replacement of a defective gene.

One strategy which enables an inserted gene to be targeted to its correct position in the genome is called *homologous recombination.* This has been demonstrated to occur in yeast in which recombination between incoming DNA sequences and their homologous regions in the yeast genome enables the incoming DNA to be directed to its appropriate locus (Fig. 7.8). In vitro studies using mammalian cells have shown that it is also possible to target by homologous recombination, although the frequency of this occurring is low, e.g. 1 in 100 to 1 in 100 000. The difference between yeast and mammalian cells is the smaller genome of the former which enables corresponding sequences to be more readily identifiable by incoming DNA.

Two developments have made homologous recombination an achievable goal in mammals. First, PCR allows many cells to be screened for the appropriate recombination event. This is achieved by constructing DNA amplification primers which are able to detect the creation of a novel junction formed between target and incoming DNA. DNA inserted elsewhere in the genome is not detected because the junction fragment would not be present. Although homologous recombination is a rare event, a technique with the sensitivity of PCR can detect it.

The second development has been the availability of mouse embryonic stem cells (usually abbreviated to ES cells). ES cells are pluripotential cells which can be established from an early embryo. A vector containing the gene of interest is transferred into these cells by physical means, e.g. microinjection. In the great majority of cases random insertion of DNA will occur. However, in a very few cells, the gene of interest will pair with its corresponding DNA sequence. Transferred and targeted DNA can then be exchanged by homologous recombination. ES cells which have undergone targeting are selected by PCR and then injected into the blastocoel cavity of a fertilised mouse embryo. The ES cells become incorporated into the latter which is allowed to develop in a foster mother. The chimaeric animal resulting will express the transferred gene in its appropriate location in the genome.

Work is now underway to isolate and utilise ES cells in domestic animals. In terms of human gene therapy, the approach using ES cells is an example of germline therapy and so the prohibitions mentioned earlier apply. However, on this occasion the gene to be inserted is now in its correct position and so is less likely to interfere with the function of other genes or be expressed inappropriately. Once homologous recombination is shown to be consistently successful and without complications the moratorium on germline gene therapy may be reviewed (see Ch. 10). At present, many technical problems related to homologous recombination remain to be resolved. However, the potential to target genes to their correct locus in the genome, particularly if this were to involve somatic cells, is very appealing.

Present and future directions

Over a short time frame, the number of protocols for gene therapy has increased dramatically (Fig. 7.9). There are

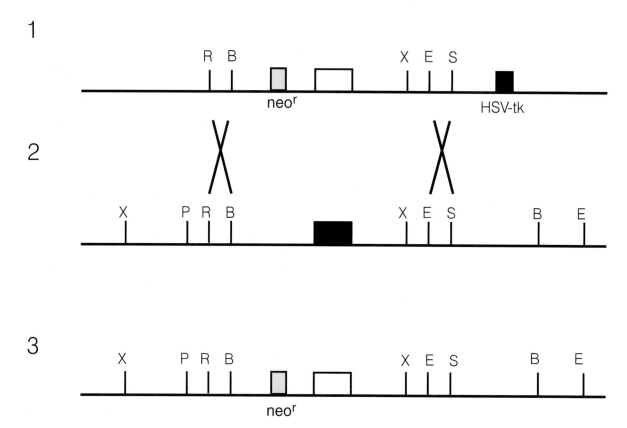

Fig. 7.8 Homologous recombination to insert a gene in its correct location.
During gene transfer, the problem of random integration into the genome could be avoided if genes were able to be targeted to their correct position and then made to replace the defective gene. An incoming DNA segment is depicted in (**1**) and the correct place into which it needs to insert is (**2**). The wild-type (normal) gene is shown as □ and the mutant as ■. R, B, X, E and S identify recognition sites for restriction enzymes. Identical restriction enzyme patterns for (**1**) and (**2**) show where the two DNA areas are the same, i.e. homologous recombination is possible at these loci. The incoming DNA segment has two additional genes which will be used to distinguish cells which have undergone homologous recombination. The genes are neo^r (neomycin resistance) and HSV-tk (herpes simplex virus thymidine kinase gene). If there is _homologous recombination_ at the corresponding DNA sequences marked by (X) the structure depicted in (**3**) will result, i.e. the mutant gene is replaced by the normal gene which also brings with it neomycin resistance but not the HSV-tk gene. Another way in which (**1**) can insert into the genome is via _random integration_. In this case the whole segment of (**1**) will be acquired since random integration occurs through the ends of linearised DNA, i.e. (**1**) will simply link to the end of (**2**) or more likely another part of the genome and so there will be both mutant gene and normal gene plus the neomycin and HSV-tk genes in tandem array. Cells which contain the neomycin resistance gene can be selected for by growing them in the presence of the drug G418 which will kill all other cells. The neo^r gene will select cells which have undergone either homologous recombination or random integration. The two options can be distinguished by a second drug (ganciclovir) which is cytotoxic to cells which contain the HSV-tk gene, i.e. cells integrating the gene in a linear fashion will be destroyed. The end result is selection for cells which have undergone homologous recombination. The cells which have integrated the appropriate gene are identified by PCR.

Number of gene therapy trials

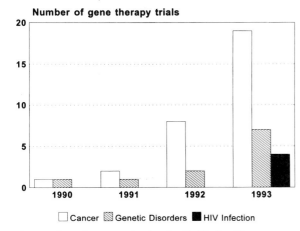

Cancer ▨Genetic Disorders ■HIV Infection

Fig. 7.9 Gene therapy protocols submitted to the US Recombinant DNA Advisory Committee (derived from Culver and Blaese 1994).

Table 7.7   Current status of some gene therapy trials for genetic diseases.*

Disease	Defective product	Target cell(s)	Status
Severe, combined immuno-deficiency disorder (SCID)	Adenosine deaminase deficiency (20–30% of SCID)	Peripheral blood T lymphocytes	Started: long-term effects being assessed; results to date promising
Cystic fibrosis	Cystic fibrosis transmembrane regulator (CFTR)	Respiratory epithelium	Both adenoviral and liposomal-mediated transfer approaches for delivery of CFTR to respiratory epithelium being assessed
Familial hyper-cholesterol-aemia	Low density lipoprotein receptor	Liver hepatocyte	Animal stage completed; human trials disappointing results to date
Gaucher disease	β glucocerebro-sidase	Reticulo-endothelial cells	Retroviral gene transfer of normal enzyme into bone marrow progenitors being assessed

* An earlier priority for gene therapy was β thalassaemia in view of its frequency and the abnormal cells being easily accessed via the bone marrow. However, the level of gene expression required as well as the elusive haematopoietic stem cell has meant that, for the present, β thalassaemia must wait until there are better technologies available. Compared to the above, a smaller number of gene therapy trials are being undertaken in genetic disorders such as α_1-antitrypsin deficiency, Fanconi anaemia, Hunter syndrome, chronic granulomatous disease and PNP (purine nucleoside phosphorylase – another cause of SCID) deficiency.

many that are about to begin. One of the significant developments which has occurred since the first edition of this book has been the change in emphasis from genetic disease to cancer and HIV. A summary of the current status of some of these trials is given in Table 7.7. Monitoring of gene therapy trials by external bodies such as the RAC (US Recombinant DNA Advisory Committee) has been intensive but as experience is gained, the guidelines have been relaxed. Further discussion on this aspect of gene therapy is to be found in Chapter 9.

Future developments will involve more in vivo strategies for gene transfer (Box 7.8). Novel approaches which allow genes to be introduced into specific regions of the body are also being considered. An example of the latter is the use of microencapsulation for treating neurological disorders (Box 7.9). Another interesting development will be in the transplantation aspect of gene therapy and involves the use of cord blood as a source of stem cells (see Chs 4, 10 for further details).

ANTIBODIES AND VACCINES

Monoclonal antibodies

Structure and function
Antibodies are proteins made by higher vertebrates. These proteins form a defence against foreign tissues, cells and organisms. Antibodies comprise two heavy and two light chains (Fig. 7.10). The part of the antibody which binds to the recognition sequence on the target (called the antigen) is designated the Fab portion. The remainder is the Fc segment which defines a number of the antibody's properties. The ability of each antibody to recognise a *single unique antigen* is utilised in many diagnostic strategies.

Antibodies are conventionally produced by immunising animals, such as rabbits or goats, with an antigen. The antiserum produced contains a mixture of antibodies in terms of both their Fab and Fc components. The problem of heterogeneity resulting from the traditional immunisation protocol was overcome in 1975 with the development of monoclonal antibodies (Fig. 7.11). Thus, antibodies of a single antigenic specificity and type could be produced in an unlimited amount and with defined activities. Monoclonal antibodies have made a major impact as diagnostic tools in clinical medicine. In the longer term, the therapeutic potential of monoclonal antibodies will hopefully become significant.

Box 7.8 Targeting genes

For in vivo gene transfer to become an effective option in gene therapy it will be necessary to ensure targeting to the correct cell occurs. At present this is possible by: (1) *insertion targeting* – the transferred gene is selectively taken up by the appropriate cells – or (2) *expression targeting* – any cell can take up the gene but only those which normally express it are able to do so following transfer. Insertion targeting can be obtained by using viral-specific tropism, e.g. the adeno-associated virus attaches to cells of the respiratory tract. Alternatively, viral vectors can have inserted into them ligands which will bind with appropriate receptors on cell surfaces, e.g. transferrin. Another way to ensure selectively is via antibody fragments attached to the virus, e.g. an antibody which will bind to α-fetoprotein – a possible target in hepatoma. Perhaps a more 'physiological' approach will involve expression targeting. To follow this route will require a thorough understanding of a gene's regulatory elements, the transcription factors which bind to it and how these can be manipulated. For example, a complex gene transfer vector could be made which comprises: (1) a viral vector, (2) a gene which expresses following exposure to the substance PSA (prostate specific antigen) and when this gene expresses it activates (3) a cytotoxic drug, and (4) a safety component to this expression targeting vector could be the gene HSV-tk (the thymidine kinase gene from herpes simplex virus 1). The expression vector described above would work in the milieu provided by prostate cells (both normal and malignant). In this environment the cytotoxic drug would be activated and local tissue damage would occur. In effect this would produce a 'genetic prostatectomy' in the case of prostate cancer. A safety control mechanism ('suicide gene') is built into the above construct since exposure to ganciclovir would activate the tk gene and so destroy the cells which had taken up the vector and its three genes.

Medical applications

Diagnostic uses of the monoclonal antibodies include radioimmunoassays (RIAs), enzyme immunosorbent assays (EIAs), flow cytometry and in situ hybridisation for histological immunotyping of tissue sections. When linked to radionuclides, monoclonal antibodies have the potential to function as markers for detecting cancer, including secondary deposits, and for monitoring progress of treatment.

In therapeutic terms, monoclonal antibodies are useful as an in vitro means to purge bone marrow of residual neoplastic cells prior to transplantation. Alternatively, monoclonal antibodies to tumour-specific antigens can be linked to effector molecules such as radionuclides, drugs or toxins. In theory, this combination enables in vivo targeting of specific therapy to a localised region thereby minimising treatment side-effects.

Box 7.9 Attempting to replace a missing gene or growth factor in the central nervous system using the physical gene transfer approach of microencapsulation

At the molecular level there are interesting developments related to genes and neurological disorders such as Alzheimer disease and amyotrophic lateral sclerosis (motor neurone disease). Another line of research is investigating nerve growth factors which could be used in neurological disorders or to repair neurones following damage. How could this knowledge be applied to practical use? Specific problems associated with gene therapy in the central nervous system include the necessity to cross the blood–brain barrier, the desirability of being able to localise expression and the availability of a mechanism which allows the introduced gene to be switched off. One way to meet these requirements is with microencapsulation.

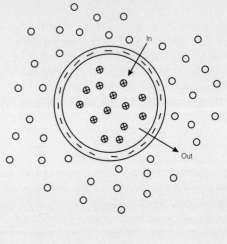

O = surrounding host cells

⊕ = transduced allogeneic cells

✗ = ——●■▲——
     tk ✱

In the diagram, the host nerve cells are indicated by open circles. Allogeneic nerve cells grown in culture are transduced with a vector (x) containing a gene of interest, e.g. a nerve growth factor (indicated by * in the diagram), the thymidine kinase gene (tk) and a promotor (●). The transduced cells are placed in a microcapsule, i.e. they are surrounded by a semipermeable barrier which allows nutrients in (transduced cells can survive) and cellular products out (i.e. growth factors produced can reach the patient's nerve cells). The microcapsule can be implanted intrathecally. Production of the nerve growth factor can be turned on and off if the promotor is inducible or there is a steady output if the promotor is constitutive. If no longer required, the transduced cells can be destroyed by giving the patient ganciclovir. If there is a problem with the microcapsule and it starts to leak, the allogeneic cells are exposed to the patient's immune system and destroyed.

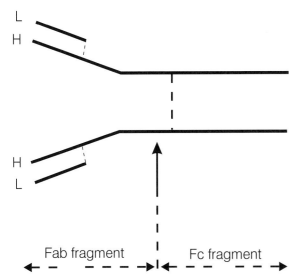

Fig. 7.10   Structure of an antibody (immunoglobulin).
Components are two heavy chains (H) and two light chains (L) which are held together by disulphide bonds (– – –). Papain breaks an immunoglobulin molecule into two pieces (Fab and Fc). The Fab portion of the antibody binds to antigens (foreign materials). The Fc portion is important since it contains the receptors which allow the antibody to bind to macrophages and B lymphocytes. The Fc portion also determines the class of immunoglobulin, i.e. IgG, IgM, IgD, IgA or IgE.

Three problems have prevented monoclonal antibodies from having effective in vivo therapeutic activity. These are the immunogenicity of monoclonals, the ability of these antibodies to penetrate target sites and the satisfactory linking of effector compounds to the monoclonals. Immunogenicity of monoclonals simply reflects the murine source of this product. Thus, a mouse-derived monoclonal antibody will have a short survival time in humans since it is foreign and so induces an immunological response against both its Fc and Fab segments. The murine-specific Fc component of the monoclonal antibody is in fact of limited value because it can only weakly recruit human effector elements which are particularly useful in the antibody response. For example, human IgG subclass 1 demonstrates better antitumour activity than mouse and other human IgG subclasses since it can more readily induce cell-mediated killing. Solutions to the problems described are now being sought with genetically engineered monoclonal antibodies.

'Humanised' monoclonal antibodies
The antigenicity of monoclonal antibodies can be reduced by taking the mouse gene for the antibody and replacing portions of its heavy and light chains with gene segments which are human in origin. Thus, the antigen recognition portion remains murine but the remainder (including the Fc component) is human. The chimaeric monoclonal formed is less antigenic but still retains some antigenicity which is directed to the Fab component.

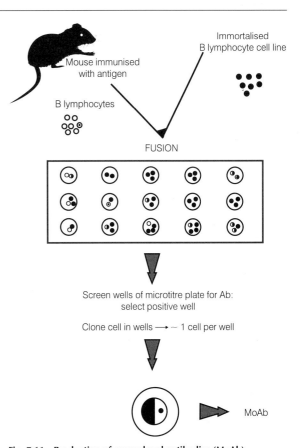

Fig. 7.11   Production of monoclonal antibodies (MoAb).
An immunised mouse will produce a polyclonal antibody response. Of the many lymphocytes involved in this response (○) there will be one which produces a specific antibody that is required (◎). To isolate the latter, the mouse's antibody-producing B lymphocytes are fused with an immortalised B cell line (●). In the wells after fusion are found many combinations: mouse B lymphocytes, immortalised B cells and various fused cells called hybridomas (indicated by half-filled circles). Enrichment for hybridomas is possible by using a special medium to grow these cells. Hybridomas in the wells are then screened for the antibody being sought. A positive well is found. This may contain a number of hybridomas, but only one is likely to be the correct one. A single (monoclonal) hybridoma is isolated by serial dilutions until there is only one hybridoma per well. This cell line then becomes an unlimited source of the monoclonal antibody.

This type of antibody is known as an *anti-idiotype antibody*.

More sophisticated refinements at the DNA level can be undertaken to replace murine components involved in anti-idiotype antibody formation. At the same time, the Fc portion coding for human IgG subclass 1 can be incorporated into the 'humanised' monoclonal antibody to enhance its effector activity. Clinical trials using 'humanised' monoclonal antibodies are being conducted. Results would suggest that the genetically engineered antibodies are less immunogenic and have a longer survival time. However, therapeutic results remain disappointing.

Improving tissue penetration

Since an antibody is a relatively large molecule it cannot easily cross the cell's membrane. Chemical cleavage of antibodies by pepsin or papain enables separation of the Fab and Fc portions, thereby reducing the size of the antibody and so enhancing tissue penetration. However, chemical proteolysis is technically difficult and there is considerable batch variability.

Genetic engineering is again proving useful in overcoming such problems since it is possible to alter the DNA sequence to place a premature stop codon between the Fab and Fc components thereby making a similar structure to that produced by chemical cleavage. The advantage of the rDNA-derived agent over the chemically derived substance is a more precise and uniform product which does not require extensive purification beyond that which would be necessary for any rDNA-derived product.

Enhancing effector function

A number of agents can enhance the monoclonal antibody's ability to kill cells. They include cytotoxic drugs and toxins such as ricin, genetically modified pseudomonas exotoxin and diptheria toxin. To bind these compounds to the monoclonal antibodies by current technology requires the antibody to be chemically modified. This is difficult to accomplish without interfering with the affinity of the monoclonal antibody for its target. Furthermore, chemical modification steps are not easily standardised and so batch variability occurs. The in vivo use of 'immunotoxins' as they are known, is associated with side-effects such as capillary leak syndrome, myalgia and hepatoxicity.

Biotechnology companies are now approaching these problems with strategies based on genetic engineering. For example, more efficient binding sites for the effector molecules can be created as part of the primary structure of the 'humanised' monoclonal antibody. Segments of the immunotoxin which induce non-specific tissue binding, e.g. carbohydrate, can be prevented from forming by using prokaryotic expression systems which are not sophisticated enough to glycosylate a peptide (see Ch. 2, Fig. 2.23 and compare with the following discussion on production of the hepatitis B vaccine). Future approaches might entail the fusion of the antibody and effector genes so that both are expressed as the one substance. The potential applications for the monoclonal antibody and its alterations through genetic engineering promises some exciting therapeutic developments.

Vaccines

The success stories involving rDNA-derived therapeutic drugs summarised in the first section of this chapter have not been matched in vaccine production. With the exception of the hepatitis B virus vaccine, which is described in more detail below, the results with rDNA vaccines have been disappointing. This is despite the considerable financial input driven in part by the urgency to find a vaccine for AIDS. Commonly used and effective vaccines such as poliomyelitis, measles and rubella are, with few exceptions, composed of *live* (infectious) and *attenuated* (nonpathogenic but immunogenic) organisms. As such they have proven to be highly effective, relatively cheap and so affordable by most communities. Other conventional vaccines are made up of *inactivated* (killed) microorganisms (e.g. Salk poliomyelitis vaccine) or the vaccines comprise one or more antigenic components (i.e. *subunit* vaccines) such as are found with influenza and recombinant hepatitis B.

Despite the efficacy of the above vaccines, it should be noted that a number would never have reached clinical use with the stringent licensing regulations now in force. For example, the oral poliomyelitis (Sabin) vaccine can revert on rare occasions to the wild-type (neurotoxic) strain and so produce poliomyelitis (see Ch. 5). Subacute sclerosing panencephalitis is a very rare neurological complication following infection with the measles virus including some vaccine-derived strains. It can be fatal or lead to permanent neurological sequelae including mental retardation. It is unlikely that the two vaccines mentioned would have been marketed with these potential risks in today's litigation-conscious community. Modern production techniques require more stringent quality control steps during manufacture as well as better assessment of toxicity. In terms of standardisation and quality control, rDNA technology has a lot to offer.

Hepatitis B vaccines

In 1982, a hepatitis B vaccine became available. The source for this vaccine was plasma from known chronic hepatitis B carriers. In this circumstance, stringent purification and inactivation procedures became mandatory. Thus, the vaccine was expensive and the amount which could be produced was limited by the availability of infected plasmas. The vaccine was not well received by the public in view of the theoretical risk that other viruses, e.g. HIV, might be transmitted despite the inactivation processes undertaken. Because of these problems and the importance of hepatitis B virus as a cause of liver disease (see Chs 5, 6) a rDNA-derived vaccine was released in 1987.

The recombinant hepatitis B virus vaccine is a subunit vaccine directed to the surface antigen (HBsAg) of the virus. The expression vector required for this vaccine has to be relatively sophisticated since the viral surface protein coat is glycosylated. Thus, either yeast-derived or mammalian expression systems are necessary. The former has been used. Purification steps are therefore required to remove contaminating yeast proteins.

The recombinant vaccine has now been used extensively and confirmed to be very effective, although a question mark remains about its immunogenicity compared to the plasma-derived product. If the recombinant vaccine proves to be less immunogenic it may be the result of a different glycosylation pattern produced by the yeast host or the segment of the surface antigen which is being expressed. A number of trials have shown that neonates and children who are at risk for hepatitis B develop a comparable protection time frame (3–5 years) when vaccinated with either vaccine. The longer term consequences of the vaccine's antigenicity remain to be determined.

Despite what has been mentioned, the recombinant hepatitis B vaccine has many advantages which include: (1) the availability of an unlimited source of antigen, (2) production can be standardised more effectively, (3) there is greater flexibility with the type of structure which is produced, and (4) with time, the vaccine will become cheaper and safer. The cost is not a small consideration since the communities which have the highest carrier rate for hepatitis B are often those which can least afford expensive vaccines. With time and competition in the market place the hepatitis B vaccine's cost has reduced considerably. To date, the only successful human rDNA-derived vaccine is that for the hepatitis B virus. However, it should be noted that more successes with rDNA vaccines have been obtained in veterinary practice as well as the meat and livestock industry.

Another alternative to the three types of conventional vaccines described earlier is the use of *synthetic peptides*. These are derived from segments of the infectious agent that are considered to be highly immunogenic. In theory, this enables better standardisation since the antigens to which the immune system is stimulated are defined. Synthetic peptides would also circumvent the risk that live attenuated vaccines could revert to wild-type and so become infectious. However, the synthetic peptide approach has been disappointing. In part this reflects the complex physical conformation which may be required to promote optimal antigenic stimulation. Segments of the infectious agent involved in this complex may be discontinuous and so not represented in a limited linear peptide. There is also good evidence that immune responsiveness is dependent on genetic factors which could be HLA-related. Thus, limited antigenic exposure such as that resulting from a synthetic peptide may not be equally effective in all circumstances, whereas the whole organism in live, attenuated or inactivated forms gives a broader antigenic profile.

Novel vaccines

The hepatitis B virus has shown that the rDNA approach to vaccine production can work in humans. The potential for *innovative developments* available through rDNA technology may enable difficulties associated with vaccination for other infections to be resolved. These are illustrated by reference to AIDS and influenza.

The HIV's RNA is capable of integrating into host DNA and so it can remain latent until it is activated. Thus, the use of live attenuated virus as a vaccine poses a potential risk if the integrated (latent) form were able to become activated and produce a mutated (wild-type) strain at a later stage (a similar problem was described earlier with the polio Sabin vaccine). The other conventional approaches involving inactivated HIV-1 or subunit components are also unsatisfactory since gp120 (Fig. 5.2), an important antigenic surface envelope protein which enables the virus to attach to the lymphocyte's CD4 receptor, is subject to considerable antigenic variation. Finally, HIV may be transmitted by infected cells as well as in free virus form. Thus, intracellular virus may escape immune surveillance.

Nucleic acid (DNA) vaccines involve the direct injection of genes (in the form of naked or plasmid DNA) expressing viral proteins into a patient to produce a sustained antigenic stimulus and so generate an ongoing immune response. In animal studies these vaccines stimulate both humoral and cell-mediated immune mechanisms (similar to what occurs following vaccination with live attenuated vaccines) without integrating into host DNA. Thus, they are an alternative, but safer, approach to live viral vaccines and are better than inactivated (dead) vaccines in the breadth of the immune response they elicit. Routes of administration are flexible, e.g. parenteral, mucosal or the gene gun which delivers tiny amounts of DNA-coated gold or tungsten beads. DNA vaccines are presently being assessed in AIDS.

As well as the safety issue there is also the suggestion that DNA vaccines will be beneficial in the circumstances of changing target antigenicity. This is a problem in AIDS but is also illustrated by the influenza virus (see Ch. 5). From 1990 to 1993, the conventional inactivated influenza vaccine had as its H3N2 component the epidemic strain A/Beijing/353/89. This strain provoked the appropriate antibody response until the 1992–1993 influenza epidemic but thereafter antigenic drift produced new strains which had less cross-reactivity, e.g. A/Georgia/03/93 and so the conventional vaccine was less effective. Using a DNA vaccine it became possible to target the more *conserved internal proteins* of the influenza virus and aim for host cell-mediated responses as well as antibody responses to the more divergent surface antigens. Animal studies have shown promising results and it may turn out that DNA vaccines will prove to be useful in overcoming changes in surface antigens as a way in which a microorganism can escape the host's immune surveillance mechanisms.

Chimaeric vaccines are produced by taking the genome of a live attenuated virus (e.g. vaccinia, adenovirus or poliovirus) and genetically engineering it to express other viral antigens. These vaccines are promising future developments. For example, a number of antigens can be produced simultaneously, thereby reducing the overall costs for vaccination programmes. A disappointing trend in many communities is the increasing numbers of infants who are not being vaccinated or are inadequately vaccinated despite proven efficacy of this preventative measure. This may represent a cost consideration but is also partly due to the complexity of the immunisation schedules with multiple injections required. A chimaeric vaccine would improve the effectiveness of vaccination regimens by simplifying their implementation.

Immunostimulation from the primary viral component (e.g. vaccinia) may also be helpful in provoking an additional response to the secondary antigen (e.g. HIV). At the DNA level, it is not difficult to modify a viral genome to enable it to contain additional genes and so express a greater range of potentially antigenic proteins. What remains to be determined is the efficacy of chimaeric

vaccines, e.g. what effect will previous exposure or immunisation to vaccinia have if this is co-expressed with the HIV?

Infectious agents which are difficult or dangerous to produce by conventional culture techniques, e.g. rabies virus, could also be better developed through rDNA means. Genetic manipulation would also be useful to reduce the likelihood of reversion to wild-type strains, e.g. poliomyelitis, or to increase the antigenicity of a particular component derived from the infecting organism.

MONITORING RESPONSE TO THERAPY

Drug resistance

Resistance to chemotherapy remains a major problem in the treatment of infectious diseases and cancer. In cancer, resistance can be *intrinsic,* i.e. the cancer is unresponsive to any form of chemotherapy, or *acquired.* If acquired, initial responsiveness ultimately gives way to a drug-resistant tumour. To overcome or delay resistance, combinations of drugs rather than single agents are frequently used in chemotherapy regimens. Nevertheless, resistance still develops to a wide range of products.

There are many ways in which drug resistance can be influenced, e.g. route of administration, concentration attained at target site and so on. At the level of the tumour itself, it is becoming apparent that a number of cellular and molecular (DNA) modifications are associated with drug resistance. Changes at the molecular level include: (1) mutations which alter protein-binding affinity, (2) increased gene expression via amplification or enhanced transcription, and (3) increased efflux or decreased uptake of the drug. Another observation is that development of resistance to one class of cytotoxic agents is often accompanied by resistance to a wide range of chemotherapeutic drugs. This is known as multidrug resistance (MDR). The changes described can be illustrated by reference to the drug methotrexate and the substance P-glycoprotein.

Methotrexate

Folic acid, a key vitamin in the synthesis of DNA, must be maintained in its fully reduced tetrahydrofolate state to be active. Inhibition of the enzyme required for this (dihydrofolate reductase, DHFR) will interfere with cell growth. Methotrexate is an antimetabolite whose anticancer effect occurs through inhibition of DHFR. Mention was made in Chapter 5 of the malaria parasite which can overcome the effect of antifolate drugs such as pyrimethamine by disrupting the binding between the antifolate drugs and DHFR. At the DNA level, this occurs on the basis of point mutations at critical sites in the gene coding for the DHFR molecule (see Table 5.6).

A different mechanism is found in tumour cells which acquire resistance to methotrexate. Elevated DHFR activity,

as shown by increased expression of mRNA, occurs secondary to gene amplification. Thus, the effect of the antifolate drug methotrexate is negated by an increase in DHFR, the enzyme which methotrexate is attempting to inhibit. There are two ways in which this can occur at the DNA level. The original tumour could have comprised two populations of cells: drug-sensitive and drug-resistant. During treatment, the latter has a growth advantage and ultimately becomes the predominant population. Alternatively, a single drug-sensitive population has undergone a series of mutations one of which produces a cell with increased DHFR activity. This clone is thereafter positively selected for by continued drug treatment.

Multidrug resistance: P-glycoprotein

The observations that resistance to cytotoxic drugs could involve a wide range of seemingly unrelated compounds and that a single defect was likely to be associated led researchers to seek a gene encoding for multidrug resistance (MDR). One potential site for drug resistance to occur was the cell's membrane since it became apparent that drugs were being excluded from drug-resistant cell lines. Comparisons of plasma membranes isolated from different drug-sensitive and drug-resistant tumour cell lines identified a unique glycoprotein in the latter. This substance was called P-glycoprotein in view of its apparent function as a permeability barrier to drugs in association with multiple resistance. P-glycoprotein was detectable in a number of tumour cell lines resistant to a wide range of cytoxic agents. The initial identification of P-glycoprotein was by monoclonal antibodies. The gene coding for this substance was next cloned and characterised. A number of important observations subsequently emerged.

The gene for P-glycoprotein is known as *MDR1*. Sequence homology studies based on DNA and protein sequences place P-glycoprotein into the ATP-binding cassette transporter family, i.e. it actively extrudes a variety of compounds out of cells at the cost of ATP hydrolysis. *MDR1* is expressed in a wide range of tissues which would suggest a normal physiological role for P-glycoprotein. Drug-resistant cells demonstrate increased expression of the gene, as seen from northern blot analysis for its 4.5 kb mRNA. The reason for this became evident in Southern blots which showed gene amplification to have occurred with multiple copies, e.g. ×60, of the gene present in resistant cell lines. Confirmatory in vitro evidence followed from transfection studies in which drug-sensitive cells were able to be converted to the resistant phenotype following acquisition of additional copies of the P-glycoprotein gene.

The normal function of the gene remains to be determined. One hypothesis is that it allows the cell to extrude unwanted substances. Alternatively, P-glycoprotein is involved in some transport mechanisms required for the cell's normal functions. With respect to the second hypothesis, it is relevant to note that the structure of P-glycoprotein bears similarities to the *CFTR* gene (cystic fibrosis transmembrane regulator or cystic fibrosis gene, see Ch. 3). Both *MDR1* and *CFTR* are found on the long arm of chromosome 7. Expression can switch from one to the other within a single cell suggesting some form of coordinated regulation. Recently, another MDR gene has been found (*MDR3*) which is located 34 kb from *MDR1*. The function of *MDR3* is now being determined.

Experimental work comparing P-glycoprotein expression in tumour cell lines that have intrinsic drug resistance or acquire resistance following treatment has shown a good direct correlation between the level of the *MDR1*-specific mRNA and the chemotherapy responsiveness of the underlying cell. Thus, a way becomes available to assess drug susceptibility in cancer cells. Strategies to utilise the *MDR1* gene to reduce toxicity in normal cells following intensive chemotherapy were discussed earlier in the section on gene therapy. Another important observation to emerge from the P-glycoprotein story is that certain drugs, e.g. calcium channel blockers and cyclosporine A, also bind to P-glycoprotein and so can be used to inhibit competitively this transporter protein thereby reversing the MDR effect. Some clinical consequences which have resulted from the discovery of P-glycoprotein are illustrated by reference to the human lymphomas (Box 7.10). *MDR1* is only the beginning. Other genes are also involved in MDR and these are starting to be cloned and characterised.

Drug sensitivity

What determines the correct drug dose? There are many considerations, including absorption, distribution, metabolism and elimination. To ensure that safety considerations are foremost, manufacturers often determine a range which is therapeutically acceptable and then recommend an appropriate dosage regimen. However, what is ignored are the *pharmacogenetic* differences between individuals which can have profound effects on the drug's metabolism and so its efficacy. Since the genetic profile is more difficult to change, the way around this problem would be to alter drug dosage.

A pharmacogenetic effect in drug metabolism was first

Box 7.10 Multidrug resistance in lymphomas and P-glycoprotein

Lymphomas can be broadly classified into (i) non-Hodgkin lymphoma and (ii) Hodgkin lymphoma. Since lymphoma is usually associated with widespread haematogenous spread by the time of presentation, the treatment for this tumour involves combination chemotherapy. Lymphomas respond readily to chemotherapy but a proportion, particularly the non-Hodgkin type, will relapse. One reason for relapse is multidrug resistance and so the significance of *MDR1* in lymphoma has been extensively studied. The effect of *MDR1* in terms of the lymphoma's response to treatment is complicated by the normal cells which express this gene. Hence, measurement of *MDR1* activity is difficult. Ways in which this is done include: *mRNA quantitation* (northern blots, PCR, in situ hybridisation); *P-glycoprotein quantitation* (immunohistochemistry staining, western blots, flow cytometry) and *functional assays*, e.g. intracellular retention of chemicals. None of the assays is ideal and this has slowed down research. Notwithstanding this problem there is some support for the observation that tumours which are P-glycoprotein expressing carry with them a poorer prognosis. This would have the potential to identify those individuals who should be treated by more aggressive or alternative therapeutic regimens. Another observation to come from lymphomas and *MDR1* is that the substances which can compete with P-glycoprotein, e.g. cyclosporine A, are effective in overcoming tumour resistance but are limited by the toxic effects which they produce. Hence, modulation of *MDR1* is a potential way in which to enhance chemotherapy for lymphoma but more selective and non-toxic agents are required (from Yuen & Sikic 1994).

---

illustrated in the early 1970s by the antihypertensive debrisoquine and the anti-arrhythmic sparteine. When testing these drugs, it became evident that occasional individuals had severe side-effects which were accompanied by very high plasma levels of the drugs despite the correct dosage regimen being followed. Family studies were very useful because these showed that other members displayed a similar response in terms of side-effects and plasma levels. Thus, the inability to handle normal amounts of these drugs was considered a genetic effect and it became possible to classify the population into the normal phenotype (known as 'fast metabolisers') and those with an inadequate or 'slow metaboliser' phenotype. This trait demonstrates autosomal recessive inheritance with ~2–10% (including Caucasians, African-Americans and Asians) being of the 'slow' type.

Cytochrome P-450

Debrisoquine metabolism occurs via a liver-specific cytochrome enzyme P-450. The P-450 cytochromes form a family of ~30 proteins that are involved in oxidative metabolism of a range of drugs and chemicals. The particular cytochrome P-450 relevant to debrisoquine and sparteine is known as CYP2D6. Inability to metabolise these drugs efficiently reflects reduced to absent activity of this enzyme. Identification of at-risk individuals by conventional biochemical means is difficult. The approach requires a test dose with assessment in the urine of the drug to metabolite ratio over a period of several hours. In itself, the assessment of potential sensitivity can lead to drug-related side-effects and interpretation of results is not easy if there are other drugs being taken or the patient is ill.

The gene coding for CYP2D6 is on chromosome 22. It comprises nine exons and is 4.3 kb in size. CYP2D6 is the best understood of the P-450 cytochrome isoenzymes and is involved in the biotransformation of a wide range of *modern drugs* including antidepressants, antipsychotics, β blockers and some anti-arrhythmics. With PCR it has been possible to study the gene from patients with slow metaboliser phenotypes and identify either point mutations or, rarely, a deletion in their DNA. It is interesting to note that one individual who required excessive amounts of an antidepressant drug known to be metabolised by the CYP2D6 gene, i.e. an 'ultra-fast' metaboliser, demonstrated amplification in DNA with multiple copies of the gene present. Thus, molecular analysis opens up the possibility that drug metabolism status can be assessed by study of DNA from peripheral white blood cells. Both heterozygote- and homozygote-affected individuals would be identifiable and a prediction made about their ability to handle a wide range of drugs which are metabolised in a similar manner. The problems of drug-exposure and interpretation of the conventional urine assays would also be avoided.

The inclusion of molecular technology into clinical pharmacology will produce important new findings about drugs as well as their interactions with other compounds. Complications resulting from drug–drug interactions are well known and prominently displayed in drug package inserts. However, the pharmacological basis is not always clear. For example, a new group of antidepressants known as the **s**elective **s**erotonin **r**euptake **i**nhibitors (SSRIs) are becoming popular because of their tolerability and safety profile. One problem with these drugs is unpredictable side-effects if other antidepressants are also taken. This interaction reflects competition between drugs for cytochrome P-450 so that the binding of one product inhibits the binding of a second. Since the cytochrome P-450 isoenzymes are involved in the metabolism of a broad range of products, the potential for drug–drug interactions are considerable. As indicated above, a 'poor metaboliser' phenotype can lead to toxic effects in some cases. On the other hand, this phenotype might also prevent the formation of active metabolites and so impair a drug's beneficial effects (Fig. 7.12).

Another interesting area of research concerns the role of cytochrome P-450 in carcinogenesis. The villain in this

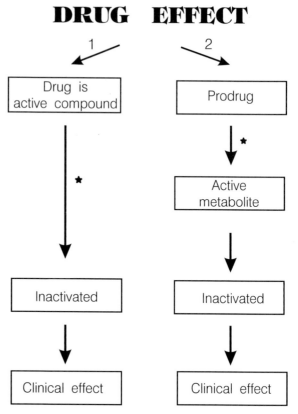

DRUG EFFECT

Fig. 7.12 Genetic alterations in drug metabolism can lead to different clinical phenotypes.
Very broadly, the effects of a drug rely on it being absorbed, distributed in various body compartments and then inactivated, usually in the liver or kidney. However, more subtle changes can be defined at the biochemical and molecular levels. (1) In this pathway a drug, e.g. a tricyclic antidepressant, requires cytochrome P-450 for its final degradation. Thus, 'slow metabolisers' (★) are prone to side-effects since the plasma level of the antidepressant drug will be inappropriately high for the dose given. On the other hand, the pathway in (2) is different since the drug, e.g. codeine – a pain reliever, works through the formation of a small amount of morphine as its active metabolite. A 'slow metaboliser' (★) in this circumstance will have the problem of less pain relief because the amount of morphine produced from codeine will be diminished.

circumstance is the 'fast metaboliser' phenotype because it would enable an individual to convert more of a procarcinogen into its active carcinogenic metabolite. A lot has been written about this potential function for CYP2D6, but definitive evidence that it is actually occurring is still awaited. Not surprisingly, subtle changes in genes such as have been proposed above will best be identified by a molecular pharmacology approach.

FURTHER READING

Recombinant DNA-derived drugs

The American Society of Clinical Oncology 1994 American Society of Clinical Oncology recommendations for the use of hematopoietic colony-stimulating factors: evidence-based, clinical practice guidelines. Journal of Clinical Oncology 12: 2471–2508

Aronson D L 1990 The development of the technology and capacity for the production of factor VIII for the treatment of hemophilia A. Transfusion 30: 748–758

Bray G L, Gomperts E D, Courter S et al 1994 A multicenter study of recombinant factor VIII (Recombinate): safety, efficacy, and inhibitor risk in previously untreated patients with hemophilia A. Blood 83: 2428–2435

Kaufman R J 1991 Developing rDNA products for treatment of haemophilia A. Trends in Biotechnology 9: 353–359

Kumar R 1995 Recombinant hemoglobins as blood substitutes: a biotechnology perspective. Proceedings of the Society for Experimental Biology and Medicine 208: 150–158

Majumdar G, Savidge G F 1993 Recombinant factor VIIa for intracranial haemorrhage in a Jehovah's witness with severe haemophilia A and factor VIII inhibitors. Blood Coagulation and Fibrinolysis 4: 1031–1033

Spivak J L 1994 Recombinant human erythropoietin and the anemia of cancer. Blood 84: 997–1004

Vose J M, Armitage J O 1995 Clinical applications of hematopoietic growth factors. Journal of Clinical Oncology 13: 1023–1035

Zuck T F, Riess J G 1994 Current status of injectable oxygen carriers. Critical Reviews in Clinical Laboratory Sciences 31: 295–324

Gene therapy

Anderson W F 1995 Gene therapy. Scientific American 273: 124–128

Blau H M, Springer M L 1995 Gene therapy – a novel form of drug delivery. New England Journal of Medicine 333: 1204–1207

Bridges S H, Sarver N 1995 Gene therapy and immune restoration for HIV disease. Lancet 1995 345: 427–432

Brugger W, Heimfeld S, Berenson R J, Mertelsmann R, Kanz L 1995 Reconstitution of hematopoiesis after high-dose chemotherapy by autologous progenitor cells generated ex vivo. New England Journal of Medicine 333: 283–316

Capecchi M R 1994 Targeted gene replacement. Scientific American 270: 52–59

Charache S, Terrin M L, Moore R D et al 1995 Effects of hydroxyurea on the frequency of painful crises in sickle cell anemia. New England Journal of Medicine 332: 1317–1374

Culver K W, Blaese R M 1994 Gene therapy for cancer. Trends in Genetics 10: 174–178

Fray R G, Grierson D 1993 Molecular genetics of tomato fruit ripening. Trends in Genetics 9: 438–443

Kay M A, Woo S L C 1994 Gene therapy for metabolic disorders. Trends in Genetics 10: 253–257

Kerr W G, Mule J J 1994 Gene therapy: current status and future prospects. Journal of Leukocyte Biology 56: 210–214

Lever A M L, Goodfellow P (eds) 1995 Gene therapy. British Medical Bulletin 51: 1–234

Martin J B 1995 Gene therapy and pharmacological treatment of inherited neurological disorders. Trends in Biotechnology 13: 28–35

Tiberghien P 1994 Use of suicide genes in gene therapy. Journal of Leukocyte Biology 56: 203–209

Antibodies and vaccines

Adair J R, Whittle N R, Owens R J 1990 Designer antibodies. In: Carney D, Sikora K (eds) Genes and cancer. Wiley, New York, p 151–161

Donnelly J J, Friedman A, Martinez D et al 1995 Preclinical efficacy of a prototype DNA vaccine: enhanced protection against antigenic drift in influenza virus. Nature Medicine 1: 583–587

McDonald W M, Askari F K 1996 DNA vaccines. New England Journal of Medicine 334: 42–45

Rabinovich N R, McInnes P, Klein D L, Hall B F 1994 Vaccine technologies: view to the future. Science 265: 1401–1404

Spooner R A, Murray S, Rowlinson-Busza G, Deonarain M P, Chu A, Epenetos A A 1994 Genetically engineered antibodies for diagnostic pathology. Human Pathology 25: 606–614

Vallera D A 1994 Immunotoxins: will their clinical promise be fulfilled? Blood 83: 309–317

Monitoring response to therapy

DeVane C L 1994 Pharmacokinetics of the newer antidepressants: clinical relevance. American Journal of Medicine 97: (suppl 6A) 13S–23S

Kroemer H K, Eichelbaum M 1995 "It's the genes, stupid" Molecular bases and clinical consequences of genetic cytochrome P450 2D6 polymorphism. Life Sciences 56: 2285–2298

Online Mendelian Inheritance in Man (OMIM) – reference number *171050 P-glycoprotein-1 [PGY1; GP170: multidrug resistance; MDR1; doxorubicin resistance] and *171060 P-glycoprotein-3 [PGY3; multidrug resistance; MDR3]. (OMIM can be accessed through the Internet – see Glossary).

Pinedo H M, Giaccone G 1995 P-glycoprotein – a marker of cancer-cell behavior. New England Journal of Medicine 333: 1417–1419

Shustik C, Dalton W, Gros P 1995 P-glycoprotein-mediated multidrug resistance in tumor cells: biochemistry, clinical relevance and modulation. Molecular Aspects of Medicine 16: 1–78

Yuen A R, Sikic B I 1994 Multidrug resistance in lymphomas. Journal of Clinical Oncology 12: 2453–2459

FORENSIC MEDICINE

INTRODUCTION

Genetic differences identifiable by protein polymorphisms have been used in forensic laboratories since the late 1960s. Initially, protein markers were based on the ABO blood groups. Subsequently, other blood groups, serum proteins, red blood cell enzymes and more recently histocompatibility (HLA) antigens have been typed. One disadvantage of protein polymorphisms has been the limited degree of variability associated with these markers. Thus, the finding of commonly occurring protein polymorphisms in two samples is of doubtful value if the probability is sufficiently high that they could represent chance events. For protein markers, the probability that coincidence could explain genetic identity lies in the vicinity of 1 in 100 to 1 in 1000. Therefore, the emphasis in the legal sense has been on *exclusion* rather than *positive identification* when two samples have been compared.

Other problems inherent in protein analysis relate to the amount of tissue required for testing and the relative ease with which proteins degrade. These considerations are particularly relevant to the scene of a crime where the ideal laboratory conditions will not be found and tissue available for analysis will more often than not be limited in amount and quality. Thus, evidence derived from protein markers is unlikely to be helpful or even available for a crime committed in the past.

In 1978, the first human *DNA polymorphism* related to the β globin gene was used to identify a genetic disease. In 1980, it was reported that **r**estriction **f**ragment **l**ength **p**olymorphisms (RFLPs), i.e. small variations in DNA detected with restriction endonucleases (see Chs 2, 3), were dispersed throughout the entire human genome. More complex and so potentially more informative DNA polymorphisms were described in 1985. These were called 'minisatellites'. Polymorphisms at the DNA level thus opened up the potential for a sophisticated approach to tissue comparisons. In the forensic situation, this would lead to an exclusion or perhaps even the positive *identifi-*

cation of an accused since the chance that a match between DNA markers was coincidental becomes highly unlikely with probabilities between 1 in 10^5 to 1 in 10^6 achieveable (contrast this to the 1 in 100 to 1 in 1000 given earlier for protein-based markers).

Not surprisingly, it soon became possible in British and North American courts of law for DNA evidence to be used in criminal and civil cases. The first such trial occurred in Bristol, England, in November 1987. DNA evidence in this particular case was crucial in providing the link between a case of burglary and rape. Today, courts all over the world have allowed, to varying degrees, the admission of DNA profiles as evidence in criminal trials and paternity disputes. In one survey, the most frequent use for DNA evidence was criminal cases involving rape (Table 8.1). An important appeal of DNA lay in its intrinsic variability so that *exclusion* was not necessarily the only option available. Thus, it became possible to aim for a unique DNA profile for each individual similar to the traditional (or dermatoglyphic) fingerprints.

Table 8.1   The types of criminal cases for which DNA evidence has been requested (from Decorte & Cassiman 1993).

Crime	Percentage*
Rape	42
Murder/homicide	37
Robbery	7
Assault	2
Kidnapping	1
Blackmail	1
Others	10

* Derived from 178 cases referred for DNA testing.

In 1989, during a pretrial hearing for a double murder case involving the *State of New York versus Castro*, DNA evidence was first seriously questioned. This led to the demonstration of suboptimal laboratory practices as well as doubtful interpretations of the statistical significance of DNA polymorphic data. Evidence based on DNA studies in this case was thereby deemed inadmissible. Subsequently, a number of other cases have had to be withdrawn by the prosecution because DNA data comprised an important component of the evidence. Cases already decided were appealed. The scientific controversies about DNA fingerprinting continued into the 1990s, particularly in respect to the significance of identical matches. Public interest reached remarkable levels with the trial of O J Simpson in the USA.

Thus, in a very short time, DNA technology has had a major impact on the judicial system, which is extraordinary given the slow pace with which the system usually moves. The initial rapid utilisation of DNA technology, particularly by commercial companies, produced problems. These reflected laboratory practices, differences in the interpretation of DNA polymorphic data, particularly in relation to minority ethnic groups, and the standard of quality assurance practised. Such problems have now been largely resolved with government and commercial laboratories foremost in these developments. Legislation enacted in many communities now requires the highest code of practice for laboratories involved in forensic DNA technology. The earlier scientific doubts about the validity of statistical probabilities using DNA fingerprints were also apparently resolved when, in 1994, an article appeared in *Nature* titled 'DNA fingerprinting dispute laid to rest' (Lander & Budowle 1994). This article concluded that the DNA fingerprinting 'war' was over and there were now no major problems preventing its use in the courts of law.

For the purpose of this chapter, the application of DNA technology in criminal cases as well as the establishment of familial relationships will be considered under the one category of forensic medicine.

REPETITIVE DNA

DNA, which comprises the 3.3×10^9 base pairs of the human haploid genome has a number of functions. Approximately 70% codes for genes or is involved in a number of gene-related activities such as regulation of expression. For example, DNA provides the signals for its own replication as well as those required for chromosomal replication, division and segregation. The remaining 30% of the eukaryote genome is composed of repetitive DNA sequences which appear to have no function. The term 'junk' DNA has been used to describe these areas although this would seem inappropriate since the distribution of such DNA is non-random in places and there remains some inter-species homology. A potential role of repetitive DNA as 'hot spots' for recombination has been proposed. This is an appealing hypothesis since the repeat sequences have no apparent coding (exon) function and so there would be less evolutionary pressure for conservation. A greater degree of mutational activity would thus be possible at these loci. In an evolutionary sense, this would be useful, e.g. for the development of new genes.

Table 8.2 Types of repetitive nuclear DNA in the human genome.
These have no known function and are divided into two classes: satellite DNA which has tandemly repetitive sequences and the interspersed repeats.

Designation	Size range	Examples	Features
Satellite DNA			
Minisatellites	1–30 kb	Probes called 33.6, 33.15, 3' α HVR	(common core)$_n$ Multilocus/single locus; Multiple repeats in tandem (VNTRs)
Microsatellites	<1 kb	(AC)$_n$ repeats	(XXX)$_n$, (XXXX)$_n$ are potentially more useful; Single locus; VNTRs
Macrosatellites	Repeat units are small in size, e.g. 171 bp, but repeated many thousand times	α satellite, satellite III DNA	Mostly in centromeres/telomeres; Multilocus; VNTRs
Interspersed repeats	~300 bp >500 bp to 10 kb	Alu repeats Kpn or L1 repeats	Interspersed repeats which are not necessarily repetitive internally or in tandem array

n = number of repeats; X = any nucleotide base.

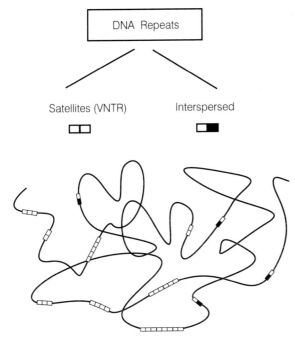

Fig. 8.1 The two classes of repetitive nuclear DNA.
Schematic representations are used for the satellite and interspersed repeat elements in DNA. The diagram shows how different numbers of tandemly inserted elements involving the satellite repeats (VNTR) can produce variability in this polymorphism. The interspersed repeats are found all over the genome but maintain a more constant size.

Repetitive DNA can be divided into two major classes. The tandemly repetitive sequences (known as *satellite* DNA) and the *interspersed* repeats. The term 'satellite' has been used to describe DNA sequences which comprise short head-to-tail tandem repeats incorporating specific motifs. These make up one-third of DNA repeats (i.e. 10% of the total genome) and are exemplified by the minisatellites, microsatellites and macrosatellites. A summary of the types of DNA repeats is given in Table 8.2 and illustrated in Figure 8.1.

Minisatellites

The first DNA polymorphisms used in the forensic laboratory were the *mini*satellite repeats. Here the common core sequence which will be repeated is *longer* than that found with the *micro*satellites (discussed below). This produces restriction fragments which are in the kilobase range compared to the microsatellite alleles which extend from 20 to 120 nucleotide bases in size. Thus, minisatellites give a much wider size range for their polymorphic DNA fragments. Because of this, Southern blotting is usually required. PCR can be used to identify some minisatellites (see p. 180).

Minisatellites are either *multilocus* (repeated throughout the genome in many loci) or *single locus* (the position of the minisatellite in the genome is fixed to one place). The

chance of finding differences between two alleles using minisatellites is very high (up to 99% in some cases) and it is also possible to use the complex polymorphic patterns arising from multiple loci to construct a unique DNA profile or 'fingerprint' for an individual. Because of the tandem array of repeat units, the minisatellite (and microsatellite) polymorphisms are examples of the type called variable *number of tandem repeats* (VNTRs).

Multilocus minisatellite VNTRs

The first minisatellites to be used in the courts of law were described by Jeffreys in the United Kingdom. These were DNA probes designated 33.6 and 33.15. The former constituted a core sequence with the motif $(AGGGCTGGAGG)_3$ repeated 18 times and the latter was a 16 base pair motif, AGAGGTGGGCAGGTGG, which was repeated 29 times. The loci detected by these two probes were dispersed throughout the genome (hence their name, multilocus VNTRs). A composite of these multiple loci produced intricate band patterns depending on how many repeats of the 'core sequence' were present. For example, in one study, probe 33.6 gave on Southern blotting of DNA digested with the restriction endonuclease *Hin*fI (a frequently cutting restriction enzyme since it recognises a four base pair sequence) an estimated 43 loci from paternal DNA and 27 from maternal DNA. The restriction fragments in the multilocus 'DNA fingerprint' clustered in the 2–4 kb range. In contrast, the single-locus VNTR patterns identified by the microsatellites produce a maximum of two DNA fragments which reflects the two alleles present. The biallelic pattern of the latter makes interpretation easier although the amount of information provided is correspondingly reduced.

Disadvantages of the multilocus minisatellite probes described above included: (1) the technical expertise necessary to get good DNA patterns; (2) the relatively large amount of DNA required for the Southern blotting studies (e.g. multilocus minisatellites may require as much as 25 times the quantity of DNA used to detect a single-locus VNTR); and (3) the difficulties which can arise in the interpretation of the large number of DNA fragments present. In particular, individual variations in the agarose gel tracks may alter the mobility of DNA as it is electrophoresed (called *band shifts*). This could be controlled, in the research laboratory, by procedures such as running multiple gels and varying the position of the samples being tested. However, the forensic laboratory, with limited amounts of DNA which are often degraded or contaminated with other sources of DNA or extraneous material capable of altering the mobility of DNA, is in a less fortunate position. Thus, it is not surprising that, apart from the relatively 'clean' samples and unlimited quantity of DNA available in family studies such as paternity disputes, the crime-related applications for the multilocus minisatellites were confined to very few experienced laboratories and special circumstances.

Single-locus minisatellite VNTRs

A second type of minisatellite VNTR identifies a single locus in the genome. There are many examples of these polymorphisms, which were first characterised in 1986–1987, throughout the genome, e.g. in association with the α globin gene locus on chromosome 16 (called 3' α HVR where HVR is an abbreviation for *hypervariable region*) or the immunoglobulin heavy chain gene locus on chromosome 14 (Fig. 8.2). The polymorphic fragments for the single-locus minisatellite VNTR are usually of a larger size than the multilocus VNTR fragments because there is a greater number of repeat units in tandem. Localisation to the one region of the genome is possible by utilising a DNA probe and stringent hybridisation and washing conditions when DNA mapping is undertaken (Fig. 8.3).

Some applications of the minisatellite VNTR

A number of commercial and government laboratories involved in DNA testing for legal purposes have utilised mixtures of single-locus minisatellite VNTRs. Each VNTR is highly informative, producing two alleles but with a *wide range of band sizes per allele*. Therefore, the chance of finding different patterns between individuals is considerably higher than that possible with an RFLP since variability with the latter will be limited to one of three *fixed* options (large/large, small/small and large/small) (Fig. 8.3). A combination of 4–6 single-locus VNTR markers gives an overall DNA profile which is very polymorphic and potentially unique to a person. In this circumstance, the probability of a chance match between individuals can be as low as 1 in 10^6 depending on the polymorphisms used and the individual's ethnic background (discussed further on p. 185).

Southern blotting analysis is usually required for minisatellites. The small amount of DNA available from a crime scene may not be limiting if filters are able to be hybridised with one VNTR probe and then rehybridised with additional DNA probes. As well as the disadvantages mentioned earlier with the multilocus minisatellites, another problem involves the testing of degraded specimens since the small molecular weight DNA present is likely to disrupt the relatively large minisatellite fragments.

Microsatellites

The more recent polymorphisms to be applied to the forensic situation are the microsatellites. These comprise small DNA polymorphisms usually <1 kb in size. The best described are the dinucleotide repeats usually involving the bases adenine and cytosine $(AC)_n$, where n (the number of repeats present) varies from 10 to 60. The microsatellites, in the way in which they are used, represent *single-locus VNTRs* because each can be made to identify one unique segment of the genome. It is estimated that the human genome contains ~50 000 of the $(AC)_n$ repeats. Thus, the value of these polymorphisms lies in their widespread distribution throughout DNA which makes them ideal for *genome mapping*. As DNA polymorphisms, they are highly informative in family studies to identify wild-type versus mutant alleles or in paternity testing (see p. 181). Microsatellites, because of their potential hypervariability, are more informative than the biallelic RFLP system but less informative than the minisatellites (see Chs 2, 3).

A technical consideration with the microsatellites is the necessity to use PCR since size differences between alleles are small. It is also necessary to utilise oligonucleotides to target a specific region in the genome so that one microsatellite locus alone is tested. On the positive side, this means that only a small amount of DNA is required as template and an automated and more rapid testing procedure is available to detect the microsatellites. More recently, microsatellites comprising a three or four base pair core (e.g. $(AGC)_n$ or $(AATG)_n$) have been described. Interpretation of gel patterns which contain amplification

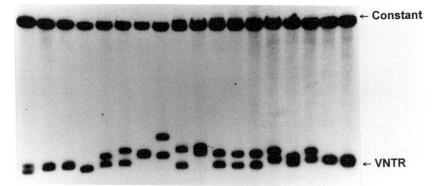

Fig. 8.2 Single-locus minisatellite VNTR patterns.
These VNTRs can have a wide range of band sizes for each of the two alleles, which increases the chance of finding unique patterns for different individuals. A Southern blot analysis of a VNTR-type polymorphism derived from the immunoglobulin heavy chain gene is illustrated. There is an upper band which is the same for all individuals, i.e. it is a constant band, and the lower two bands are polymorphic, i.e. variable in size. The latter display considerable diversity amongst a random normal Caucasian population. Since these VNTRs are single-locus changes, a maximum of two polymorphic alleles is possible.

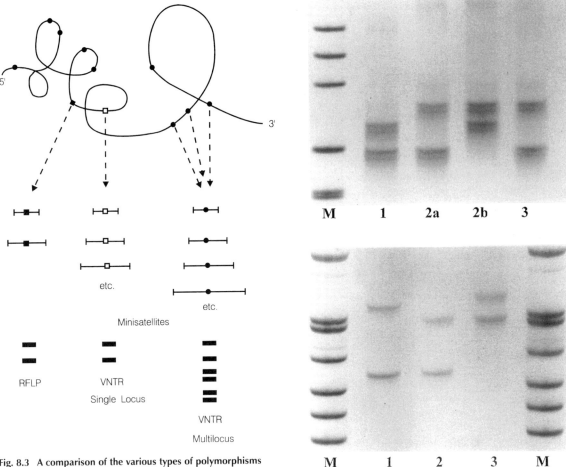

Fig. 8.3 A comparison of the various types of polymorphisms available for DNA testing.
■ Indicates an RFLP which is present at a *single* locus and will produce *two* polymorphic bands (large and small) which are of *fixed* size. Thus, the number of combinations generated by two alleles in each individual is limited to large/large, small/small and large/small. □ *Two* polymorphic bands are also obtained for the *single-locus* VNTR minisatellite but these polymorphisms are more informative because there is greater variability between the sizes obtainable for each of the two bands and so there is more chance that individuals will have different profiles (see Fig. 8.2). *Combining* a number of different single-locus VNTRs produces an even more characteristic set of markers per individual. ● is the most informative of all polymorphisms because the multilocus VNTR pattern is a composite of many VNTRs which are scattered throughout the genome. A complex DNA profile ('fingerprint') results (see also Figs 8.6 and 8.7 which show actual autoradiographic patterns for a multilocus VNTR).

Fig. 8.4 Microsatellite patterns obtained with dinucleotide and tetranucleotide repeats.
Although more dinucleotide microsatellites, e.g. (AC)$_n$, have been described, they have one disadvantage which is the frequently found 'stutter' bands thought to occur because of slippage of the *Taq* polymerase enzyme. **(Top)** In this example of a (AC)$_n$ repeat, 'stutter' bands can be seen, but it is still possible to determine which fragment has been inherited from each parent (M, molecular weight marker; 1, mother; 2a, 2b, offspring, in this case dizygotic twins; 3, father). However, the presence of 'stutter' bands can confuse the assignment of microsatellite fragments. **(Bottom)** Mis-assignment of bands or technical problems in amplifying alleles is less likely to occur with the larger microsatellites such as the tetranucleotides. The example given involves a (CTTT)$_n$ which shows very clear fragments without extraneous bands (M, 1, 2, 3, as above).

products for these is easier since the difference between alleles is now greater, i.e. from two bases to three or four. Amplification by PCR is also more reliable compared to that found with the dinucleotide repeats (Fig. 8.4).

Other repetitive DNA

Located near centromeres and telomeres are the macrosatellites, a third form of hypervariable satellite DNA.

In contrast to the microsatellites and minisatellites, the macrosatellites can be very large (e.g. megabases in size) and so PFGE may be necessary for their identification and characterisation. In practical terms, DNA polymorphisms associated with the macrosatellites are not used in forensic practice since DNA is often degraded to some extent. Thus, high molecular weight DNA essential for PFGE would be unavailable in these circumstances.

By comparison to the 'satellite' DNA repeats, *interspersed repeats* occur more frequently. These are not usually found in tandem array and are not necessarily located as multiple repeats. Therefore, they have little role to play in comparing DNA specimens. Two repetitive elements in this class are the 'Alu' repeats and the 'Kpn' repeats. The Alu repeats are so named because they frequently have a cleavage site for the restriction endonuclease *Alu*I. Alu repeats are estimated to occur every 5–10 kb of DNA. Kpn repeats (named after the restriction endonuclease *Kpn*I) are also called L1 repeats and they occur in ~1–2% of the genome and are larger in size than the Alu repeats. The distribution of Kpn repeats is consistent with mobile elements such as the transposons (DNA sequences which can replicate and insert a copy at a new location in the genome). Alu repeats have a forensic application since they are human (higher primate)-specific and so are useful in determining a specimen's origin as human or non-human.

Apart from nuclear DNA, another target for the forensic laboratory is *mitochondrial DNA* which is highly variable in a region known as the D loop. This variability predominantly results from single base changes and some length polymorphisms. Advantages of mitochondrial DNA are two-fold: the exclusive maternal origin of this DNA will facilitate the determination of family relationships and multiple (even in thousands) copies are present in each cell. For example, hair is frequently found at the scene of a crime. However, there are many potential sources for this tissue, e.g. victim, accused, police, bystanders, etc. and so *individual* hairs must be studied. Nuclear DNA (present as two copies per cell) is extractable from hair roots but not the shafts and so hair has limitations in forensic DNA testing. However, since multiple copies of mitochondrial DNA are present in the shafts, it becomes possible with PCR to type individual hairs without the necessity for roots to be present.

DNA amplification

Three properties of DNA amplification by PCR make it an ideal test in the forensic situation. (1) Minute amounts of evidentiary material left behind at the scene of the crime will provide enough template for DNA analysis. (2) DNA which has been degraded can still be amplified since only a small segment of DNA is required for primers to bind in PCR. (3) It would be possible to retest the sample at another laboratory. Minisatellites can also be identified by PCR provided the fragments are not too large, e.g. <3 kb.

Balancing the above are a number of potential problems which need particular care in the forensic situation. These include erroneous results from PCR and from contamination.

- *Erroneous results from PCR.* Does exposure to the environment with its consequent DNA-damaging effects lead to errors in PCR? Experience would now suggest that this does not occur. In other words, if amplification is possible after environmental exposure, then the end product is free of artefacts resulting from damaged DNA. The problem of errors occurring through misincorporation by the *Taq* polymerase enzyme or differential amplification of DNA sequences has been mentioned in Chapter 2. The above are not a drawback to the use of PCR in forensic practice since the test can usually be repeated. The newer types of *Taq* polymerases now available on the market are also less likely to result in these types of errors.

- *Erroneous results from contamination.* The effect of contaminating sources of DNA on PCR has already been mentioned in relation to genetic disorders and the detection of pathogens. Contamination occurs in the ideal laboratory despite strict care and the highest standards. Sources of contamination include the laboratory scientists, equipment or more frequently other amplified products. Contamination becomes an even more significant issue in the poorly controlled crime scene. For example, in the case of rape there would be a number of potential sources of DNA which could serve as a template for amplification (these sources of DNA are described in more detail on p. 183).

The eccentricities of false-positive and false-negative results attributable to PCR need to be considered both in practical terms and in designing appropriate quality assurance measures.

HUMAN RELATIONSHIPS

Human leukocyte antigen (HLA) markers

The genes which comprise the HLA complex (now more usually known as the MHC – major histocompatability complex) on the short arm of chromosome 6 are divided into three classes which occupy a physical distance of ~3500 kb. Class I genes include HLA-A, HLA-B and HLA-C and other related genes. Next to these are found the class III genes which comprise various components of the complement cascade and a number of other genes. Class II genes are called HLA-DP, HLA-DQ and HLA-DR (Fig. 8.5). Because of the very polymorphic nature of the HLA-D genes, based mainly on three hypervariable regions in the second exons of HLA-DQA1, DQB1, DRB1 and DPB1, the class II locus is particularly useful for forensic analysis. Some discussion on the functional aspects of the MHC is provided in Box 8.1.

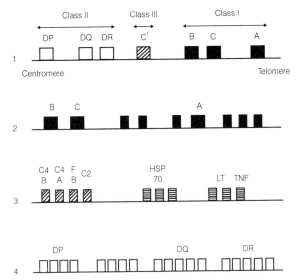

Fig. 8.5 Organisation of the MHC (major histocompatibility locus).
(1) A schematic map of HLA genes within the MHC. The three classes are shown (not to scale). C' refers to complement. (2) A more extensive map of the class I genes is given. Apart from the well defined HLA-A, HLA-B and HLA-C genes, there are a number of other class I like genes also present. (3) Class III genes are depicted – complement (components C4B, C4A, C2) and factor B (FB). Other nearby genes include a heat shock protein (HSP), two lymphotoxin genes (LT) and the tumour necrosis factor gene (TNF). (4) The various genes making up the class II region include four genes at HLA-DP, five genes at HLA-DQ, five genes at HLA-DR. In addition, there are, between DP and DQ, a number of other genes involved in the MHC's function. The close proximity of the HLA-DQ and HLA-DR genes explains why these demonstrate *linkage disequilibrium,* i.e. some DQ alleles are inherited with particular DR alleles out of proportion to their individual frequency in the population (from Abbas, Lichtman and Pober 1994).

Box 8.1 Functional significance of the MHC

The role of the class I and II genes in the immune response involves the presentation of antigens to the T lymphocyte. This is important both in recognition of self ('tolerance') and foreign antigens (e.g. the transplant situation). Class III genes are also implicated in the immune response, e.g. as components which mediate inflammation. The functions of many genes in the MHC remain unknown. Characterisation of these as well as the class I–III genes will further our knowledge of this important locus. There are over 40 human diseases which appear to involve the HLA complex. For example, class I genes are associated with haemochromatosis and ankylosing spondylitis. Class II genes are associated with autoimmune disorders such as systemic lupus erythematosus, multiple sclerosis and many others. The class III locus is associated with psoriasis. The relationship between HLA and these disorders will become better understood once the locus is fully characterised.

Tissue typing

The HLA antigens are an important set of markers for tissue typing as well as comparative studies of populations. Traditionally, HLA typing has been based on serological markers, although very subtle differences between antigens shared by individuals at the HLA-D locus have only been detectable by a 'functional' assay known as the mixed lymphocyte reaction. HLA typing is required in most types of tissue transplantation since the closer the graft is to the donor in the immunological sense, the less likely it is that graft rejection will occur. Comprehensive tissue typing can be time-consuming and despite histocompatibility at various HLA loci, examples of rejection still occur, which may in part reflect hidden antigenic determinants.

DNA typing will avoid the time-consuming assays which are otherwise necessary to obtain a comprehensive HLA profile at the protein level. The various HLA genes are extremely polymorphic which facilitates DNA analysis. Kits which allow identification of HLA determinants by oligonucleotide probes are now commercially available and in routine use in many of the tissue typing laboratories.

Family studies

In addition to assessing inter-family relationships for genetic disorders, the use of DNA markers in legal assessments of relationships within families has become increasingly common.

Paternity disputes

Protein typing for HLA and other polymorphisms still remains a valuable approach for paternity testing although DNA testing is assuming increasing importance. In this situation, fresh blood can be obtained from the various parties and analyses are conducted under optimal laboratory conditions. The polymorphic nature of the HLA markers makes them very useful in paternity studies. For example, commercial kits which utilise PCR and allele-specific oligonucleotides (see Ch. 2), enable typing for the six alleles associated with the HLA-DQA gene. The six alleles are designated A1.1, A1.2, A1.3, A2, A3 and A4. Combinations of these alleles produce 21 genotypes. One estimate suggests that there is only a 7% chance that two individuals selected at random will share the same HLA-DQA type. Whilst the variability in a six allelic system is inadequate for positive tissue identification, it provides a rapid and relatively simple DNA test which is helpful in *excluding* an individual in the case of disputed paternity or showing that a tissue sample did *not* belong to a suspect in a criminal case. The HLA-DRB1 gene is even more polymorphic. However, because of the complex nature of its band patterns compared to, for example the HLA-DQA gene, HLA-DRB1 remains particularly useful for paternity testing but less so in the crime scene where contaminating material can lead to confusing band patterns.

A combination of the HLA types and DNA polymorphisms would go one step further in estimating whether the individual in dispute was in fact the biological father. This would rely on the markers obtained, their frequency in the population and the likelihood that the combination detected could occur by chance alone. The multilocus minisatellites are even better than the HLA markers in this respect, since they produce a greater number of variable alleles which would be more likely to allow both *exclusion* as well as a more definitive assessment of the *likelihood that an individual was the biological father*. Again, the material for analysis could be prepared under optimal conditions and so the potential complicating effects of degradation or contamination on the DNA patterns, which is a vital consideration in the crime scene, become less relevant.

Determination of paternity in alleged incest is particularly difficult to resolve if conventional protein methods are used since the suspect and the related victim are bound to share a number of common markers. In these circumstances, the more highly polymorphic DNA markers are invaluable (Fig. 8.6).

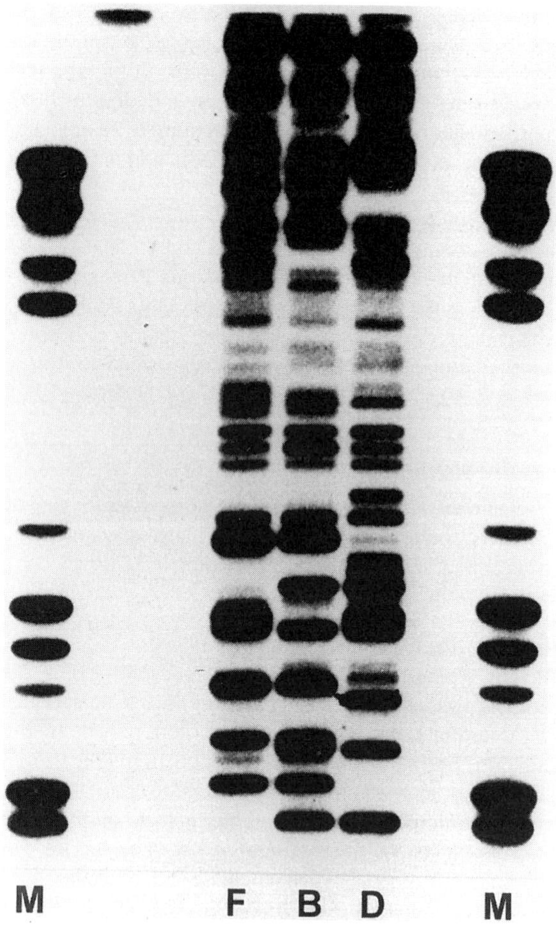

Relationships within families and individuals

An extension of the paternity situation can be found when it is necessary to confirm relationships between individual members of a family. This is illustrated in a case involving an immigration dispute with one individual being refused residence in the United Kingdom. A satisfactory conclusion to the case was possible when DNA typing with the multilocus minisatellites confirmed the individual's identity and so his right to live in the country (Box 8.2).

M F B D M

Fig. 8.6 A case of incest which is confirmed by DNA fingerprints using a multilocus minisatellite VNTR probe.
The three DNA tracks in the centre depict from right to left, the accused's daughter (D) who is the mother of a baby, the baby (B) and the accused (F) who would also be the baby's grandfather if incest were proven. The three share a number of common bands. In a *first degree relationship*, it would be expected to see on average a bandshare in the order of 62% which is what is found between father (F) and daughter (D) in this case. However, the bandshare between father and baby is 78% which shows that not only is he a first degree relative of the baby but also closely related to its mother. All paternal bands in the baby can be assigned to the putative father, proving conclusively that he is both the father and grandfather. M, molecular weight markers. (Case and photo provided by courtesy of Cellmark Diagnostics, Abingdon, UK.)

Box 8.2 DNA typing with multilocus minisatellites to confirm family links

A black African male (called X) who was born in the United Kingdom left that country to live with his father in Ghana. On return to the United Kingdom to rejoin his mother and three siblings he was refused residency since there was suspicion that a substitution in the form of a nephew or an unrelated male may have occurred. The father was not available to be tested and in fact there was the possibility of non-paternity involving 'X'. Protein markers confirmed that the woman (called 'M') who claimed to be the mother of 'X' was related to him. However, it could not be excluded that she was in fact his aunt. Minisatellite probes described previously (33.6 and 33.15) were able to define 61 distinct fragments in DNA from X. These were all found in the putative mother or the father (the latter's pattern was inferred from DNA profiles obtained from the three available siblings). This confirmed that X was related to the family since DNA fingerprints are seldom shared amongst unrelated individuals. 25 fragments in X were shown to have come from M and so there could be little doubt that M was the mother of X. On the basis of this evidence X was granted residency (from Jeffreys et al 1985).

TISSUE IDENTIFICATION

Evidentiary samples

At the scene of a crime there may be stains (e.g. blood or semen on the victim's clothing) or other tissues (e.g. skin, hair under the victim's fingernails) which require identification and characterisation (Table 8.3). The first question which might need to be answered is whether these samples are human in origin. As indicated earlier, the Alu repeat sequences can resolve this problem. The next step is to utilise DNA from evidentiary samples to build DNA profiles. These will then be compared to DNA patterns from the suspect(s). In the crime scene, it is important to realise that there are likely to be a number of DNA sources, e.g. the victim him/herself, third parties or the environment. The potentially complicating issues facing the forensic scientist are well illustrated by the evidentiary samples obtainable in a case of rape. In this situation, DNA can come from: (1) the victim in the form of blood, body tissues or secretions (vaginal, anal or oral in origin) and bacteria; (2) one or more assailants; (3) semen from earlier voluntary intercourse; (4) animals or bacteria from the crime scene; and (5) also the possibility that the source of DNA was 'planted' by a third party, including the victim.

DNA testing remains feasible in the complex circumstances just described. DNA originating from microorganisms or other animals does not usually cross-hybridise with human-specific DNA probes although faint bands attributed to bacterial or non-human products have been reported by some laboratories. DNA obtained from sperm is more 'robust' when it comes to isolation procedures. Therefore, laboratory protocols can be designed to utilise this property and enhance the isolation of sperm DNA at the expense of DNA from other tissues. Thus, the level of the contaminants can be reduced or even excluded. The dilemma arising from multiple human DNA sources can be approached by comparisons. Blood or tissue from the victim is obtained to identify his/her DNA band patterns which are then 'subtracted' from the overall profile. The contribution from an innocent third party (e.g.

sexual partner) can be treated in the same way. DNA patterns from evidentiary samples can then be compared to those obtained from potential assailant(s). From these comparisons it might be possible to identify specifically the accused(s) as the source of the DNA stain or semen.

Alternatively, blood from the victim may have spilled or splashed onto an assailant's clothing or the crime scene, e.g. a car. DNA isolated from the blood spots will subsequently provide important evidence connecting the victim with the individual wearing the clothing or the crime scene (Fig. 8.7, Box 8.3).

Practical applications

Evidence based on DNA testing will be used by the prosecution to confirm a link between the victim and the accused. On the other hand, DNA evidence could turn out to be more beneficial to the defence if DNA patterns excluded a match. An accused who is on trial because of evidence obtained from an eye-witness may request DNA

Table 8.3   The types of biological samples referred for DNA testing (from Decorte & Cassiman 1993).

Specimen	Percentage (178 cases)
Mixed (semen + blood or epithelial cells, hair)	48
Stains (blood, semen)	41
Hair	3
Saliva (cigarette butts, envelopes, stamps)	3
Other (bones, teeth, tissue)	5

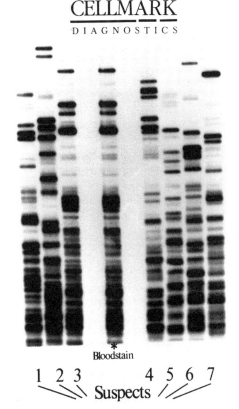

Fig. 8.7 A scene of crime DNA fingerprint comparing a blood stain with DNA from seven suspects.
Suspect number three gives a positive match since bands shared between the suspect and the blood stain are identical. (Case and photo provided by courtesy of Cellmark Diagnostics, Abingdon, UK.)

Box 8.3 DNA typing to establish a link between a homicide victim and the crime scene

In July 1991, in Vancouver, British Columbia, a 29-year-old caucasian woman was murdered and her body incinerated. When the body was discovered it was so badly burnt that identification was impossible. More relevantly, DNA could not be isolated from the ash and charred remains. Due to a number of chance events and good police work, a suspect was soon apprehended. Although denying the crime, his car was found to have blood stains, blood-stained clothing (which did not belong to the accused), as well as an empty gasoline container. The deceased's identity had been established through dental charts since the only part of her body which had been spared to some degree were the teeth. It was then suggested that some impacted wisdom teeth might serve as a source of DNA. These were removed and indeed this turned out to be the case since the impacted teeth had been protected from the heat by the jaw bone in which they were imbedded. DNA from the teeth and the crime scene (i.e. the car belonging to the accused) was examined by Southern blotting and five VNTR minisatellite polymorphisms. Perfect matches were obtained. It was estimated that based on the Royal Canadian Mounted Police caucasian database, the chance that this had occurred by chance was 1 in 540 million. Faced with this evidence the accused admitted the crime and was convicted (from Sweet and Sweet 1995).

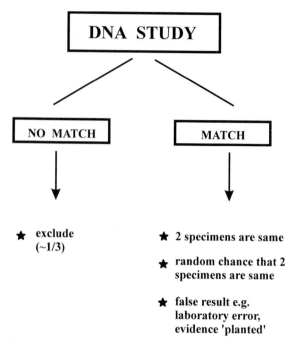

Fig. 8.8 Outcomes from a DNA comparison in the forensic situation.
DNA results comprise *one* piece of evidence in the forensic case. They provide a very powerful argument if they *exclude* an individual. On the other hand, a *positive* match can be interpreted in a number of ways and it is the function of the laboratory, expert witnesses and the court to determine which of the three options is the most likely.

testing as the only means by which his/her innocence can be proven. In one US case, the accused was actually acquitted because the prosecutor failed to test for a DNA match! DNA fingerprinting will save time in police investigations since suspects can be quickly excluded if their DNA profiles are different. Despite being acquitted of a crime, an individual will suffer humiliation and possible stigmatisation following arrest and trial. This can be averted if DNA testing avoids the arrest in the first place. Two experienced forensic laboratories (the US Federal Bureau of Investigation and the British Home Office) have found that DNA testing allowed a suspect to be excluded in approximately *one-third* of cases (Fig. 8.8).

DNA analysis has a role to play in the identification of human remains. For example, the availability of parental DNA samples might allow identification of a body when conventional means (physical appearances, dermatoglyphic fingerprints, dental charts) have been unsuccessful. Thus, dissimilar DNA profiles will exclude a relationship and similarities in DNA polymorphisms will confirm that the deceased is likely to be or is related (Box 8.4). The availability of DNA fingerprints for military personnel would be of practical assistance in the combat situation when identification of a dismembered body may be required.

Communication and education

To avoid prejudicing juries with scientific facts, pretrial hearings are usually undertaken to assess the admissibility of new scientific evidence which in this case is the DNA fingerprint. At these hearings, the judge evaluates the evidence before it is presented to the jury. Expert witnesses from both sides are called and the evidence considered in the light of various opinions. One important feature of DNA fingerprints is the way in which the data are imparted to the judge and jurors who are unlikely to have much knowledge of an already complex subject. For example, a claim by the prosecution that a DNA match between the accused and blood obtained from the victim's clothing represents a 1 in 10^6 chance of a random event is very persuasive evidence. However, an equally crucial component to this evidence is the requirement to explain to the judge and later the jury how the test was done and what are its potential drawbacks, including the method used to assess the statistical probability of a random match. In this respect it is essential that the expert witnesses have both a scientific and practical knowledge of the subject so that they can impart this information in a *meaningful* way. A good lawyer will understand the advantages and disadvantages of the technology and use this information to his/her client's benefit. This is particularly relevant to the adversar-

Box 8.4 Distinguishing individual skeletal remains in mass graves by DNA typing

As indicated in Box 8.3, teeth are important evidentiary material in forensic cases since they are more resistant to postmortem degradation as well as extremes of environmental conditions. Teeth are also easy to transport and serve as a good source of DNA. Comparisons of antemortem dental records with skeletal remains provide useful means to identify individuals even in a mass grave. Despite the enlightened times we live in, there are many reports of human rights abuses leading to mass murder. In more affluent societies, dental records may be decisive in determining the identity of individual victims. However, in the less affluent communities, which are more likely to be involved in these tragedies, dental records are unlikely to be available. An alternative mode of identification in this case might be DNA. To assess the value of DNA typing to investigate skeletal remains from mass graves, a team investigated a report from the Guatemalan government that 10 years earlier, a civil patrol had interrogated and then removed 12 men from a village. The men were never seen again although what appeared to be two large graves were found nearby. Because of fear of reprisals, the villagers had said nothing for 10 years. Subsequently, a team of international forensic experts visited the graves and exhumed 12 skeletal remains. DNA was isolated from teeth and compared to hair samples from maternal relatives of some of the 12 missing men. DNA markers sought were from mitochondrial DNA (and so because of maternal inheritance it was only necessary to compare these markers with female relatives). Mitochondrial rather than nuclear DNA was also chosen for study since the former has many thousand copies per cell compared to the two copies of nuclear DNA. Thus, small yields from the degraded human tissues could still provide an adequate DNA profile. Unlike the story described in Box 8.3, the final results did not provide conclusive data linking all the victims to the reported missing men. However, the study showed the potential for DNA technology in this example of a human tragedy (from Boles et al 1995).

ial court system (judge and jury) where lawyers can influence the jury more than is possible in the inquisitorial (judge alone) courts.

Forensic laboratories

Technology

The potential for DNA testing produced an early rush to set up DNA forensic laboratories which were frequently commercially based enterprises. Tests usually employed were the single-locus VNTR probes since these were technically less demanding than the multilocus probes and were considered to be less likely to produce artefacts on the basis of the material examined. To approach the degree of variability possible with one multilocus VNTR, laboratories obtained composite DNA profiles which were based on polymorphic patterns from four or more of the single-locus VNTR probes.

It should again be noted that the same DNA methods which are routinely used in the research or hospital diagnostic laboratories can be technically more demanding in the forensic situation. In the latter, DNA is frequently tested under suboptimal conditions. For example, the stability of DNA will depend to some extent on the way it has been maintained or stored. This cannot be controlled at the crime scene. The older the sample or the longer it has been exposed to the atmosphere (particularly ultraviolet light, moisture or high temperature), the more degraded the DNA becomes. Artefacts or technical problems resulting from contamination or degradation have led to over 40% of evidentiary samples in the US and UK courts being considered inconclusive on the basis of DNA tests.

Not surprisingly, a number of problems emerged in relation to DNA tests in forensic cases. *Laboratory standards* have at times been inadequate with band shifts being ignored or controlled for by 'correction factors'. Band shifts are a particular problem in forensic practice since they not only represent artefacts from the procedures themselves (e.g. agarose gel variation described earlier, p. 177) but they also result from contaminated or degraded material.

A second difficulty reflected the inability to *measure accurately* the sizes of restriction fragments, particularly large ones found with the minisatellite VNTRs. Related to this was the problem of distinguishing closely co-migrating band fragments. The US Federal Bureau of Investigation attempted to overcome this by defining a number of standard size ranges (called 'bins') so that alleles falling within each 'bin' were treated as the one marker. The widths chosen for individual 'bins' (and so the potential for identifying two different bands as the same) could also be used when calculating the probability that two bands were identical.

Population comparisons

A third issue which continues to provoke discussion, is the validity of *population data* against which the DNA polymorphisms from evidentiary samples are being compared. In the *State versus Castro* example, the laboratory reported that a DNA match between a blood stain found on the accused and blood from the victim had a 1 in 10^8 probability of occurring by chance alone. However, the comparisons to derive the chance association were considered invalid for a number of reasons including the fact that they had not been made against an ethnic group to which the accused belonged, i.e. Hispanic.

One advantage of DNA testing is its ability to show identical patterns (genotypes) between two samples being compared and then to take this one step further and demonstrate that the genotypes are unique, i.e. the probability of them being present in another individual is insignificant. In this respect, it is essential to know the frequency in the population of the various markers which make up the genotype. This forms the basis for the statistical calculation that the two specimens are derived from the same source. The approach described makes a number of assumptions, e.g. there is random mating and the alleles for the multiple VNTR markers segregate independently of each other, i.e. there is no linkage disequilibrium.

However, random mating may not be occurring in ethnic or minority groups within a community. The presence of linkage disequilibrium can be detected by sampling a large population, but does this approach exclude linkage disequilibrium in minority ethnic groups? Would comparative studies be affected adversely in the case of the latter occurring? Does a single allele present for one marker represent homozygosity for that marker, an additional null allele which has not been typed or two alleles which cannot be distinguished? These are some of the questions which have been asked by scientists and the courts.

The answers have been slow in coming since population geneticists themselves have not expressed uniform opinions on the validity of the population comparisons which have been made. In an attempt to produce uniformity, a 3-year study by a US National Research Council (NRC)

committee recommended that a 'ceiling principle' be used so that in calculating likelihoods either the highest frequency found in 10–15 ethnic databases or a conservative 5% was used. In this way any single locus could, at the most, contribute a 200:1 odds. The concept of a 'ceiling principle' provoked considerable debate, was not used in European courts and has now been abandoned because of the controversies and because there are now sufficient DNA databases with which to make direct comparisons. For example, a large US city might require databases comprising polymorphism statistics for Black African, Caucasian, Hispanic, American Indian and Oriental populations.

Despite all the rhetoric about regional and ethnic factors and their potential effects on the frequencies of DNA markers, some population geneticists now take the view that the differences these make to the final calculations are minimal. For example, a probability of 10^{-7} might be reduced to 10^{-6} because an ethnic group is involved. A ten-fold or even one hundred-fold difference in probability should not in itself be sufficient to convict or acquit a defendant. In the normal course of events, a court will assess the reliability and accuracy of the evidence coming from a witness without having an objective cut-off value. A similar approach should be taken with DNA fingerprinting, which should be considered as another piece of the total evidence.

FUTURE DEVELOPMENTS

Laboratory

Quality control and proficiency testing

The problems created by incorrect application or interpretation of molecular techniques in forensic medicine have provided a useful example of how utilisation of this technology must be appropriate and scientifically based. Following on from the *State versus Castro* case there are now more stringent legal requirements to ensure that laboratories involved in DNA typing are able to maintain the highest standards which include regular quality assurance and proficiency testing.

Initially, a quality control programme was difficult to organise since many of the commercial forensic laboratories utilised their own specific single-locus VNTR probes which were protected by patents. Therefore, direct interlaboratory comparisons were a practical problem. This became less of an issue as microsatellite markers and PCR technology were implemented. Today, various externally based testing programmes involving the use of unknown samples give the courts of law an indication of a laboratory's performance in DNA typing.

Another reason for complexity in quality assurance reflects the types of evidentiary samples provided. For example, some specimens are fresh, others have been

exposed to the atmosphere for variable periods of time. The different methods used to store the evidentiary sample once it had been collected might also be another variable in the DNA patterns obtained. Finally, there is the problem of contaminating DNA and the overall effect this will have on the test results. An experienced forensic laboratory will need to know how the above variables might influence the testing procedures and if necessary have the appropriate control samples or data should these be required.

Technology

The increasing potential for automation, e.g. the incorporation of fluorescent dyes into PCR products and then sizing of bands by computer-based means will provide a more objective approach to DNA forensic typing. The 'gold standard' for DNA typing is, in fact, sequence determination. It is likely that, in future, forensic laboratories will routinely utilise this as a way to compare DNA patterns, particularly as automation becomes cheaper and more readily available (goals of the Human Genome Project).

An important consideration for the defence and prosecution is the option to allow retesting of the evidentiary material to confirm the findings and/or enable the test to be performed in a 'neutral' laboratory. Stains or tissue samples will usually yield a small amount of DNA. The requirements for minisatellite assessment will, in most cases, deplete the sample. DNA evidence obtained from a single laboratory is less than ideal since confirmation will not be possible. In these circumstances, there is even greater need for good quality assurance to guarantee the appropriate laboratory standards. Again, this issue has become less of a problem with the increasing use of PCR. In the UK, a positive match following testing with seven single-locus VNTRs is followed by the taking of a further specimen from the accused. DNA from the second sample is tested in another laboratory.

The problem of band sizing has been resolved to some extent by the 'bin' approach, the use of PCR and automation. However, novel approaches may still become possible so that band sizing is avoided completely. One such method, presently under evaluation, involves variant repeats in minisatellites (also called 'digital DNA typing'). The variability in these DNA markers relies on the differing nucleotide composition which can be found within a minisatellite repeat of the same length. Internal variation

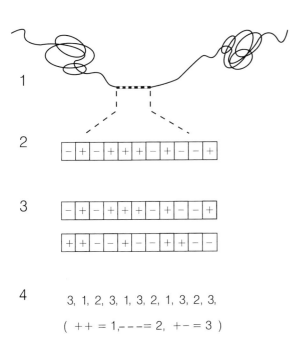

4   3, 1, 2, 3, 1, 3, 2, 1, 3, 2, 3,

(++ = 1,--- = 2, +- = 3)

Fig. 8.9 Minisatellite variant repeats producing a 'bar code' for locus identification.
Also known as 'digital DNA testing', this is based on the variation present in a single VNTR minisatellite. **(1)** Shows a minisatellite locus with 11 repeat units in total genomic DNA. **(2)** The individual locus can be amplified by PCR. The 11 repeat units within the locus can be characterised on the basis of whether they contain (+) or do not have (−) a restriction enzyme recognition site for HaeIII. **(3)** Patterns described in (2) become more complex if both alleles are studied. **(4)** A ternary code can be derived on the basis of both alleles having a HaeIII site (code = 1), neither allele having the restriction site (code = 2) and one of two alleles with the HaeIII site (code = 3). The complexity of the ternary code becomes unlimited as more minisatellite loci are characterised. The bar code does not involve sizing of DNA fragments.

with these markers avoids the necessity to size DNA fragments (Fig. 8.9).

Courts and society

In the matter of DNA fingerprinting, the courts and scientists demonstrate an interesting dichotomy. The courts are conservative and prefer to deal with precedents. On the other hand, scientists are constantly striving to develop new techniques and in the field of molecular medicine, the changes seem to occur on a daily basis. Some adjustment, perhaps by both parties, will be required to ensure that this does not interfere with the long-term development and utilisation of DNA fingerprinting.

During 1995, it became legal in the UK for law enforcement agencies to take DNA in the form of hair roots or buccal samples (i.e. non-intimate samples), by force if necessary, from those convicted of serious crimes. The aim is to establish a national DNA database from such offenders. This requirement, to a varying degree, applies in a number of States in the USA. For example, in Virginia *all* convicted of a crime, no matter how serious, have DNA collected and stored. In some States, DNA is released if the person is found to be innocent, in others DNA is handled in the same way as conventional fingerprints and kept. Some successes have already been reported with these databases including the identification of rapists and murderers and the ability to link different crimes. Despite the concern of civil libertarians, it is very likely that DNA repositories will become commonplace. The ethical and privacy issues emerging from DNA databases are considerable and will be discussed further in Chapter 9.

Dermatoglyphic fingerprints, first introduced in the 1890s, are an integral component of the legal system. Whether an individual's fingerprints are in fact unique to that person was never initially determined by the rigorous scientific standards which are demanded today. There remains a lot to be learnt and developed in relation to DNA fingerprinting which will play an important role in the forensic laboratory of the future.

FURTHER READING

General

Berghaus G, Brinkmann B, Rittner C, Staak M (eds) 1991 DNA – technology and its forensic application. Springer-Verlag, Berlin
Decorte R, Cassiman J-J 1993 Forensic medicine and the polymerase reaction technique. Journal of Medical Genetics 30: 625–633
Lander E S, Budowle B 1994 DNA fingerprinting dispute laid to rest. Nature 371: 735–738

Repetitive DNA

Housman D E 1995 DNA on trial – the molecular basis of DNA fingerprinting. New England Journal of Medicine 332: 534–535

Human relationships

Abbas A K, Lichtman A H, Pober J S 1994 Cellular and molecular immunology. Saunders, Philadelphia, p 96–114

Jeffreys A J, Brookfield J F Y, Semeonoff R 1985 Positive identification of an immigration test-case using human DNA fingerprints. Nature 317: 818–819

Pena S D J, Chakraborty R 1994 Paternity testing in the DNA era. Trends in Genetics 10: 204–209

Tissue identification

Boles T C, Snow C C, Stover E 1995 Forensic DNA testing on skeletal remains from mass graves: a pilot project in Guatemala. Journal of Forensic Sciences 40: 349–355

Cohen J, Stewart I 1995 Beyond all reasonable doubt. Lancet 345: 1586–1588

Sweet D J, Sweet C H W 1995 DNA analysis of dental pulp to link incinerated remains of homicide victim to crime scene. Journal of Forensic Sciences 40: 310–314

Weir B S 1995 DNA statistics in the Simpson matter. Nature Genetics 11: 365–368

Future developments

Jeffreys A J, MacLeod A, Tamaki K, Neil D L, Monckton D G 1991 Minisatellite repeat coding as a digital approach to DNA typing. Nature 354: 204–209

McEwen J E 1995 Forensic DNA data banking by state crime laboratories. American Journal of Human Genetics 56: 1487–1492

9

ETHICAL AND SOCIAL ISSUES

INTRODUCTION

Ethics can be defined as: (1) a system of moral principles by which human actions and proposals may be judged good or bad, right or wrong; (2) rules of conduct recognised in respect of a particular class of human actions (Macquarie dictionary).

The fundamental principles of good medical care are also the basic principles of ethics: *beneficence* or 'do good' and *non-maleficence* or 'do no harm'. These have become very significant particularly with the rapid advances in medical research which have occurred since the Second World War. In 1964, the Declaration of Helsinki made recommendations which were meant to guide physicians in their conduct of biomedical research on human beings. These were intended to improve diagnostic and therapeutic procedures which of necessity would sometimes be combined with professional care. The original Declaration of Helsinki, although reviewed and amended by a number of World Medical Assemblies, the last being in 1989, remains essentially the same.

Key principles of ethics for guiding the practice of medicine (including molecular medicine) are beneficence, autonomy, confidentiality and justice. *Beneficence* refers to the obligation to do good as well as the avoidance or removal of harm. The ethical principle of *autonomy* arises from respect for the individual and the recognition that the person has the right to truthful information about his/her clinical condition, his/her options for treatment and the opportunity to participate in the treatment decision. If the process is experimental, the subject needs to understand this and consent freely. Professional *confidentiality* and *privacy* are important ethical principles related to autonomy. Confidentiality and privacy, as will be illustrated, can lead to difficult dilemmas in relation to DNA diagnostic testing. The principle of *justice* comes into consideration in the distribution of resources. Included in this is equal access to health services and the avoidance of unnecessary or wasteful procedures. In terms of research, justice would require useful, high-quality work and the application of new knowledge in a way which is most valuable clinically but not necessarily the most profitable.

Because of the significance of genes and their genetic information, developments in molecular technology have attracted the attention of the public, the media and have been monitored by a number of statutory bodies and committees. Input into these committees from laypersons has enabled more broadly based assessments of protocols and options. To date, there have been few instances of unethical behaviour by physicians and scientists in the area of genetic engineering. Careful monitoring needs to be continued to ensure that this is maintained. On a broader front, communities have followed established and accepted principles for the conduct of genetic engineering. However, there are a number of examples which illustrate how unethical practices can develop in medical and scientific research. There is no reason why the above would not apply to genetic engineering if the advantages to be gained were perceived to be worthwhile. The risk of this happening would be less if the community were both well informed and involved in decision making.

Training in ethics for health professionals and scientists does not comprise a significant component of many curricula. This is unfortunate since situations which have both ethical and social implications are likely to develop, particularly in the field of molecular medicine. For example, an expanding pool of knowledge about human DNA is already placing some pressure on available resources. The ethical and social implications of this will only increase as the Human Genome Project gains momentum. It has also been stated that the Human Genome Project, as an example of scientific endeavour, is being matched in many communities by shrinking resources which will be further depleted to meet the costs of this research.

The innovative developments which are possible through genetic engineering need to be balanced by their potential ethical and social implications. Professional knowledge and expertise carry with them the responsibility for the careful weighing of *risk* and *benefit* for the individual. Therefore, while the degree of benefit may not always be predictable, at least there should be no harm to emerge. The predictable risk should never outweigh a possible benefit. A number of issues which have potential ethical and social implications are summarised in this chapter. They illustrate the present situation as well as making predictions about the future. As has been shown consistently in this book, the advances in this field will be substantial. Accompanying these developments will be questions of ethics and the effects of these changes on society as a whole.

CLINICAL PRACTICE

DNA testing

DNA has a certain versatility which makes it ideal for genetic testing. On the other hand, this can lead to unnecessary or inadequate testing which produces more harm than good. The advantages of DNA for diagnostic testing include: (1) human tissue for study is easy to obtain since DNA is identical in *all cells* and at *all times* during development; (2) DNA diagnosis does not require a particular gene to be identified before its presence can be sought; and (3) the range and complexity of human diseases which can be detected will continue to expand as more DNA probes are described.

Presymptomatic ('predictive') DNA testing

Unlike many other genetic disorders, the implications of DNA testing in Huntington disease have been studied in great detail and by various health professionals. From this work, it is possible to identify the 'good and bad' in DNA diagnostic testing. Until DNA tests became available in 1986, the individual who was at risk for Huntington disease had to wait until his/her fourth decade or longer before the first clinical features became apparent. In the meantime, he/she would have firsthand experience of the inexorable clinical deterioration which occurs in this disorder because a parent or close relative would have been affected.

Huntington disease was first localised to the short arm of chromosome 4 in 1983, although the gene itself was not isolated for another decade. However, even without the actual gene, an *indirect* approach to presymptomatic diagnosis became available. Linkage analysis enabled the mutant phenotype to be predicted within the context of a family study. From this, the known a priori risk of 50% for an offspring of an affected individual could be lowered to 1–5% or raised to 95–99% depending on whether the individual had inherited a DNA marker which co-segregated with the normal or Huntington disease clinical phenotype. The risk estimates given included a conservative 1–5% error rate which reflected the recombination potential for the DNA markers in linkage with the putative Huntington disease gene. With the isolation of the causative gene and ability to detect a common mutation in this gene, DNA testing progressed to the next level, i.e. *direct* DNA predictive testing for Huntington disease.

The positive side to DNA testing in Huntington disease is that the known risk for an individual can be assessed at any time in his/her life. With this knowledge, informed decisions about lifestyle and other key issues such as reproduction can be made, e.g. prenatal diagnosis has become an option, whereas previously at-risk couples often elected to have no children or adopt or undergo in vitro fertilisation (IVF) by donor insemination. In some cases, the replacement of uncertainty with a DNA diagnosis, whether it was favourable or not, enabled the individual to adapt and lead a better quality life.

However, there are also negative aspects to DNA testing. For linkage analysis, it is essential to study a number of family members. Key individuals are particularly important since it is these who allow linkage between a clinical phenotype and a DNA marker to be established. In the context of Huntington disease, a key individual is one who has unequivocal evidence for the disorder. Less helpful is the family member who is at an age that a confident diagnosis of 'not affected' is possible although never a certainty (see Ch. 3).

Apart from the resource-intensive nature of linkage analysis, there are a number of ethical issues which must be considered in undertaking a family study, particularly one which may need to be extended to include distant relatives. Confidentiality will invariably become a concern. Even the first step, which involves construction of a pedigree, is undertaken without the consent or knowledge of family members. Disclosure of the pedigree may lead to key individuals being coerced by others into supplying information or giving blood. This could reach the courts of law with an uncooperative relative being threatened by legal action. At present, the example given above produces no major dilemmas since the unwilling participant would not be compelled by the courts to give blood or have his/her privacy infringed by the enforced disclosure of personal details. Nevertheless, it should be kept in mind that some communities have already legislated for compulsory blood testing in certain circumstances, e.g. blood alcohol levels after road accidents. While the two situations bear little resemblance to each other, it would not be

less traumatic and increases the number of disorders detectable since DNA can be isolated with relative ease from chorionic tissue. The above, as well as the increasing availability and commercialisation of DNA diagnostic kits, has meant that the decision whether or not to undergo prenatal diagnosis and so perhaps terminate a pregnancy is being faced by an increasing number of couples. In this respect, it is relevant to note that some governments, e.g. the US Congress, have expressed concern that the proliferation of laboratories offering DNA genetic testing is placing, in some circumstances, *unnecessary* pressure on individuals to make decisions about their genetic status. There can be additional pressures and complications if there is a third party, i.e. a fetus in the present context.

The ethical and social implications of terminating a pregnancy have not been discussed, but these will also

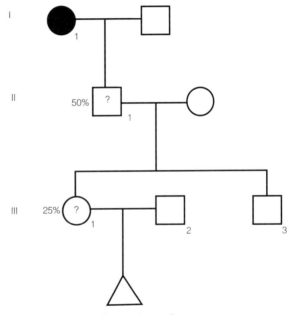

inconceivable in the future, particularly after DNA testing has become more firmly established as a routine procedure, that a court might be convinced to create a precedent on the grounds that good for the majority (the individual and his/her family) should outweigh what might appear to be 'trivial' concerns to one other person.

There are ethical dilemmas which arise as a consequence of too much information becoming apparent following DNA testing. Some examples are given in Box 9.1. Another issue which remains to be resolved in Huntington disease is *when* to test for this disorder. DNA predictive testing is usually undertaken for an individual who is at 50% risk, i.e. a first-degree relative has Huntington disease. However, there are cases when the risk remains at 25% because the at-risk parent does not want to be tested, examined neurologically or acknowledge that he/she is at risk. An ethical dilemma occurs when the person at 25% risk is tested and shown to carry the Huntington disease mutation. In this situation, the genetic status of other family members is revealed (Fig. 9.1).

Prenatal diagnosis

The increasing scope for presymptomatic testing is being matched by the number of prenatal diagnoses which are being undertaken. Prior to the availability of chorionic villus sampling, prenatal detection of genetic diseases was only possible during the second trimester of pregnancy. Delayed results meant a difficult termination of pregnancy should this be requested. Today, first trimester diagnosis by chorionic villus sampling makes termination of pregnancy

Fig. 9.1 DNA testing for Huntington disease.
An individual with Huntington disease is depicted as I₁. Her son (II₁) has a 50% a priori risk for Huntington disease. Her granddaughter (III₁) has a 25% a priori risk. In this situation, DNA testing is not usually undertaken for the latter because her risk has not reached 50%, i.e. her father tests positive or he develops symptoms. The dilemma arises if the granddaughter wishes to know whether she carries the mutation for Huntington disease (e.g. she is considering pregnancy) *and* the asymptomatic father refuses to be seen by a neurologist or undertake a DNA test or for whatever other reason does not want to be involved. Even without his participation, DNA testing in the granddaughter can be undertaken. A negative result in her does not alter the father's overall risk. However, a positive result will mean her father *must* have Huntington disease and it will increase the risk for her brother (III₃) from 25% to 50%. Does she tell the brother this? Will it be possible to prevent the information that the father is affected from reaching him? If either or both siblings are affected, they will no longer be able to seek help and support within the family framework because this would establish the father's genetic status.

influence a couple's final decision. In this respect it is sobering to read what individuals who have the severe form of β thalassaemia think about prenatal diagnosis. In one study, all stated that they wished to have children of their own. The great majority indicated their willingness to undergo prenatal diagnosis for β thalassaemia should their partner also be a carrier. Less than half indicated that, taking into consideration the constraints placed on their life, they were still happy to have been born. The remainder expressed a wish that their parents should have undergone prenatal diagnosis and termination of pregnancy if necessary. Thus, there is no easy solution from the patients' perspective. The dilemmas facing the couple can be even more difficult since there is often no first-hand experience of the genetic disorder being tested.

Research and DNA testing

Does the researcher who is studying the molecular basis of genetic disorders have specific responsibilities? As indicated previously, early studies of Huntington disease suggested a single mutation for this disorder. This is now confirmed to be correct and so linkage analysis, apart from the risks of non-paternity or recombination, was an appropriate way in which to undertake presymptomatic testing for 10 years before the actual gene was found. However, a different story occurred with adult polycystic kidney disease which involves two distinct loci with the second (albeit rare locus) only becoming apparent 3 years after DNA testing for this disorder had started (see Ch. 3).

A difficult question arising from DNA testing is when does the research phase become a routine clinical diagnostic service? This has recently been highlighted by the finding of the *BRCA1* gene and its implications for DNA testing in breast cancer. Unlike *APC*, the gene for familial adenomatous polyposis, which has now reached the stage when it can be considered an accepted DNA test for this form of bowel cancer, the same cannot be said for *BRCA1*. Policy statements from learned professional bodies have been produced to indicate that until further knowledge about *BRCA1* becomes available, DNA testing in this circumstance is still within the realm of *research*.

The problem of when the research phase ends and routine DNA testing starts will be asked more frequently as other DNA probes become available. Pressure from peers or consumer groups will make it more difficult to postpone changing what is a research study into the more definitive diagnostic service. In these circumstances, the researcher has the onerous task of ensuring that data (comprising laboratory, counselling and educational components) are of sufficiently high standard that appropriate long-term decisions, within the framework of a research study, will be possible. This will frequently also involve administrative and political decisions about the allocation of resources.

Counselling

The emphasis today is on *non-directive* counselling. Thus individuals seeking advice because of a family history of genetic disease or a couple in the prenatal diagnosis situation will need to make their own decisions based on the information they are given. However, the rapid advances in molecular genetics can assist and at the same time complicate the counselling process.

Information gained from molecular genetics can make the counselling process more difficult because it provides some data, but not enough for the couple to make a decision. For example, a question frequently asked is how severely affected an offspring with a particular genetic disorder will be. In β thalassaemia, the molecular basis for the milder, non-transfusion-dependent form called 'thalassaemia intermedia' has been determined in some cases. However, other factors are also involved. Has knowledge based on the molecular pathology of this disorder reached the stage that the co-inheritance of β thalassaemia with other genetic changes (e.g. an increase in fetal haemoglobin or co-existing α thalassaemia) will enable a confident prediction of milder severity to be made? The answer is 'no'. A similar dilemma will be described in cystic fibrosis (see p. 194). Therapeutic options, particularly those related to molecular medicine, are also changing. For example, gene therapy in cystic fibrosis and the availability of an oral chelating agent in thalassaemia will become future treatment considerations. These should improve the outlook for both disorders and need to be considered by prospective parents.

Greater demands are being placed on counsellors. There will be few who alone can provide a comprehensive counselling session based on a wide range of issues including the clinical consequences, recent advances in molecular medicine and potential future developments. The complexities associated with DNA testing, in particular, require a team approach, since it is unlikely that a single individual will have the breadth and depth of knowledge required to understand all genetic disorders and then convey this information to the consultand in a way which will be meaningful.

The importance of a team approach in dealing with clinical disorders with a genetic component became reinforced with the *BRCA1* gene and the realisation that DNA testing for the more common cancer predisposition genes was 'just around the corner'. In this situation, the patient who is at risk should have access to a range of health professionals (e.g. oncologist, surgeon, geneticist, counsellor, nurse) who provide, within the appropriate environment, the information which will assist in decision making.

Justifiable criticism has been levelled at commercial DNA-testing laboratories which make available a DNA result without considering whether or not the test is necessary or without the resources which would allow the test's implications to be understood by the patient or health professionals. This could be resolved if the provision of

DNA testing were clearly linked to counselling. A worrying development in the cystic fibrosis DNA story was the announcement in late 1994 that a UK-based laboratory would provide DNA testing for four cystic fibrosis mutations *direct to the public*. All that was required was a mouth wash sample, an information sheet and a fee to cover the DNA test and additional advice in some circumstances. What response this has generated from the public, what the long-term effects of this type of testing are and who will be tested are issues which remain unclear. Whether the laboratory will provide the counselling and educational components of DNA testing is also unresolved.

Population screening

The utility of DNA testing strategies, particularly the potential for PCR to test many samples quickly and relatively cheaply, has meant that widespread screening of the population becomes a practical consideration. As more genes are sequenced, the number of mutations identifiable by PCR will increase. There are many types of screening programmes which could be implemented based on DNA analysis (Box 9.2).

Two contrasting examples of *selective screening* programmes targeted to at-risk populations and started before the DNA era illustrate the potential advantages and disadvantages of this type of testing. *Tay–Sachs disease* is a fatal neurodegenerative disorder of childhood. It is inherited as an autosomal recessive trait. Since the early 1970s, individuals at risk for Tay–Sachs have been screened and counselled. The incidence of Tay–Sachs disease has been reduced without the problems which developed following the implementation of population screening for *sickle-cell disease*, another autosomal recessive disorder. Sickle-cell disease leads to considerable morbidity and mortality, although the ultimate outcome is not entirely genetic in origin as environmental factors are important. The US-based sickle-cell screening programme, which was also started in the early 1970s, was targeted to the at-risk black population. The initial version of this programme produced more harm than good. Results led to a lowering of self-image, overprotection by parents and discrimination. The discrimination came from employers, insurance companies, health insurers and potential spouses. What were some of the differences between these two programmes?

One important consideration in the success of Tay–Sachs screening was that the target group comprised individuals of Jewish origin who had, in contrast to the black population, better *educational* opportunities. Thus, the necessity for counselling and public education to explain the significance of mass screening and the results obtained are important ethical considerations. Another contrast between the two programmes was the close community consultation and participation undertaken prior to genetic testing for Tay–Sachs. Because of the problems associated with sickle-cell screening, changes were made to the programme, e.g. removal of the legal compulsion to screen

and improved counselling/education facilities. These enabled more successful testing to be pursued.

The modern screening dilemma can be illustrated by cystic fibrosis. DNA amplification by PCR enables the common ΔF508 defect to be detected in peripheral blood or more accessible tissues such as buccal cells present in a 10 ml mouth wash specimen. As summarised in Chapter 3, there are over 600 mutations which produce cystic

---

Box 9.2 Population screening strategies

The term 'screening' can have many different meanings. To government, it usually suggests population testing and a large financial committment. To the individual family it means that at-risk members are being tested for some trait. It has been suggested that the latter should not be called 'screening' because of its focused and limited nature. However, 'screening' will be used as a general term and its exact meaning defined as required. There are a number of different screening strategies possible.

- *Selective screening* – individuals at increased risk for a genetic disorder are targeted. DNA testing is very useful in this circumstance since a mutation known to be present in a family can be sought. Alternatively, DNA polymorphic markers enable a linkage approach, within the constraints of a family study, to be followed.
- *Mass screening* – the testing of entire population groups usually looking for a common disorder. An example of this is cervical cancer checks. To be effective, a large percentage of women over the age of 45–50 must participate and the test needs to be sufficiently sensitive and specific to make the programme cost-effective.
- *Epidemiological surveys* – mass screening-type approaches to ascertain the prevalence and incidence of a particular disorder. Important ethical considerations in setting up screening programmes include the ability to do good by early detection, i.e. the disorder is potentially serious and there is effective treatment available. Privacy and confidentiality issues will arise since the information gathered could be acquired by third parties.
- *Occupational screening* – DNA screening in the workplace which can have two different aims. The first identifies individuals at greater risk for industrially related complications. For example, the combination of α_1-antitrypsin deficiency in a worker and a dusty workplace would be more likely to lead to chronic lung disease. The second utilises DNA technology to detect DNA damage which is related to the workplace. This would provide objective data in terms of cause and effect. Although screening at the workplace has merit, a potential criticism would be that emphasis is being placed on exclusion of the at-risk worker rather than attempting to make the workplace a safer environment.

Table 9.1 **Geographical distribution of the ΔF508 mutation of cystic fibrosis (Cystic Fibrosis Genetic Analysis Consortium 1994).** There are over 600 mutations which produce cystic fibrosis. Mutations other than ΔF508 occur less frequently and will not be detected in a screening programme for ΔF508.

Continent[*]	ΔF508 mutation in cystic fibrosis chromosomes (%)
Northern Europe	70
Southern Europe	55
North America	66
Middle East	28
Africa	68
Australasia	75
All continents	66

*Ethnic heterogeneity within continents is a further consideration in assessing the local frequency of the ΔF508 mutation.

fibrosis, although the ΔF508 is the most common amongst individuals of Northern European origin. Other mutations are much less frequent (Table 9.1). Debate continues as to the value of *mass population screening* in contrast to testing individuals or families at risk, i.e. *selective screening*.

The ethical and social issues raised include the use of a test which will not detect all those who are affected. For example, if only the ΔF508 mutation is sought, false-negative results in couples from a population with a frequency for the ΔF508 mutation of 70% will be 0.51 (1 − (0.7 × 0.7)), i.e. approximately half the couples will not be informative by this approach. The detection of the less common mutations (some of which are only present at a 1–2% frequency in the population) will add to the workload but not increase substantially the information to come from the screening programme.

Even if laboratory facilities are available, major efforts directed towards genetic counselling and public education would be required to ensure that those tested fully understood the implications of the results. The financial resources to carry out a mass screening programme would be enormous. In view of this and the inability to detect all mutations with *present technology*, most national genetic societies have recommended limited or selective screening of groups who are at higher risk than the general population.

Additional problems which need to be resolved before embarking on a widespread cystic fibrosis screening programme include the uncertainty of the mutations in respect to disease severity. Thus, counselling in a number of instances will be difficult and incomplete. An example can be seen from the IVF programmes which have shown that some cases of infertility in otherwise healthy males are due to cystic fibrosis (see Box 3.5). If, in this circumstance, the female partner carries the ΔF508 mutation, offspring

conceived by IVF using sperm isolated from the infertile male have a 50% chance of being compound heterozygotes for ΔF508 and one of the mild defects carried by the infertile male. The phenotype that this produces will be difficult to determine since without IVF-based technology infertile males would not have transmitted that particular mutation.

Many of the *newborn screening programmes* have now successfully incorporated DNA analysis for cystic fibrosis into their testing protocols (see Ch. 4). This has streamlined the laboratory's work and has reduced anxieties in families because the number of newborns required to be recalled for definitive testing is reduced. However, the advantages will need to be balanced with the growing cohort of children in the population who have been identified as carriers of the cystic fibrosis mutation. What effect, if any, this information will have on the way they develop will be important to document. Finally, in all examples of population screening, data storage would have to be both accessible and secure to ensure confidentiality.

Therapeutic developments (including the potential for gene therapy) have improved the affected individual's quality of life and survival and would need to be considered in assessing the value of mass screening for what was once a uniformly fatal disorder in early adult life.

Discrimination

To the individual, DNA screening has the advantage of detecting potentially reversible problems early or allowing him/her a longer disease-free period by preventing exposure to industrial toxins which are particularly noxious to an individual with his/her genetic constitution. On the other hand, the disadvantages are that discrimination, loss of privacy, loss of health benefits and loss of employment can follow. Some hypothetical examples of how this might occur are given by Alpers and Natowicz (1993).

Both employers and insurance companies are becoming increasingly aware of the potential for DNA screening or testing. There are a number of reasons given to justify this, e.g. the comment by the employer that failure to identify those at risk will unfairly discriminate against others in the workplace. The insurance company, which is a business, will point to the duty owed to the other policyholders and shareholders. For example, a UK House of Commons committee was informed by representatives of the insurance industry that providing life insurance for the ~5% of individuals presently excluded because of pre-existing medical conditions would at least double the cost of current premiums.

The potential consequences which would follow if insurance companies required compulsory DNA testing have been discussed on many occasions and remain a sensitive issue in molecular medicine. At present, insurance companies base their policies on actuarial calculations related to the likelihood of death, loss or damage. In terms of life insurance, those who fall into high-risk groups are usually able to be insured, although at a much higher premium. The medical examination, the medical history and in some cases a compulsory test for HIV are accepted as part of the prerequisites for obtaining life insurance. However, the potential to utilise the DNA or genetic makeup of an individual has opened a Pandora's box. Governments have responded in a number of ways, e.g. a form of universal insurance cover which would have no preconditions or the complete ban of genetic testing when it comes to insurance matters. The US government in 1995 legislated to define an individual's genetic status as a disability in some circumstances. Thus, it would be discriminatory for an employer to use this information in employing or dismissing an employee. The legislation does not apply to the private insurance companies who have generally kept a 'low profile' in respect to DNA testing. In the UK, the insurance industry has had to take a more active stance since it has been given a period of time to establish appropriate guidelines to cover the use of genetic information for the purpose of life insurance or the government would move to legislation. In that country, the insurance industry has started a dialogue with the relevant health professionals.

A number of key issues need to be resolved. For example, will DNA testing comprise another component of the actuarial assessment or will it be used to deny insurance or make the 'loaded' premium unattainable? For example, a family history of Huntington disease will, at present, place an individual who wants to take out life insurance in the at-risk category and the premium will be modified accordingly. However, how would the insurance company react to the following two scenarios: (1) prior to insurance being obtained the at-risk individual voluntarily undertakes DNA testing and is shown to carry the mutant gene and (2) following the insurance premium being negotiated, DNA testing identifies the mutant gene. It is presumed that in both the above hypothetical situations, the finding of a normal gene will place the person in the general population risk category.

Testing children

One important advantage of DNA testing is that it can be undertaken at any age. However, this can lead to an ethical dilemma when otherwise healthy children are tested. In these circumstances, informed consent is given by their parents. Why are children being tested? An acceptable medical indication is a childhood disorder or an adult-onset disorder for which early intervention will improve the prognosis. Non-medical indications which have been identified by consumer groups include lifestyle planning decisions such as choosing the appropriate environment in which to bring up a child with a genetic disorder. Dubious indications include the options for planning of future educational, career or reproductive decisions by the *child*. It is often difficult to know what component reflects the parental wishes, what directly relates to the child and

whether it would be more appropriate to test at a later age when the child him/herself would be in a better position to make an informed choice.

In the end, it is the parents acting as the child's guardians who will make the request for DNA testing. Whether this is to relieve anxiety on their part or a legitimate medical indication or a combination of both can be difficult to assess. However, relief of parental anxiety must be balanced by the problems of DNA testing mentioned already and the potential harm to a child whose disclosed genetic status may lead to an unnecessary change in the way he/she is allowed to develop.

To avoid difficult decisions in these circumstances, DNA programmes, for example that for Huntington disease, will not test individuals unless they themselves are able to give informed consent, i.e. they have attained the age of 16–18 years. The Huntington disease example is relatively straightforward since this adult-onset disease has no known treatment which can influence its natural history. On the other hand, more debate has occurred in respect to familial hypertrophic cardiomyopathy, another autosomal dominant disorder which *usually* manifests in adult life (see Ch. 3 for further discussion on the molecular genetics of this disorder).

One view held in respect to testing children for familial hypertrophic cardiomyopathy is that presymptomatic DNA testing will relieve anxiety for the parents and the child if the underlying mutation is excluded *or* ensure that those identified as carrying the mutation are more effectively followed and encouraged not to participate in the competitive sports which are associated with sudden death. The opponents to this view point out that no medical intervention has yet been shown to alter the natural history of the disorder and the harm coming from early knowledge that the child carries a 'defective' gene could have a negative effect on that individual's subsequent development. In this debate it is interesting that the proponents for DNA testing are more likely to be the clinicians (e.g. cardiologists, paediatricians) who deal directly with affected individuals while the opponents for testing are the clinical geneticists who are taking a more global view of the situation. Informed debate and objective research are now required to enable the health professionals to achieve a consensus so that the decision-making process by those at risk is not dependent on the bias of an individual or a professional group.

Guidelines are helpful in complex cases, but flexibility needs to be maintained to deal with those situations where DNA testing will be beneficial to the child. In the context of a research protocol, the testing of children can occur but due consideration will need to be made since the process is more complex than that involving the mature adult who can consider the implications for him/herself. A series of conclusions and recommendations relating to the issues of DNA testing in children has come from a report of the Clinical Genetics Society of the UK (Box 9.3).

Tissue storage

Human-derived tissues are stored in a number of ways. Storage may relate to 'routine laboratory practice', e.g. the keeping of residual tissue sections, blood samples or even blood spots present on Guthrie cards obtained from a neonatal screening programme. On the other hand, storage may be more formal as a DNA bank where immortalised cell lines enable tissues to be kept and retrieved for an indefinite period of time (see Ch. 3).

Tissue storage has produced additional ethical and legal considerations since DNA technology has shown that even archival samples are able to be tested. A trivial but recent example of this was the suggestion that the long-standing dilemma whether the US President Abraham Lincoln had Marfan syndrome could be resolved through a PCR study of tissue fragments and blood spots taken from clothing worn at the time of his assassination. This became possible when the DNA sequence related to fibrillin, the gene for Marfan syndrome on chromosome 15, was published. Opinions about the ethical aspects of such a study were divided. One group considered that there were no legal and ethical issues involved. Another questioned the ethics of this proposal.

Formal guidelines have been established to define the rights of the individual who has had DNA banked. How the material is used, the anticipated condition of the material and mode of storage are clearly stated. Ownership of the material remains with the depositor. Nevertheless, dilemmas arise when material has been stored for one purpose but then another DNA test becomes available and the stored material (the depositor may be deceased at this time) would be helpful in defining the genetic status of other family members.

It is interesting that in a recent review of laboratories which undertook DNA banking, it was noted: (1) the amount of DNA banking was increasing, and (2) policies in relation to the banked material were deficient, particularly in academic institutions. Apart from the policies issue, the discrepancy between what is undertaken in commercial companies (driven in some part by the legal consequences if there is an error) and the academic institutions is a disturbing observation which needs to be carefully monitored and corrected if confirmed.

Ownership of biological material, e.g. frozen embryos, may also come into dispute when the involved partners have separated and there is conflict as to what should be done with the stored sample. Another practical example would be the Guthrie card. In many communities, the State has legislated that 'Guthrie spots' will be taken and these are used to screen for a number of medical conditions. Could blood taken from an infant for the purpose of neonatal screening be used to test for other genetic diseases? The good that this can bring to the individual and/or the community needs to be balanced with the person's right to privacy. A further question is whether the Guthrie card could be subpoenaed by a court of law to resolve a legal matter. In another survey, but this time of laboratories which dealt with Guthrie cards, the same lack of uniformity and policies mentioned earlier in terms of DNA banking was observed. Some of the laboratories had accumulated many Guthrie cards over the years and so had a very large potential bank of tissue from which DNA could be isolated.

Genetic registers and databases

The keeping of registers is an integral part of a genetics service. Registers come in various shapes and sizes. The compilation of a pedigree is one form of a register. As indicated previously, many individuals identified on the pedigree are not aware of its existence or have not given permission for their inclusion. In fact, to acquire informed consent in these circumstances is both difficult and in itself an intrusion of an individual's privacy. A further extension of the genetic register is the availability, in a central database, of a list of names or identities of individuals who have a particular type of genetic disorder. The significance of this in providing information for health planning or to assist other family members is balanced by the potential for unauthorised disclosure of such data. The privacy issue is particularly significant when third parties, e.g. employers, insurance companies or the courts of law, may gain access to this information. In these circumstances, the ethical principles of autonomy and confidentiality outweigh the considerations that third parties may derive benefit or harm from such information. However, this can be interpreted in different ways and a court of law may consider the reverse holds in a particular situation.

DNA fingerprints provide a unique profile for an individual. It is likely that in future they will be used in place of the dermatoglyphic patterns which are now an acceptable routine procedure in criminal cases (see Ch. 8). However, unlike dermatoglyphics (which would not be universally available for all individuals in a community), there will be alternative sources of DNA for most, if not all, members of that community. These would have been obtained for specific purposes, e.g. 'Guthrie spots' during newborn screening, or inadvertently through routine blood counts or typing specimens from blood donors, but could now be used (or misused) for other purposes.

Quality assurance

The rapid advances in molecular medicine and the negative effects these can have on laboratory practices have been illustrated by suboptimal standards practised in some forensic laboratories (Ch. 8). An important ethical issue (and a legal one) in laboratory and clinical practice includes the obligation to ensure high standards. Quality assurance programmes have been implemented in many clinical and laboratory situations to monitor outcomes. On the other hand, the very rapid developments in molecular technology, particularly in respect to diagnostic laboratories, have managed in a number of cases to overtake the implementation of formal quality assurance programmes. Deficiencies in this respect may not reflect a primary reluctance by the laboratory to participate in such a programme but may result from the intense pressure to start a new diagnostic test because a probe or DNA sequence is available. Thus, expansion to increase the *quantity* of testing can take priority over the *quality* of the results. In the circumstance of increasing knowledge, as will be observed in molecular medicine for some time, a balance is essential to ensure that data provided by a laboratory are of the highest standard possible given the resources available. DNA diagnostic 'kits' will be helpful in respect to quality assurance, but this will come at an extra financial cost.

Future issues

All the ethical and social implications of DNA testing raised earlier have centred on single-gene disorders. In these circumstances, diagnosis is relatively straightforward and the resulting phenotypes are more or less predictable. However, in the not too distant future, the more complex *multifactorial* and *somatic cell disorders* will become recognisable at the DNA level. Thus, coronary artery disease, neuroaffective disorders, dementias and cancers will have their underlying DNA components defined. On the positive side, there is the potential to understand more fully the DNA/DNA and the DNA/environment interactions which are necessary to produce the various clinical phenotypes. This will facilitate the implementation of more effective preventative and therapeutic measures. The negative components include the increasing gap which will emerge between what is known about many diseases and what can be done to treat them. There will also be the potential for misuse of this information by the individual, the State, industry or third parties.

Taking multifactorial genes one step further will be the DNA characterisation of genes involved in the individual's fundamental make-up including physical features, behavioural and other components such as intelligence. This knowledge is presently a long way in the future. There is also an 'informal agreement' in the scientific community that normal traits should not be the subject of genetic diagnosis. It would be interesting to see if this decision holds once the genetic components for these characteristics start to be defined. Even if scientific curiosity can be kept in check, the pressure from the public to measure these normal traits will be immense. A type of precedent, the IQ test, already exists.

The education of both health professionals and the public in the many far-reaching aspects of molecular medicine is an important priority for the future. Educational processes, particularly at school and tertiary levels, are required to ensure that the full medical, ethical and social implications of new developments relevant to genetic engineering can be appreciated by the majority and debate is not distorted by vocal minority groups. Informed community input involving ordinary human wisdom is required to play a role in the direction or application of research and clinical aspects of molecular medicine.

Gene therapy

The issue of germline versus somatic cell therapy has already been raised in Chapter 7. At present, the former is proscribed. This reflects the inevitable consequence of germline manipulation, i.e. any changes which result will be transmitted through the germ cells to offspring in subsequent generations. On the other hand, somatic cell gene therapy is considered similar to other accepted medical procedures, e.g. manipulations required for bone marrow or organ transplantation. Ultimately, it is the *risk:benefit ratio* which determines whether somatic cell therapy is medically and ethically acceptable in individual cases. Thus, a potentially life-threatening disorder for which there is no effective treatment or where the available treatment is associated with significant complications, e.g. bone marrow transplantation, is the first prerequisite. Once the somatic cell therapy approach can be justified as technically and therapeutically acceptable, it is allowed to proceed after protocols are reviewed by the appropriate monitoring bodies and a follow-up process is set into place. The various steps mentioned above have already taken place in the example of adenosine deaminase deficiency. Informed consent, since the affected individual was a 4-year-old child, was provided by her parents (see Ch. 7).

Somatic cell gene therapy, particularly that involving the retroviral vectors, has not been associated with complications to date. However, many protocols have been utilised in situations where the patient's clinical state was terminal and death, as a consequence of the underlying disorder,

soon followed. The long-term effects of gene therapy need careful monitoring, e.g. what is the risk in humans that retroviruses can produce cancer, is there any 'leakiness' from the somatic cells so that the introduced foreign gene can reach the germline and how beneficial is gene therapy? To answer these questions will require effective, long-term surveillance of patients treated in this way. Similarly, the potential, if any, for germline or enhancement therapy may need to be revisited in the near future (see Ch. 10). The many options made available by gene therapy will inevitably produce protocols requiring careful consideration by ethics committees.

RESEARCH

Pursuit of knowledge

The pursuit of knowledge for the sake of knowledge is becoming more difficult in today's goal-driven research, with investigators frequently encouraged to strive for a marketable end-product. While shrinking resources are in part the reason for this type of rationalisation, the trend has the long-term potential to stifle creative or innovative work. Another way for this to occur is through bureaucratic interference. In many communities, governments provide a significant proportion of the research funds. Increasingly, the governments are now demanding a greater input into the types of work being undertaken. Thus, in subtle ways the State is able to influence the direction that research will take. While in some ways this is a reasonable request on the part of government, it has the potential to make research politically driven and can suppress the creative individuals who have made significant contributions in the past but will have difficulty thriving in the increasingly regulated environment.

Similarly, there would be more than one Nobel prize winner who would not be competitive on current research funding allocations which are based on a strict process of peer review. On the positive side, peer review ensures that there is some form of justice when it comes to distribution of grants and ensures that the money has been spent 'wisely'. On the negative side, the type of research proposals submitted are frequently designed with short-term goals in sight so that the next round of peer review has more chance of success.

The molecular medicine era is one of the most exciting in the modern history of medicine. Many developments have occurred at a time when there have been significant shifts in the way in which research is being undertaken. The free-thinking or *hypothesis-driven* research is giving way in some circumstances to *goal-driven* activities which have attracted funding because of perceived commercial benefits. The Human Genome Project illustrates another approach in which a mammoth undertaking will be resolved by technological blitzkrieg. Some purists do not

perceive genome mapping as anything other than data collection because it does not follow the traditional avenues in research. However, there is no doubt that information coming from the technology-driven Human Genome Project will produce very useful and practical information. Times are changing rapidly. The effects that these shifts in strategies or emphasis will have on society through future developments in molecular medicine will take time to assess.

Patents and intellectual property

The medical and scientific community is strongly encouraged to derive benefits from the patenting of important discoveries. Until the molecular era, patents were obtained for therapeutic substances or technologies. The concept of patents has been important because it has given private industry the impetus to invest money in research and development. More recently, genes and their products have become the subject of patents. Thus, DNA diagnostic kits utilising a particular DNA sequence for amplification by PCR will incur royalty payments to cover the use of PCR and perhaps the patented DNA sequence. Initially, the patenting of genes or DNA sequences raised concern that there would be a reduction in dissemination of information throughout the scientific community. To date, this has not occurred to any significant extent. However, a research group at the US National Institutes for Health (NIH) surprised the scientific community by filing patents for over 6000 anonymous human brain-derived DNA sequences in 1991. These 'genes' were isolated from a brain cDNA library and their uniqueness demonstrated by sequencing a segment of the cDNA and showing on DNA database searches that the sequences were not present in the databases. Thus, they represented unique DNA segments (called ESTs – expressed sequence tags) which, since they came from a cDNA library, were likely to be segments of genes with, as yet, unknown function.

The controversy arising from the patenting of anonymous cDNA clones reflects the philosophy behind a patent, i.e. a novel idea or invention which has some utility. Since the above cDNA clones have no known function their utility is difficult to assess. The opinion held by many scientists and governments is that genes, humans and components of humans are not 'inventions' and so are inappropriate for patents. Groups on both sides of the Atlantic were drawn into the NIH controversy which was eventually defused when the patent applications were withdrawn in 1994. However, commercial biotechnology companies continue to pursue this option to patent anonymous DNA segments in the anticipation that eventually they will prove to be significant. The validity of this argument will eventually be tested in the courts of law or patent offices.

Late in 1994, the NIH became involved in another dispute when a patent for the *BRCA1* gene was filed by the University of Utah and Myriad Genetics Inc., a private biotechnology company. The NIH considered that its contribution to the isolation of the breast cancer gene was sufficient to require that it be included in the patent. The NIH filed a competing application. After a relatively short negotiation period, it was agreed that the patent should include the three key groups. It would be interesting to speculate whether this amicable solution would have occurred if the third party had been an investigator in a small institution rather than the NIH with its considerable resources and influence. Some have also argued that individuals from the very large pedigrees which were essential for the isolation of the *BRCA1* gene should not be excluded from any financial gains. The two examples highlight potential ethical issues in scientific practice directly related to DNA technology. This will only be the beginning if matters concerning patents, in particular what can and cannot be patented and how knowledge will be pursued and shared, are not resolved.

GOVERNMENT

Allocation of resources

The rapid advances in medical knowledge are starting to produce ethical dilemmas even in the most affluent of societies. Health and welfare policies in communities are highly variable ranging, from complete coverage of each individual by the State to a 'user-pay' system. Advocates for the different systems consider theirs is the best and perceived deficiencies are resolved by 'fine tuning'. Hence, many of the systems have developed chronic problems centred on the conflicting demands yet limited resources and the public's perception of what rights are intrinsic to life within that community.

Information about the human genome, the expanding DNA testing opportunities and alternative therapeutic regimens through gene therapy will put further pressure on the health system. The exponential growth in knowledge of the single-gene disorders has been impressive but is still in its infancy for the more complex polygenic and multifactorial conditions. The issues of what are priorities will need to be addressed. Since many of the present-day systems are already stretched in terms of resources it will be interesting to see what priority molecular medicine is afforded by governments. In some cases, the pendulum might swing too far the other way and urgent current problems, e.g. the aged, the intellectually disabled, might be further disadvantaged by the promises that molecular medicine will provide a panacea for all problems *and* function as a reservoir for attracting money through its biotechnology potential. Only informed debate will ensure that the allocation of resources for the many future priorities is appropriate and fair.

In the environment of conflicting demands, health planners are demanding more accountability and outcomes based on value for money. In this context the moral principle of justice applies to ensure that access is equitable. On the other hand, to achieve value for money and efficiency, it is essential that the other principles of autonomy, confidentiality and privacy are not discarded or manipulated because they add an extra cost.

Consumer pressure and lobbying will also play an important role in the future progress of molecular medicine. Already there exists a strong lobby group which has well-established negative views about gene technology. In some ways this has been beneficial to the development of molecular medicine because it has ensured that lobby groups from the other extreme have not been able to coerce politicians into promoting unsavoury aspects of gene technology. However, it is essential that the public continues to be educated through exposure to information which is factual and understandable. The more sensational media headlines, e.g. *'Infidelity – it may be in your genes'* have been less helpful in this respect.

Embryo research

The ethical, moral and social issues concerning termination of pregnancy have, with recent technological developments, become relevant to manipulation or research into the embryo or pre-embryo. The pre-embryo is defined as the stage of development from fertilisation until the product of conception becomes implanted in the uterus (~14 days after fertilisation). In this circumstance, a key issue is what constitutes an individual and when during development does an individual become a distinct entity? There are two commonly held views in this respect. An individual becomes a discrete entity at the time of fertilisation. In this case, embryo research would be forbidden since it would have the potential to interfere with the individual's right to life. Another opinion is that an individual cannot exist until about 14 days after conception by which time the primitive streak is being formed. Prior to that time it is possible for the embryo to split and so produce identical twins, i.e. it is not a distinct entity. Based on the above and a number of other arguments, research involving the embryo is allowed in a number of countries until 14 days post-conception. There are additional safeguards imposed, e.g. research is only permitted in the case of spare embryos obtained at IVF which would otherwise be destroyed and there is a prohibition on transfer of embryos to the uterus of any species once they have been genetically manipulated.

Monitoring committees

The formal monitoring processes involving gene therapy are illustrated by the USA's Federal Drug Administration (FDA) and the RAC (Recombinant DNA Advisory Committee). The FDA's role is in the development of safe and effective biological products from their initial investigational phase to commercial production. The RAC is a committee of the NIH and considers gene therapy proposals which originate from federally funded projects or institutes. Both the FDA and the RAC are critical for the effective implementation and evaluation of gene therapy in North America and, indirectly, in other countries. The FDA has a primary role in the therapeutic product while the RAC considers the scientific value of the technology, including potential ethical issues. Public scrutiny of gene therapy proposals to the RAC is possible.

Two broad groups will become involved in gene therapy and hence require approval from the FDA and the RAC (and equivalent bodies in other countries). The first will be the *biotechnology company* whose ultimate survival will depend on a marketable product. The second is the *academic institution* which is also encouraged to think 'commercially' but at the same time has more scope to deal with fringe issues or look at lesser commercial priorities, e.g. a gene therapy cure for haemophilia B is unlikely to come from the private sector since this form of haemophilia is much less common than haemophilia A and so demand will not be sufficient to allow recovery of development costs.

On the other hand, to complete the full FDA approval process is costly and unlikely to be met by many academic institutions. In this circumstance, three scenarios are possible: (1) the academic institution must find a commercial partner, (2) the work will not be possible, or (3) the product used is inferior since drastic cost-saving measures will be required. In this circumstance, the partnership option would seem the most appropriate but this will be offset by the differing priorities. In acknowledgement of the poten-

tial conflict, the NIH has provided funding for a limited number of core gene therapy vector facilities. These will enable the academic institutions access to FDA-approved materials with which to undertake research programmes in gene therapy. In this circumstance, government intervention has reduced the potential conflicts which have indirectly developed from a very thorough reviewing process in the field of gene therapy.

At a more local level, a number of committees are usually in place to monitor activities related to genetic engineering. They include the committees which deal with matters involving human and animal experimentation and the committees which are predominantly responsible for monitoring the biosafety issues involving protocols which utilise recombinant DNA. From time to time ad hoc committees are formed to review and examine topical or sensitive matters. Membership of monitoring committees is usually designed to ensure that there is an appropriate mix of professionals, consumers, ethicists and other interested parties. The work done by these committees, which is often voluntary and very time-consuming, has been a key factor in the smooth progress of molecular medicine to date.

FURTHER READING

General

Andrews L B, Fullarton J E, Holtzman N A, Motulsky A G (eds) 1994 Assessing genetic risks: implications for health and social policy. National Academy Press, Washington

Charlesworth M 1993 Bioethics in a liberal society. Cambridge University Press, Cambridge

Richards J R, Bobrow M 1991 Ethical issues in clinical genetics. Journal of the Royal College of Physicians of London 25: 284–288

Walters W A 1991 Human reproduction: current and future ethical issues. In: Walters W A (ed) Clinical obstetrics and gynaecology. Bailliere Tindall, London, vol 5

Clinical practice

Alpers J S, Natowicz M R 1993 Genetic discrimination and the public entities and public accommodations titles of the Americans with Disabilities Act. American Journal of Human Genetics 53: 26–32

Bird T D, Bennett R L 1995 Why do DNA testing? Practical and ethical implications of new neurogenetic tests. Annals of Neurology 38: 141–146

Clinical Genetics Society: report of a working party on the genetic testing of children (Clarke A, chairman). Journal of Medical Genetics 31: 785–797

Collins F S 1996 BRCA1 – lots of mutations, lots of dilemmas. New England Journal of Medicine 334: 186–188

Cystic Fibrosis Genetic Analysis Consortium 1994 Population variation of common cystic fibrosis mutations. Human Mutation 4: 167–177

Fost N 1993 Genetic diagnosis and treatment: ethical considerations. American Journal of Diseases of Children 147: 1190–1195

Harper P S 1995 Direct marketing of cystic fibrosis carrier

screening: commercial push or population need? Journal of Medical Genetics 32: 249–250

Hersch S, Jones R, Koroshetz W, Quaid K 1994 The neurogenetics genie: testing for the Huntington's disease mutation. Neurology 44: 1369–1373

Hudson K L, Rothenberg K H, Andrews L B, Ellis Kahn M J, Collins F S 1995 Genetic discrimination and health insurance: an urgent need to reform. Science 270: 391–393

Huggins M, Bloch M, Kanani S et al 1990 Ethical and legal dilemmas arising during predictive testing for adult-onset disease: the experience of Huntington disease. American Journal of Human Genetics 47: 4–12

McEwan J E, Reilly P R 1994 Stored Guthrie cards as DNA 'banks'. American Journal of Human Genetics 55: 196–200

McEwan J E, Reilly P R 1995 A survey of DNA diagnostic laboratories regarding DNA banking. American Journal of Human Genetics 56: 1477–1486

O'Reilly A 1995 The consumer's perspective. In: R J Trent (ed) Handbook of prenatal diagnosis. Cambridge University Press, Cambridge, p 230–247

Ponder B A J 1994 Setting up and running a familial cancer clinic. In: Ponder B A J (ed) Genetics of malignant disease. British Medical Bulletin 50: 732–745

Rennie J 1994 Grading the gene tests. Scientific American 270: 88–97

Report of the Committee on the Ethics of Gene Therapy (Clothier report) 1992. Her Majesty's Stationery Office, London

Ryan M P, French J, Al-Mahdawi S, Nihoyannopoulos P, Cleland

J G F, Oakley C M 1995 Genetic testing for familial hypertrophic cardiomyopathy in newborn infants: clinicians' perspective. British Medical Journal 310: 856–859

Schiliro G, Romeo M A, Mollica F 1988 Prenatal diagnosis of thalassemia: the viewpoint of patients. Prenatal Diagnosis 8: 231–233

Statement of the American Society of Human Genetics on genetic testing for breast and ovarian cancer predisposition 1994. American Journal of Human Genetics 55: i–iv

Statement on use of DNA testing for presymptomatic identification of cancer risk 1994. Journal of the American Medical Association 271: 785

Research

Pompidou A 1995 Research on the human genome and patentability – the ethical consequences. Journal of Medical Ethics 21: 69–71

Thomas S M, Davies A R W, Birtwistle N J, Crowther S M, Burke J F 1996 Ownership of the human genome. Nature 380: 387–388

Government

Kessler D A, Siegel J P, Noguchi P D, Zoon K C, Feiden K L, Woodcock J 1993 Regulation of somatic-cell therapy and gene therapy by the Food and Drug Administration. New England Journal of Medicine 329: 1169–1173

Modell B, Kuliev A M 1993 A scientific basis for cost-benefit analysis of genetics services. Trends in Genetics 9: 46–52

10

THE FUTURE

HUMAN GENOME PROJECT

It is estimated that the human genome consists of approximately 80 000 genes located on the 23 pairs of chromosomes. To date only about 3000 of these genes have been mapped to specific loci. A list of some genetic disorders for which DNA markers are available and their chromosomal locations is given in Table 10.1. The potential to map and sequence the 3.3×10^9 base pairs which make up the human haploid genome was first discussed in 1986–1987. Benefits from such a project would flow to many areas including clinical medicine, biological research and biotechnology. Following extensive scientific and public consultation, the US Congress launched the Human Genome Project in 1988. The project officially commenced in October 1990 and is due to finish 15 years later in the year 2005 at an estimated cost of 3 billion dollars. It was anticipated that genome work would be conducted in many international laboratories.

The project is described as making available the source book for biomedical science in the 21st century. Coordination and central planning were initially undertaken by the US National Institutes of Health and the US Department of Energy. Today there are many involved bodies, e.g. US National Center for Human Genome Research, HUGO (Human Genome Organisation – an international group which fosters collaboration and information transfer between 'genome laboratories') and various genome groups which have been formed to encourage and coordinate this activity at the national level, e.g. UK HGMP (UK Human Genome Mapping Project).

Components

The Human Genome Project has a number of programmes, which are summarised in Table 10.2. The first involves the construction of comprehensive genetic and physical maps of the human genome. Genetic maps have been described earlier and are obtained by using polymorphic DNA markers in family linkage studies. The most useful polymorphisms in terms of genome mapping are the microsatellites, of which there are now several thousands covering all chromosomes. Details of the microsatellites are being systematically stored in the Genome Database (GDB) sponsored by the Howard Hughes Medical Institute and the Johns Hopkins University (Baltimore). Linkage between a DNA marker and a specific phenotype is often the first step from which the relevant gene can be isolated and characterised.

The distance between markers on a genetic map is defined in terms of a centiMorgan (cM) with 1 cM equal to ~1 Mb (megabase). At present, genetic maps are available for most loci in the human genome. The ultimate aim of the project is to refine this so that the entire genome is covered by DNA markers which are 1 cM apart. Each of the DNA markers generated will need to have an unambiguous identity and for this the concept of sequence tagged sites (STSs) has been proposed. This means that sequencing of DNA markers will be required. Each marker will then be identified by the part of its sequence that is unique to that marker. Thus, data emerging from the work of different groups and the use of different technologies can be easily accessed and compared. The technology to undertake this is available but the amount of work required is considerable.

From a genetic map it will be necessary to construct physical maps so that the distance between the DNA markers can be determined in absolute terms (i.e. base pairs, kb or Mb) and a restriction map can be constructed for each locus. This is a mammoth undertaking since it will be necessary to have entire regions of the genome characterised in terms of overlapping clones. Generating random clones in the form of libraries is a routine procedure. The difficult task is ordering them into large

Table 10.1   An abridged list of genetic disorders for which there are DNA markers available.*

Only ~3000 of the proposed 80 000 human genes have been mapped to specific chromosomes and about half of these have been cloned.

Chromosome	Disease (chromosome arm)
1	Phaeochromocytoma, cutaneous malignant melanoma (p), Charcot Marie Tooth type 1B, familial hypertrophic cardiomyopathy, Gaucher disease (q)
2	Hereditary non-polyposis colon cancer (p), Waardenburg syndrome, Ehlers Danlos syndrome IV (q)
3	Von Hippel–Lindau, Long QT syndrome, hereditary non-polyposis colon cancer (p), glycogen storage disease IV (q)
4	Huntington disease, achondroplasia (p)
5	Familial adenomatous polyposis (q)
6	Spinocerebellar ataxia I, haemochromatosis (p)
7	Cystic fibrosis, Long QT syndrome, obesity (q)
8	Langer–Giedion syndrome (q)
9	Friedreich ataxia, hereditary fructose intolerance, hereditary haemorrhagic telangiectasia, acute hepatic porphyria (q)
10	Multiple endocrine neoplasia 2A & 2B, megacolon, medullary thyroid carcinoma (q)
11	β Thalassaemia, sickle-cell disease, familial hypertrophic cardiomyopathy, Long QT syndrome, Beckwith–Wiedemann syndrome (p), multiple endocrine neoplasia I, Wilms tumour, ataxia telangiectasia, hereditary angioedema (q)
12	von Willebrand disease (p), phenylketonuria (q)
13	Retinoblastoma, Wilson disease (q)
14	Familial hypertrophic cardiomyopathy, α_1-antitrypsin deficiency (q)
15	Prader–Willi syndrome, Angelman syndrome, Marfan syndrome, Bloom syndrome, familial hypertrophic cardiomyopathy, Tay–Sachs disease (q)
16	α Thalassaemia, adult polycystic kidney disease, tuberose sclerosis, familial mediterranean fever, HbH mental retardation syndrome (p)
17	Miller–Dieker syndrome, Li Fraumeni syndrome (p), breast and ovarian cancer, neurofibromatosis 1 (q)
18	Familial amyloid neuropathy (q)
19	Myotonic dystrophy, malignant hyperthermia, familial hemiplegic migraine (q)
20	Creutzfeldt–Jakob disease (p), Fanconi anaemia (q)
21	Early onset familial Alzheimer disease, familial amyotrophic lateral sclerosis 1, Down syndrome locus (q)
22	Neurofibromatosis 2, DiGeorge syndrome (q)
X	Duchenne muscular dystrophy, retinitis pigmentosa, Norrie disease (p), haemophilia A, B, agammaglobulinaemia and severe combined immunodeficiency, fragile X, Menke disease (q)
Y	XY sex reversal (SRY gene) (p)

*p = short arm and q = long arm of the chromosome. Additional disease loci are not excluded in some of the above examples. A more comprehensive list is found in McKusick's *Mendelian Inheritance in Man*, or an annual summary published in the *Journal of Medical Genetics*.

overlapping sets called 'contigs'. At present, contigs stretch over a relatively small distance (perhaps a few hundred kilobases). In the first 5 years of the project it was proposed to increase the contig distance to 2 Mb. DNA markers identified by sequence tagged sites will be located approximately 100 kb apart. Subsequently, the aim of the project is to generate cosmid or yeast artificial chromosome (YAC) contigs which span the entire genome.

Another goal will be to improve the current methods for DNA sequencing and encourage the development of novel approaches. Defined regions of the genome will be sequenced but the actual sequencing of the entire genome

Table 10.2   Components of the Human Genome Project and the goals for the first 5 years (1991–1995)[a].
The project is planned to cover the period 1990–2005 and was originally grouped into three 5-year periods.

Programme	Goals for the first 5 years[a]
1. Mapping and sequencing the human genome	a) Construct genetic maps with DNA markers at 2–5 cM distances b) Construct physical maps with 100 kb spaced STSs c) Sequence small 1–2 Mb regions of interest in the genome d) Develop existing technology and encourage innovative approaches to sequencing
2. Mapping and sequencing genomes of model organisms	a) Construct genetic, physical maps of the mouse genome. Sequence key regions of the mouse genome b) Complete sequencing of the genome of *E. coli, S. cerevisiae* and *C. elegans*; continue sequencing and mapping of *Drosophila*
3. Gene identification[b]	Identify genes for humans and model organisms and place these on the maps
4. Informatics: data collection and analysis	a) Develop software and database designs to support large scale collection, storage, distribution and allow ready access to databases b) Develop new methods for interpretation and analysis of genome maps and DNA sequences
5. Ethical, legal and social considerations	Research issues of privacy, confidentiality, stigmatisation, discrimination, equity of information, professional and public education
6. Research training	Create training posts particularly in interdisciplinary sciences related to genome research, provide training courses
7. Technology transfer	Develop closer liaison with industry and the private sector. Facilitate technology transfer into and out of centres of genome research
8. Outreach	Encourage a flexible distribution system so that results and developments are quickly transferred to the community or potential users

[a]Because of significant progress in the first 3 years, the goals of the project were revised in 1993 to cover the period 1990 to 1998 and are listed here. [b]Gene identification was not explicit in the initial 5-year plan but now has been clearly identified as an important aim.

will be deferred until technology enables it to become more efficient and less costly (e.g. one goal would aim to reduce the cost for each base pair sequenced to less than US$0.5). It is highly likely that when the time comes for the entire genome to be sequenced the methodology used will be of a form not yet described! The development of high speed robotic work stations will be an integral component of this phase.

A separate programme focuses on mapping and sequencing of model organisms. This has two ultimate aims. First, model organisms will provide less complex genomes to facilitate technology development. Second, the models will enable comparative studies to be made between the human genome and those from non-human sources. Information generated from these comparisons will be essential for an understanding of the evolutionary processes, how genes are regulated and the aetiology of some genetic disorders (as an example, see the discussion on Imprinting in Ch. 3). The model organisms chosen include:

- Bacteria: *E. coli, Bacillus subtilis,* two species of mycobacteria
- Yeast: *Saccharomyces cerevisiae*
- Simple plant: *Arabidopsis thaliana*
- Nematode: *Caenorhabditis elegans*

- The fruit fly: *Drosophila melanogaster*
- Mammal: the mouse.

Another programme involves informatics and will address methods to store data generated by the project, i.e. genome maps and DNA sequences. A considerable amount of software development will be required in terms of data storage, data accessibility and data analyses.

The Human Genome Project started despite considerable criticism and controversy. This has ranged from economic considerations to those who do not approve of 'shot-gun' approaches to research. Therefore, an additional programme was added to consider the ethical, legal and social (abbreviated to 'ELSI') implications of the project. 3% of the total budget has been set aside to research issues such as privacy and confidentiality (e.g. who will have access to an individual's genetic makeup), stigmatisation or discrimination (e.g. what insurance companies or employers might do with the information generated from the project), what adverse effects might result from knowledge of the genome's sequence and so the potential to predict health or disease in an individual. The importance of educating the public and professionals about the project and its implications will also be considered.

A separate programme will set out to train, at the pre- and post-doctoral levels, individuals who will have a good knowledge of genome research methodologies. Skills resulting from this will not only be in the area of molecular biology but will include computer science, physics, chemistry, engineering and mathematics. Interdisciplinary approaches to training and skill acquisition will be encouraged. Training and short courses in defined areas of the project will be supported. This programme will, in the long term, be beneficial to industry and the private sector.

Other programmes will concentrate on technology transfer and outreach. These components of the project alone are considered to justify the money being spent since they will lead to technological developments which will have widespread use in research, industry and the practice of medicine. Development of novel methodologies will be strongly encouraged. Technology transfer to medicine, industry, the private sector and the community at large will be rapid so that discoveries from the project can be developed to their full potential. Direct funding of private companies will be made available to expedite needed developments.

Consequences

If all programmes and goals are successful, the Human Genome Project will have far-reaching effects. Information obtained about single-gene disorders will generate both good and potentially adverse effects. For example, a lot will become known about genetic diseases but the gap between what is known and what can be done in the therapeutic sense will widen. The cystic fibrosis and Huntington disease examples have illustrated some of the advantages and disadvantages to emerge from DNA technology. However, the consequences of single-gene disorders will appear to be modest when information from the project starts to flow over into the multifactorial and somatic cell diseases. Information about human chromosomes will also be obtained as part of the project. Since a large proportion of morbidity in utero and in the neonatal period can be attributed to chromosomal abnormalities,

further knowledge on how chromosomes divide and replicate will be extremely valuable. Funding required for the project will be enormous. Priorities in terms of resource allocations will be many and could lead to ethical, social and political dilemmas.

Completion of the last segment of DNA to be sequenced will mark the end of the project but the beginning of much hard work directed to 'decoding'. At this stage it will be necessary to determine the functional significance of many of the genes or DNA sequences, e.g. regulatory elements. This will be particularly challenging in complex genetic traits for which there are polygenic effects.

Progress

There have been some interesting developments since the project started in 1990. An impressive and perhaps unexpected early success has come from the mapping and sequencing of model organisms. It is expected that a number of these will be completely sequenced by the mid or late 1990s. Additional organisms, e.g. *Helicobacter pylori* and *Haemophilus influenzae*, that were not part of the original project, have now been completely sequenced. Data from the model organisms have identified new genes and provided information of direct relevance to the human genome.

The important goal of reducing the cost of DNA sequencing to about US$0.5 per base is likely to be realised in the late 1990s, although the sequencing throughput has been disappointing in relation to the human genome. No novel methods have been described. Unless this changes, one estimate is that at the present rate it could take 5000 machine years (e.g. 10 centres each with 100 machines working full time for 5 years) to complete the human genome sequence. Nevertheless, the wealth of information coming from the model organisms has strengthened the resolve to sequence the entire 3.3×10^9 bases in the human genome. Some ways in which this could be achieved with present methodology have been proposed (see Gibbs 1995).

DEVELOPMENTAL BIOLOGY

One of the exciting advances in molecular medicine relates to the way in which the body develops and how genes are involved in these changes. Following on from an understanding of the normal molecular processes has come knowledge of malformations and their underlying mechanisms. The success of this work has depended on basic biological research utilising animal models such as the fruit fly *Drosophila melanogaster*, the mouse and more recently the zebrafish. The interspecies conservation of the important developmental genes has meant that the equivalent ones in humans were able to be identified and charac-

terised. The signficance of this work was acknowledged with the 1995 Nobel Prize for Physiology or Medicine (see Ch. 1). In both vertebrates and invertebrates, two families of genes and a number of others are important in development.

Homeobox (hox) genes

Mutations in the body form of *Drosophila* which caused a part of the body to be replaced by a structure normally found elsewhere were shown in the early 1980s to involve a number of genes called homeotic genes. All vertebrates, including humans, have four homeotic gene complexes located on different chromosomes. In *Drosophila*, it has been possible to show that the physical arrangement of the homeotic genes in these complexes is identical to the order in which the genes are expressed along the anteroposterior axis of the embryo during development, i.e. the more 5′ a gene the more posterior is it expressed in the developing body. In mammals, the homeotic genes are known as *hox* genes (*hox* derives from *homeobox*) and are also believed to specify cell identity along the anteroposterior axis of the embryo.

A conserved DNA sequence is found in all homeotic genes. It is called the homeobox and is about 180 bp in size. The 60 amino acid sequence encoded by the homeobox has DNA-binding properties and so the homeoproteins are transcription factors. Thus, this class of genes can regulate the expression of many other genes. Comparative DNA analyses have shown that homeotic genes derived from common ancestral genes and their subsequent divergence reflects the morphological complexity of the organism in which they are found. For example, insects have a single cluster of the homeotic genes, in *Drosophila* there are two and in vertebrates the number is four. In humans, the homeobox clusters (*HOX 1–HOX 4*) are found on chromosomes 7p, 17q, 12q and 2q, respectively. An amazing observation is that in all species the genes remain aligned in the same relative order as in *Drosophila*. The conservation between the genes is so high that vertebrate genes can replace their invertebrate counterparts in transgenic *Drosophila* embryos.

To understand the role of the homeoboxes in development, *Drosophila* or animals with mutations, spontaneous or created by recombinant DNA means, are being studied. The latter approach has provided some important evidence to suggest that *hox* is the mammalian equivalent of the *HOM* genes since structural deformities (of the head and neck) result (see also animal models, p. 208, which illustrates what happens when the mouse *hox-1.5* gene is inactivated). Despite the identification of these highly conserved genes, the search for natural mutants has been less fruitful. This may reflect the fact that abnormalities in *hox* are only expressed as an abnormal phenotype when both alleles are inactivated (in contrast see the dominant effect from *PAX* genes). Thus, the exact role that the *HOX* genes play in normal human development remains obscure and will require further characterisation at the molecular level.

Paired-box (pax) genes

More recently, another conserved DNA sequence has been characterised in mice and other species as divergent as worms and humans. This is called the paired box. The relevant genes are known as *pax* (paired box). In the human, there are nine of these genes dispersed over many chromosomes. An 128 amino acid sequence in *pax* is conserved in mammals and *Drosophila*. Like the homeobox, this sequence has the properties of a DNA transcription factor. Some of the *pax* genes also contain homeobox sequences. A number of natural mutants involving *pax* produce clinical problems in animals, including humans.

An interesting developmental disorder in the mouse is called *Splotch*. This affects neural crest-derived components leading to the development of spina bifida and dysmorphic features, including white spots on the body. It has now been shown that *Splotch* is the result of a 32 base pair deletion in the mouse's *pax-3* locus. The corresponding genetic locus in the human is *PAX3*. Mutations affecting this gene have been identified in the human autosomal dominant disorder called Waardenburg syndrome. In this condition there is deafness and pigmentary disturbance. The two tissues involved are both of neural crest origin. Thus, a genetic defect has highlighted the role that a conserved gene plays in development of the neural crest in humans. The *Splotch* animal model also becomes available to study normal neurological development including the abnormality which leads to spina bifida.

Two other developmental defects in humans have been associated with *PAX* genes (Table 10.3). This contrasts with the *hox* genes (see p. 208) and may be explained by the dominant nature of the *pax* mutations which express if one of the two alleles is mutated. Another interesting observation about *pax* is that abnormal function, which has occurred following a chromosomal translocation, can lead to tumour formation. A number of the *PAX* genes are implicated in the development of cancer. The *PAX* 2, 5 and 8 group is particularly interesting in this respect. Considerably more information on *pax* is sure to come in the future.

Table 10.3 PAX genes and associated disorders in humans (from Stuart & Gruss 1995).
A family of nine genes which can affect development and induce tumour formation. Structural similarities enable the genes to be grouped into four classes.

Gene	Location	Loss of function	Gain of function
PAX1	20p11	–	–
PAX9	14q12	–	–
PAX3[*]	2q35	Waardenburg syndrome	Rhabdomyosarcoma
PAX7[*]	1p36	–	Rhabdomyosarcoma
PAX2	10q25	Kidney, retinal abnormalities	Wilms tumour
PAX5	9p13	–	Glioblastoma, medulloblastoma
PAX8	2p12-q14	–	Wilms tumour
PAX4[*]	7q22-qter	–	–
PAX6[*]	11p13	Aniridia	–

*These genes contain both a paired box and a homeobox.

Other genes

Many of the DNA-binding proteins utilise zinc to stabilise their functional domains. In these proteins there is binding of two cysteine and two histidine molecules through a zinc ion. The segment formed is finger-like, hence the name zinc finger genes. Apart from being DNA transcription factors and so capable of influencing the expression of a number of genes, the role played by these genes in development remains to be fully determined. One autosomal dominant genetic syndrome known as Grieg cephalopolysyndactyly (cranial and hand abnormalities) is caused by a defect involving the zinc finger gene *GLI3*.

Gap junctions are considered important for normal development of the heart. One gap junction gene known as *connexin43* has been shown in a transgenic knock-out study and in children with complex heart malformations to be involved in aetiology. It is proposed that in this circumstance there is failure in cell-to-cell communication affecting the developing embryo and so leading to the complex developmental cardiac malformations which result. Clearly, there will be many more genes found which play a role in the normal growth process. In the area of developmental biology, molecular analysis of the function of *hox*, *pax* and other related genes will allow further understanding of how human development is programmed.

ANIMAL MODELS

Unlike humans, animals can be manipulated experimentally and genetically. Therefore, animal models for human diseases have been extensively developed. These models arise spontaneously or can be induced following exposure to chemicals or other substances. They provide a means by which the natural history of a disorder can be followed progressively over a number of generations. Therapeutic options can also be tested prior to human trials. As discussed previously in Chapter 3 (genetics of hypertension) some of the more complex multifactorial diseases in humans may best be resolved by research using a genetic breeding approach coupled with DNA analysis.

Traditional

Over the years, inbred strains, particularly the laboratory mouse, have been the mainstays for studies involving a wide range of human disorders. Inbred mice are produced by repeated sister–brother matings over about 20 generations. The end result is a syngeneic mouse which will be identical (e.g. homozygous) at every genetic locus and identical to other mice of the same strain. Another type of inbred mouse is the congenic one. Although derived from one strain, selective breeding allows this animal to have genetic material from a second strain at a single locus.

The naturally derived animal models provide considerable information but they have limitations, e.g. the mutation may not be representative of that which is found in the human disorder. More importantly, there are many diseases for which a suitable animal model does not exist. Recombinant DNA approaches provide a means by which new animal models can be created or existing ones manipulated specifically to test the function of genes.

Transgenic

Transgenic mice illustrate a strategy to develop animal models for human disorders. They are produced by microinjection of DNA into the pronucleus of a fertilised oocyte and are useful for disorders associated with an abnormal output, e.g. an excess or a mutant form of a protein. Although the gene of interest is not inserted into its correct position in the genome, it still remains possible to add new genes which can function in vivo. Thus, expression of the mutant transgene will produce the clinical phenotype. An extension of this is the transgenic mouse which has been created by gene knock-out. This involves homologous recombination between an introduced mutant gene and the corresponding wild-type gene. The normal gene is thus replaced by one with a known mutation. In this way gene function can be inhibited or the effect of a specific mutation observed.

Embryonic stem cells (ES cells) have been critical to the development of knock-out transgenics. Since ES cells are totipotential, they can be genetically manipulated and then reintroduced into the blastocyte of a developing mouse to produce a chimaera. Foreign DNA which has become integrated into the germline of the chimaera will enable the gene to be transmitted to progeny. Appropriate matings will produce homozygotes containing the transgene (Fig. 2.24). ES cells allow a gene to be targeted to its appropriate locus and so replace its normal wild-type counterpart by homologous recombination (Fig. 7.8). Using this approach, a better understanding of genetic inheritance or disease pathogenesis becomes possible (Fig. 10.1, Boxes 10.1, 10.2).

The utility of knock-out studies to define the function of unknown genes is illustrated by the mouse *hox-1.5* gene which was inactivated by homologous recombination. Homozygous mutants for this defect developed a pheno-

Box 10.1 The cystic fibrosis transgenic mouse

A number of mouse models of cystic fibrosis have been created by targeted disruption of the animal's *cftr* (cystic fibrosis *t*ransmembrane *r*egulator) gene. Mice that are homozygous for this defect demonstrate the same electrophysiological defects as seen in the abnormal human intestinal and respiratory cells, i.e. the affected mice have an absent cyclic AMP-activated chloride secretory response. The cystic fibrosis transgenic mice will enable a prospective assessment of tissue changes in cystic fibrosis as well as providing models to test therapeutic substances more objectively, e.g. the effect on chloride transport could be used as a measure of efficacy. An unanswered question in the cystic fibrosis story is why the mutant gene is so common in many populations. A number of hypotheses have been proposed. One which overlaps with the central role of chloride in cystic fibrosis suggests a survival advantage that heterozygotes would have against the chloride and fluid loss which occurs following infection with cholera. Some evidence for this hypothesis has now come from transgenic mice which show that homozygous mutant and heterozygous animals have acquired resistance to the diarrhoea induced by the cholera toxin. Since the *CFTR* gene is critical to intestinal chloride and fluid secretion, it is plausible that a mutation in the gene provides a selective advantage in an environment exposed to *Vibrio cholerae*. Like the thalassaemia story, this advantage only becomes a liability in the homozygous-affected individual.

Box 10.2 Prion diseases

Prions (proteinaceous infectious particles) were first suggested as potential agents in causing spongiform encephalopathies in the early 1980s. The characterisatic features of these inevitably fatal disorders is a long latency period (30 or more years in humans) leading to progressive decline in cognition and motor function. Naturally-acquired animal models include scrapie and more recently the highly publicised 'mad cow disease' (BSE). Human examples include kuru, Creutzfeldt–Jakob disease and Gerstmann–Straussler–Scheinker disease. The protein responsible for this disorder is known as PrP (prion protein) and exists in two forms. That which is normally present is PrP^c (cellular) and the mutant is PrP^{Sc} (for scrapie). A change in conformation from normal to mutant protein is the underlying basis for prion disease. Thus, it represents an 'infectious' agent which does not contain nucleic acid. Evidence for the PrP connection came with the cloning and characterisation of the relevant gene (*PRNP*) and the finding that patients with Creutzfeldt–Jakob or Gerstmann–Straussler–Scheinker disease had mutations in this gene. Definitive proof and interesting findings followed the creation of several transgenic mice which were able to overproduce or underproduce the PrP protein. These studies confirmed the infectivity of the PrP^{Sc}, showed that PrP is not essential for normal development (a future approach to treatment could involve interfering with the mutated *PRNP* gene) and suggested that there may be other prion-induced diseases since some PrP^c overproducing transgenic mice developed a disorder affecting muscles and the peripheral nervous system.

type similar to the human DiGeorge syndrome, i.e. absent parathyroid and thyroid glands with defects of the heart, major blood vessels and cervical cartilage. In this manner, a systematic study of the various homeobox and related DNA sequences by gene knock-out using homologous recombination will enable their role in development to be determined. A similar approach will be useful to determine the significance of candidate genes in the multifactorial disorders (Table 10.4).

PREDICTIONS

Technology

An important expectation of the Human Genome Project is that, like the US space programme, the technological developments which occur as the project progresses will have far-reaching effects both in and outside of medicine. Automation is one development which has progressed, particularly in relation to DNA sequencing and microsatellite analysis. The ultimate aim would be to provide cheaper and simpler techniques which could be used in a greater number of laboratories. In the long term, it would not be surprising to find DNA sequencing rather than specific mutation analysis as the method of choice for the diagnosis of genetic defects.

In the meantime, screening for point mutations should become less laborious as more effective strategies are described. A promising one being evaluated is an enzyme belonging to the class of 'resolvases'. This enzyme has the potential to identify single base mismatches in relatively large segments of DNA which would be a great advantage over SSCP, a technique which is only effective with small fragments of DNA. Important future developments will also include the increasing availability of DNA diagnostic kits which have non-radiolabelled DNA or RNA probes. These will be user-friendly and so applicable to many clinical laboratories. The development of newer methods to amplify DNA or the removal of the contamination problem

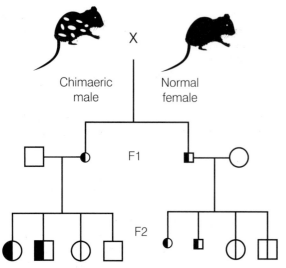

Fig. 10.1 Replacement of a wild-type gene by a mutant one in transgenic mice.
Homologous recombination allows an inserted gene to be targeted to its correct position in the genome. Using ES cells the gene can be integrated into the germline for future transmission to progeny. A male chimaeric mouse with one mutant insulin growth factor 2 gene (*igf2*) is made by homologous recombination using ES cells. This mouse is crossed with a normal female. Progeny from the union are normal in size or small (indicated by large and small symbols). The small progeny result when the only normal *igf2* gene has been maternally transmitted; the normal sized progeny result when offspring have two normal copies of the *igf2* gene or the only normal gene present is paternal in origin. This can be explained by imprinting, i.e. to be active the *igf2* gene requires paternal transmission (from DeChiara et al 1991).

with the present techniques will see greater use of DNA amplification across many areas including clinical practice, research and industry.

It is not inconceivable that DNA amplification will become so user-friendly that testing at the bedside or the local family physician's office will become commonplace.

Somatic cell genetics

Somatic cell genetics now joins the single-gene, polygenic, multifactorial and chromosomal disorders as a distinct group of genetic disorders. The role played by changes in the genetic material of somatic cells is well exemplified by cancer, e.g. the familial adenomatous polyposis example described in Chapter 6 in which a series of mutations affecting somatic cells enables progression from a localised tumour to a malignant, metastasising cancer.

The potential for mutation at the somatic cell level has recently been highlighted by a novel molecular defect which produces anticipation (increasing severity or earlier age at onset of a genetic disease in successive generations). In the case of Huntington disease, the fragile X syndrome and myotonic dystrophy (discussed in Ch. 3) it has been shown that anticipation is associated with a heritable and

Table 10.4 Some animal models of human disorders.
Both spontaneously arising and genetically engineered animal models are important for an understanding of some human diseases.

Human disease model	How produced and information which can be obtained
Cystic fibrosis	Homozygous ΔF508 and examples of other *cftr* mutations (non-human genes are written in lower case) have been produced by gene knock-out studies.
Ulcerative colitis	Knock-out of mouse G protein α_{12} (G – guanine nucleotide binding) produces a model for ulcerative colitis and provides further information on the function of G proteins.
Alzheimer disease	A mutated form of the *APP* gene implicated in Alzheimer disease has been used to make transgenic mice. This produces an excess of the β amyloid protein and features of Alzheimer disease. The mouse model confirms the association between *APP*, β amyloid and Alzheimer disease.
Hypertension	A transgenic rat with increased expression of a mouse renin gene develops hypertension. The significance and mechanisms involved are being investigated.
Unstable triplet repeats	A transgenic mouse expressing the *SCA1* gene with an associated normal or expanded $(CAG)_n$ develops, in the latter, a clinical disorder similar to spinocerebellar ataxia type 1.
Down syndrome	Irradiated mice were screened for chromosomal abnormalities. A translocation of a segment of mouse chromosome 16 (homologue to human chromosome 21q) was identified. Mice were bred to produce trisomy. The animal model shows some features of Down syndrome.
α Thalassaemia	Mice deleted for the two α globin genes were created by gene knock-out. Homozygous-affected mimic Hb Bart's hydrops fetalis and are being used to study potential gene therapy strategies for treatment.
Skin	Transgenic mice producing a mutant keratin gene develop features of the human skin disorder epidermolysis bullosa simplex. The potential contribution of oncogenes such as *ras* and *fos* in the development of skin tumours is being investigated by making transgenics with mutant genes.
Congenital heart diseases	Gene knock-out of a *connexin* gene in mice has produced a model for pulmonary stenosis.

unstable DNA nucleotide triplet which can increase in size from one generation to another. The triplet repeat demonstrates both somatic instability as well as instability following inheritance through one of the parental germ cells. This novel mechanism for instability remains to be fully characterised.

Somatic cell changes associated with ageing, autoimmune disease and congenital malformations will become fruitful areas of research.

Multifactorial disorders

The important and common medical problems in humans such as heart disease, dementia and diabetes have both genetic and environmental components in their aetiologies. Knowledge of the former will lead to improved means to detect those who are at risk. The environmental contributions to the multifactorial disorders will be understood better if their effects on DNA can be identified and characterised. In practical terms this will enable more effective and focused public health prevention programmes. It should also become possible to target therapeutic options to the specific interactions which are occurring between genes and environment.

At present, a major barrier to progress at the molecular level is the use of strategies such as linkage analysis which are powerful tools for the single-gene disorders but less effective when the genetic contributions to disease are subtle or polygenic. More sophisticated computer analysis programs have been suggested as the way to progress in these complex traits. The blunt force of data accumulation which will become possible as the Human Genome Project advances may, in the longer term, provide the answers. Alternatively, novel approaches will be required to investigate more effectively these complex but important public health issues.

Information about normal physiological processes, e.g. memory, will be obtained as by-products of research into the complex polygenic and multifactorial traits.

Ethical, social and legal issues

Mention of the ELSI components of the Human Genome Project has already been made. However, it is useful at this stage to predict what particular ethical, social and legal issues will develop from an ever-increasing knowledge base of DNA.

An important ethical consideration is privacy. This will have already involved a limited number of individuals and their families who have undergone DNA testing for the diagnosis of genetic disorders. It will become a more widespread concern as the gene database expands and affects a larger number in the community. Similarly, the utilisation of genetic information by government or industry will need a delicate balancing act to ensure that good is not outweighed by the harm that indiscriminate use of personal information can have for an individual and his/her family.

A second and increasingly important ethical consideration in the future is resource allocation. The topic of health care and its cost is high on most political agendas. The additional pressures which will be exerted by molecular medicine as 'new discoveries' occur will require careful

and honest consideration so that the potential versus actual good does not remove health dollars from what are otherwise effective but less high profile medical practices.

A professional issue relates to scientific practice and the role of patents and intellectual property. This topic has been actively discussed and the pros and cons defined. In this aspect of molecular medicine it will be important that the economic advantages do not outweigh the logical and sensible mix which should be possible between academic and private sector interests.

On the legal front, the developments in DNA fingerprinting since 1985 have been rapid. No doubt better strategies for DNA amplification and fingerprinting will become available. For example, the internal variability being measured by the digital DNA typing patterns may be more useful than comparisons between individuals within populations. The issues of privacy are no longer theoretical discussion points since a number of communities have passed laws which make it compulsory for convicted criminals to have DNA fingerprints taken. The safeguards to optimise use of DNA fingerprinting but, at the same time, not compromise the civil liberties of the individual will need to be carefully monitored.

A challenge in terms of the social implications of molecular medicine will be the ability to educate the public. This is essential to ensure that future political decisions are based on wide community support and not small but vocal lobby groups. In this endeavour, a key target group will be the school population. Already some changes to curricula are being made to ensure that an appropriate knowledge base is developed to 'demystify' DNA.

In all matters related to education, a more proactive role by those with a broader understanding of molecular medicine will be required.

Therapeutics

The present ways to produce recombinant DNA drugs will appear to be primitive in comparison to future strategies. Knowledge about active sites, methods to improve bioavailability and delay or evade resistance will enable multifunctional drugs to be designed and then produced in large and affordable quantities by recombinant DNA technology.

In terms of gene therapy, the somatic cell and cell marking aims depicted in Figure 10.2 are now being assessed in clinical trials. The one clinical success with somatic cell therapy has been the treatment of adenosine deaminase deficiency. Future successes are awaited. Some controversy has emerged suggesting that clinical trials may have moved too fast before the basic in vitro and in vivo laboratory studies had been adequately examined. More effective gene transfer protocols and vectors are required before gene therapy will realise its expected potential. In the meantime, the clinician and scientist will have to avoid the pressures to rush into a premature clinical trial. Success with cell marking has been more notable and this technique has allowed the in vivo surveillance of cells to determine the role they play in a number of physiological and pathological processes.

The ability to target a gene to its correct locus will no doubt reopen the discussion on germline gene therapy in humans. The role of enhancement gene therapy, e.g. changing normal traits or characteristics, is not a consideration at present. However, there is a fine line between giving growth hormone to make an individual taller for cosmetic reasons and, at some future time, administering an exogenous gene for the same purpose. Those who take the 'slippery slope' view would suggest that the move from replacing growth hormone genes to genes for other normal traits might not be that large a step. There are interesting challenges ahead! Only the health professional who is well educated in molecular medicine will make a positive contribution to this debate.

Exciting prospects for the future include the use of cord blood cells as sources of stem cells for gene transfer and the potential for fetal gene therapy in utero. DNA-derived vaccines will be improved by genetically engineering vectors that carry a number of antigens. These will provide greater opportunities for cost-effective vaccination programmes to be affordable by more of the world's growing population. Therapeutics will be an important area to which molecular medicine will make significant contributions.

The therapeutic options will only be limited by the imagination of the scientist.

Molecular medicine team

One of the significant changes in internal medicine in the past few decades has been the movement to specialisation. As part of this trend the team approach to treatment of human disorders has developed. For example, the improving survival associated with cystic fibrosis has been attributed to the comprehensive management which becomes possible in special cystic fibrosis clinics. In this environment, patients have improved access to medical and ancillary services, e.g. physiotherapy, genetic counselling.

The team approach is particularly relevant to molecular medicine since the rapid and complex developments in the field are occurring at a rate which makes it very difficult for any single health professional to maintain a comprehensive knowledge base. The expectations from the community, the power of advocacy and lobby groups, the access to information (newspapers, television and increasingly the Internet) has meant that not infrequently the individual health professional has less control over what he/she would consider to be optimal care, e.g. an individual wants to know his/her cystic fibrosis carrier status since there is a vague family history of this disorder. In privately orientated health care programmes, the individual who has the money can find out this information, but is it necessary or justified? The dilemma involving the conflict between basic research versus clinical practice in gene therapy has been mentioned earlier.

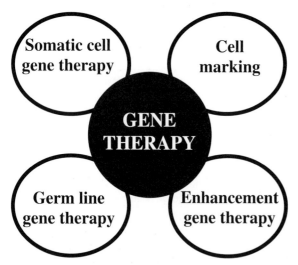

Fig. 10.2   Potential goals in gene therapy.
The two non-controversial arms of gene therapy are somatic cell gene therapy and marking studies. Although experimental, these forms of gene therapy are considered comparable to other new treatment regimens. More controversial are germline gene therapy and enhancement therapy which are presently proscribed in medical and scientific communities.

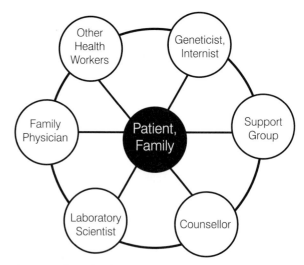

Fig. 10.3   The molecular medicine team.
A suggested team which could be assembled to deal with a molecular medicine-based clinical problem. The relative contributions from the components which make up the outer circle would depend on the underlying disorder. The primary care (family) physician and the community support groups play a key role as the health professionals and lay-persons who, in the longer term, will have the closest contact with patients and their families.

In the complex situations which can arise from the many developments in molecular medicine, it is essential that health professionals continue to be educated on an individual basis but perhaps, more effectively, as part of a team. This is well illustrated by recommendations which have been made in respect to clinics for familial cancer.

The implementation of the team approach, an example of which is given in Figure 10.3, will be critical to the future development of molecular medicine.

FURTHER READING

General
Mueller R E, Young I D 1995 Emery's elements of medical genetics, 9th edn. Churchill Livingstone, Edinburgh

Human Genome Project
Berks M and the *C. elegans* genome mapping sequencing consortium 1995 The *C. elegans* genome sequencing project. Genome Research 5: 99–104

Boguski M S 1995 Hunting for genes in computer data bases. New England Journal of Medicine 333: 645–647

Collins F, Galas D 1993 A new five-year plan for the U.S. Human Genome Project. Science 262: 43–46

Dujon B 1996 The yeast genome project: What did we learn? Trends in Genetics 12: 263–270

Gibbs R A 1995 Pressing ahead with human genome sequencing. Nature Genetics 11: 121–125

Guyer M S, Collins F S 1993 The Human Genome Project and the future of medicine. American Journal of Diseases of Children 147: 1145–1152

Hoffman E P 1994 The evolving genome project: current and future impact. American Journal of Human Genetics 54: 129–136

Juengst E P 1994 Human genome research and the public interest: progress notes from an American science policy experiment. American Journal of Human Genetics 54: 121–128

McKusick V A, Amberger J S 1994 The morbid anatomy of the human genome: chromosomal location of mutations causing disease. Journal of Medical Genetics 31: 265–279

Olson M V 1995 A time to sequence. Science 270: 394–396 (the 20 October 1995 issue of *Science* is a special genome issue)

Developmental biology
Britz-Cunningham S H, Shah M M, Zuppan C W, Fletcher W H 1995 Mutations of the *connexin43* gap junction gene in patients with heart malformations and defects in laterality. New England Journal of Medicine 332: 1323–1329

Epstein C J 1995 The new dysmorphology: Application of insights from basic developmental biology to the understanding of human birth defects. Proceedings of the National Academy of Sciences USA 92: 8566–8573

Gehring W J, Hiromi Y 1986 Homeotic genes and the homeobox. Annual Review of Genetics 20: 147–173

Lalwani A K, Brister R, Fex J et al 1995 Further elucidation of the genomic structure of *PAX3*, and identification of two different point mutations within the *PAX3* homeobox that cause Waardenburg syndrome type I in two families. American Journal of Human Genetics 56: 75–83

Lawrence P A, Morata G 1994 Homeobox genes: their function in Drosophila segmentation and pattern formation. Cell 78: 181–189

McGinnis W, Kuziora M 1994 The molecular architects of body design. Scientific American 270: 58–66

Nusslein-Volhard C 1994 Of flies and fishes. Science 266: 572–574 (the 28 October 1994 issue of *Science* has a special segment on developmental biology)

Read A P 1995 Pax genes – *Paired* feet in three camps. Nature Genetics 9: 333–334

Stuart E T, Gruss P 1995 PAX genes: what's new in developmental biology and cancer? Human Molecular Genetics 4: 1717–1720

Animal models
Burright E N, Clark H B, Servadio A et al 1995 *SCA1* transgenic mice: a model for neurodegeneration caused by an expanded CAG trinucleotide repeat. Cell 82: 937–948

Capecchi M R 1994 Targeted gene replacement. Scientific American 270: 52–59

Chisaka O, Capecchi M R 1991 Regionally restricted developmental defects resulting from targeted disruption of the mouse homeobox gene *hox 1.5*. Nature 350: 473–479

DeChiara T M, Robertson E J, Efstratiadis A 1991 Parental imprinting of the mouse insulin-like growth factor II gene. Cell 64: 849–859

Dzau V J, Gibbons G H, Kobilka B K, Lawn R M, Pratt R E 1995 Genetic models of human vascular disease. Circulation 91: 521–531

Gabriel S E, Brigman K N, Koller B H, Boucher R C, Stutts M J 1994 Cystic fibrosis heterozygote resistance to cholera toxin in the cystic fibrosis mouse model. Science 266: 107–109

Justice M J, Jenkins N A, Copeland N G 1992 Recombinant inbred mouse strains: models for disease study. Trends in Biotechnology 10: 120–126

Prusiner S B 1995 The prion diseases. Scientific American 272: 48–57

Reeves R H, Irving N G, Moran T H et al 1995 A mouse model for Down syndrome exhibits learning and behaviour deficits. Nature Genetics 11: 177–184

Rubin E M, Barsh G S 1996 Biological insights through genomics: mouse to man. Journal of Clinical Investigation 97: 275–280 (a series of articles from January to May 1996 dealing with 'Molecular medicine in genetically-engineered animals')

Spiegel A M 1995 G protein gene knockout hits the gut. Nature Medicine 1: 522–524

Predictions
Bayer R, Stryker J, Smith M D 1995 Testing for HIV infection at home. New England Journal of Medicine 332: 1296–1299

Blaese R M, Culver K W, Miller A D et al 1995 T lymphocyte-directed gene therapy for ADA-SCID: initial trial results after 4 years. Science 270: 475–480

Coutelle C, Douar A-M, Colledge W H, Froster U 1995 The challenge of fetal gene therapy. Nature Medicine 1: 864–866

Crystal R G 1995 Transfer of genes to humans: early lessons and obstacles to success. Science 270: 404–410

Fulginiti V A 1993 Genetics: The quiet revolution in science and medicine: implications for research on child health issues, education of health professionals, and the new preventive and curative medicine. American Journal of Diseases of Children 147: 1139–1141

Friedmann T 1996 Human gene therapy – an immature genie, but certainly out of the bottle. Nature Medicine 2: 144–147

Kohn D B, Weinberg K I, Nolta J A et al 1995 Engraftment of gene-modified umbilical cord blood cells in neonates with adenosine deaminase deficiency. Nature Medicine 1: 1017–1023

Krishnan S, Haensler J, Meulien P 1995 Paving the way towards DNA vaccines. Nature Medicine 1: 521–522

Leiden J M 1995 Gene therapy – promise, pitfalls, and prognosis. New England Journal of Medicine 333: 871–873

Miller H I 1994 Gene therapy for enhancement. Lancet 344: 316–317

Ponder B A J 1994 Setting up and running a familial cancer clinic. In: Ponder B A J (ed) Genetics of malignant disease. British Medical Bulletin 50: 732–745

Wivel N A, Walter L 1993 Germ-line gene modification and disease prevention: some medical and ethical perspectives. Science 262: 533–538

GLOSSARY AND ABBREVIATIONS

GLOSSARY

Allele-specific oligonucleotides **(ASOs)** Oligonucleotides which are constructed with DNA sequences homologous to specific alleles. Two ASOs can be made which differ in sequence at only one nucleotide base thereby distinguishing a mutant allele with a point mutation from its corresponding wild-type allele (see Fig. 2.19).

Alleles Abbreviation for allelomorph and means alternative forms of the same gene.

Allogeneic From one person to another who is genetically dissimilar but of the same species (see autologous, heterologous).

Alu repeats The most common interspersed repeat (~300 bp in size) in human DNA accounting for about 5% of total DNA. So named because it is cleaved by the restriction enzyme *Alu*I.

Amino acids The building blocks of proteins. Each amino acid is encoded by a nucleotide triplet (see codon, Fig. 2.1, Table 2.1).

Amniocentesis Aspiration of amniotic fluid during pregnancy.

Amplification Multiple copies of a DNA sequence.

Aneuploid Any chromosome number that is not an exact multiple of the haploid number (23 in humans). Examples of aneuploidy include the presence of an extra copy of a single chromosome, e.g. trisomy 21 (Down syndrome), or the absence of a single chromosome, e.g. monosomy (as found in Turner syndrome 45,X) (see ploidy).

Anneal Formation of double-stranded nucleic acid from single-stranded forms.

Antibody More correct term is immunoglobulin. A protein produced by higher vertebrates following exposure to a foreign substance (called an antigen). The Y-shaped antibodies bind to antigens and neutralise them. Antibodies can be polyclonal or monoclonal in origin (the latter are

derived from a single cell and so all antibodies are identical) (see Fig. 7.10).

Anticipation Increasing severity or earlier age at onset of a genetic disease in successive generations (see Fig. 3.18).

Antisense Antisense DNA is the non-coding strand of DNA. It functions as the template for mRNA production which then contains the sequence present on the sense strand. Antisense RNA or antisense oligonucleotides have sequences which are complementary to mRNA and so interfere with the latter's function (see Figs 2.1, 7.6).

Apoptosis A gene-directed cellular activity in which cellular products result in self-destruction of the cell. The definition of apoptosis as a form of programmed cell death is also used, although not universally accepted (see Ch. 6, Fig. 6.10).

Assortive mating Sexual reproduction in which the pairing of mates is not random, i.e. members of a particular group which are more (less) likely to mate with other members of that group produce positive (negative) assortive mating.

Attenuated virus A virus which has become less pathogenic following passage outside its natural host.

Autologous From the same person (see allogeneic, heterologous).

Autosomal disease Disease which is the result of an abnormality affecting the 22 pairs of autosomes (non-sex chromosomes).

Bacteriophage, 'phage' A virus which infects bacteria (see clone, vector).

Base pair A measurement of length for DNA. Includes a nucleotide base with its complementary base, i.e. adenine (A) would bind to thymine (T) or cytosine (C) to guanine (G) (see complementary, Fig. 2.1).

Bioinformatics   See informatics.

Candidate gene   A gene which would be a good starter to initiate a search for the genetic basis of an inherited disorder of unknown origin, e.g. the myosin genes in muscle disorders.

Cap   Post-transcriptional change to the 5' end of the growing mRNA molecule in which a modified nucleotide (4-methylguanosine) is added. Has a functional role since it is recognised by ribosomes as the initiation signal for protein synthesis.

Carcinogen   Physical or chemical agent which induces cancer.

Carrier   An individual who is heterozygous for a mutant allele which causes a genetic disorder in the homozyous or hemizygous states.

Cell cycle   The timed sequence of events occurring in a eukaryotic cell between mitotic divisions. Divided into M (mitotic), S (DNA synthetic), G_1 and G_2 (gap or pause phases) and G_0 (resting phase). The times for each component differ between cell lines (Box 6.1).

CentiMorgan (cM)   Distance between DNA loci as determined on a genetic map. One cM distance indicates two markers are inherited separately 1% of the time. In terms of the physical map, 1 cM is very approximately equal to 1 Mb (see megabase). Name is derived from T H Morgan.

Centromere   The heterochromatic constricted portion of a chromosome where the chromatids are joined (see heterochromatin, telomeres, Fig. 2.22).

Chimaera   An individual composed of a mixture of genetically different cells. A chimaera is distinguished from a mosaic on the basis that the cells in a chimaera are derived from different zygotes, e.g. transgenic mouse formed by the embryonic stem cell approach (see mosaicism, transgenic, Fig. 2.24).

Chorionic villus sampling (CVS)   Biopsy of the chorion frondosum during pregnancy to obtain a source of fetal tissue for prenatal diagnosis (see Figs 4.1, 4.2).

Chromosome walking   Directional movement along a chromosome. Used in positional cloning to reach genes. Overlapping clones (contigs) are important for chromosome walking. Genetic distances will allow an assessment of the distance from target DNA. Physical maps, e.g. pulsed field gel electrophoresis, will provide more accurate distances for the walk. A faster walk is possible by constructing jumping libraries (see Figs 3.7, 3.8).

Clone   Refers to identical cells or molecules with a single ancestral origin. To clone a gene means to take a single gene or part of a gene and isolate it from the remainder of genomic DNA. The cloned gene can then be produced in unlimited amounts (see functional cloning, positional cloning, Figs 2.9, 2.10).

Codon   Three adjacent nucleotide bases in DNA/RNA that encode for an amino acid (see Table 2.1).

Complementary   The specific binding between the purine–pyrimidine base pairs of double-stranded nucleic acid. Thus, adenine (purine) will covalently bind to thymine (pyrimidine) and guanine (purine) to cytosine (pyrimidine) in a 1 to 1 ratio (see base pair, Fig. 2.1).

Complementary DNA (cDNA)   DNA which is synthesised from a mRNA template. The enzyme required for this is reverse transcriptase (see Figs 1.1, 6.2).

Complementation group   The term 'cistron' or 'gene' has also been used for complementation groups. This terminology reflects the method used to distinguish whether two mutations which give the same phenotype are different alleles of the same gene or different genes with a similar function. The procedure involves cell fusion studies and the effect of this on a measureable characteristic. Correction of this parameter in the hybrids (i.e. complementation) suggests different complementation groups, i.e. genes, and vice versa.

Compound (heterozygote)   An individual with two different mutant alleles at a locus.

Concordance   Both members of a twin pair demonstrate the same phenotype or trait (see discordance).

Congenic   Inbred strains that differ from one another in a small chromosomal segment, cf. syngeneic inbred strains which are identical except for sexual differences (see Fig. 5.8).

Congenital   Present at birth.

Conservation (DNA)   The finding that a DNA sequence is present in a wide range of phylogenetically distant organisms suggests functional significance since it is unlikely that during evolution a region of DNA would have remained unaltered unless it had a specific and important function, e.g. it is a gene. The *ras* proto-oncogene illustrates this since it is conserved in organisms as divergent as humans and yeast.

Constitutional (cells)   Cells which would be representative of an organism. In DNA testing for loss of heterozygosity, examples of constitutional cells which would provide a baseline for the DNA polymorphisms include lymphocytes (if the cancer is non-haematological) or fibroblasts which could be obtained from a skin biopsy.

Constitutive (genes)   Genes which are expressed following interaction between a promotor and RNA polymerase without additional regulation. Also called 'housekeeping' genes since they are often expressed in all cells at low levels. In contrast are inducible genes, e.g. metallothionein, which expresses following exposure to some heavy metals.

Consultand   The person seeking or referred for genetic counselling (see also proband).

Contigs   Overlapping clone sets which represent a continuous region of DNA.

Contiguous gene syndromes   A group of disorders which have malformation patterns often in association with mental retardation and growth abnormalities. The clinical heterogeneity found in these disorders may reflect the involvement of a number of physically related but otherwise distinct genes. Examples are given in Table 3.12, see also Figure 3.24.

Cordocentesis   The way in which to obtain a fetal blood sample by ultrasound-guided umbilical vein puncture.

Cosmid   Derived from plasmids but contains *cos* sites from phage lambda. Used as a vector for cloning DNA.

CpG islands   Regions of 1–2 kb containing a high density of hypomethylated cytosine residuals associated with guanine. CpG islands are frequently found at the 5′ end of genes (see methylation).

Cyclins   Families of interacting proteins involved in the regulation of the cell cycle. So named because their levels are cell-cycle dependent (see Box 6.1, Fig. 6.9).

Cytokines   Proteins (but not antibodies) which are released by some cells in response to contact with an antigen, e.g. interleukin-2 (IL-2). Cytokines function as intracellular mediators, e.g. generation of immune response seen with IL-2 (Fig. 7.3).

Decoding   Identifying the function of a gene from its DNA sequence.

Deletion   Loss of a segment of DNA or chromosome (see interstitial deletion, microdeletion, Figs 2.5, 2.6).

Diploid   The chromosome number found in somatic cells. In humans this will be 46, i.e. twice the number present in the germ cells (see haploid, Fig. 2.21).

Discordance   Members of a twin pair do not demonstrate the same phenotype or trait (see concordance).

Disomy   See uniparental disomy.

Dizygotic twins   Twins (fraternal) produced from two separate ova fertilised by different sperms (see monozygotic twins, Box 3.9).

Dominant   A genetic disorder is said to have dominant inheritance if the mutant phenotype is produced when only one of the two normal (wild-type) alleles at a particular locus is mutated (see recessive, Fig. 3.10).

Dominant-negative effect   Inactivation of one of the two tumour suppressor gene loci can produce what appears to be a dominant effect if the mutant protein inhibits the normal product from the remaining normal allele.

e antigen   Hepatitis B virus e antigen (HBeAg) – a part of the core antigen of the hepatitis B virus (HBcAg) which is secreted into the serum through cellular secretion pathways. HBeAg correlates strongly with infectivity.

Electroporation   The use of a pulsed electric field to introduce DNA into cells in culture.

Embryonic stem cells (ES cells)   Totipotential cells that can be cultured from the early embryo. ES cells can be induced to remain undifferentiated in culture. Foreign DNA is transfected into these cells which are then microinjected into blastocysts of developing embryos. A chimaera is produced. If chimaerism also involves the germ cells it will be possible to breed mice which are heterozygous or homozygous for the foreign gene (see Fig. 2.24).

Enhancer   DNA sequences which have the following properties: (1) they increase transcriptional activity, (2) they are effective even if inverted in position and (3) they operate over long distances (see Box 1.4).

Env gene   Encodes for the envelope protein of a retrovirus (see Fig. 5.2).

Epigenetic   Changes which affect the phenotype but not genotype. The changes may have been inherited but they do not involve an alteration in genetic information, e.g. imprinting.

Episome   Plasmid or plasmid-like extrachromosomal DNA which has the ability to integrate into the host's chromosome.

Euchromatin   Non-condensed, light-coloured bands following staining to produce G (Giemsa)-banding of chromosomes. More likely to contain transcriptionally active DNA (see heterochromatin, Fig. 2.21).

Eukaryotes   Organisms ranging from yeast to humans which have nucleated cells.

Exon   That segment in a gene which codes for a polypeptide and is represented in the mRNA (see intron).

Expressed sequence tag (EST)   Unique fragment of a gene expressed in human tissue.

Expressivity   The severity of a phenotype. Variable expressivity is a feature of autosomal dominant disorders.

Familial   A condition which is more common in relatives of an affected individual than in the general population (e.g. breast cancer).

Fingerprints   Dermatoglyphic fingerprints: derived from the ridged skin patterns of the fingers. DNA fingerprints: obtained from multilocus minisatellite DNA polymorphisms (see minisatellites, satellite DNA, Figs 8.6, 8.7).

Five prime (5′)   The 5′ position of one pentose ring in DNA is connected to the 3′ position of the next pentose via a phosphate group. The phosphodiester–sugar backbone of DNA consists of 5′–3′ linkages and this is the direction that the nucleotide bases are transcribed (see Fig. 2.1).

Flanking (markers, DNA)   DNA markers on either side of a locus: DNA sequences on either side of a gene.

Fluorescence *in situ* hybridisation (FISH)   Non-isotopic method to label DNA probes for in situ hybridisation. The ability to utilise multiple fluorochromes in the same reaction increases the utility of this procedure. The resolving power of FISH is further enhanced if interphase chromosomes are studied.

Footprinting   Technique which identifies sites where there is protein bound to DNA. This complex then becomes resistant to degradation by nucleases.

Frameshift mutation   A mutation in DNA such as a deletion or insertion which interferes with the normal codon (triplet base) reading frame. All codons 3′ to the mutation will have no meaning. For example, the triplets GGT-TCT-GTT code for amino acids glycine, serine and valine, respectively. A deletion of one nucleotide, e.g. a G of the GGT, would disrupt the reading frame, to give GTT-CTG-TT, etc. The protein product will terminate when a new stop codon is reached.

Functional cloning   Cloning strategy in which knowledge of a gene's product (function) is used to clone the gene (see clone, positional cloning, Fig. 2.14).

G418   Neomycin analogue which kills cells unless they are neomycin-resistant or carry the gene for neomycin resistance (Figs 7.2, 7.8).

Gag gene   Group specific antigen – encodes core protein for a retrovirus (see Fig. 5.2).

Ganciclovir   Prodrug which can be phosphorylated to its active metabolite by thymidine kinase from the herpes simplex 1 virus (HSV-tk). The active metabolite causes cell death by inhibiting DNA synthesis (Fig. 7.8, Box 7.6).

Gaucher disease   One of the lysosomal storage diseases due to a deficiency of the enzyme β glucocerebrosidase.

G-banding   G (for Giemsa)-banding is a commonly used procedure to identify chromosomal bands in a karyotype. Spreads of cells in metaphase are treated with trypsin and then stained with Giemsa (see Fig. 2.21).

Gene   A sequence of DNA nucleotide bases which codes for a polypeptide.

Gene therapy   The transfer of genetic material (DNA/RNA) into the cells of an organism to treat disease or for research purposes.

Genetic engineering   Colloquial term for recombinant DNA technology: the experimental or industrial applications of technologies which can alter the genome of a living cell.

Genetic map   An indirect measure of distance, constructed by determining how frequently two markers (DNA polymorphisms, physical traits or syndromes) are inherited together. Distances in genetic maps are measured in terms of centiMorgans (see physical map).

Genome   The total genetic material of an organism, i.e. an organism's complete DNA sequence.

Genotype   The genetic constitution of an organism. In terms of DNA markers it refers to the genetic constitution of alleles at a specific locus, e.g. the two haplotypes (see haplotype, Fig. 3.25).

Germ cells   Cells which differentiate early in embryogenesis to form ova and sperm.

G proteins   Abbreviation for guanine-binding proteins. They play an important role in relaying messages from the cell surface to the nucleus. Act by binding GTP (guanosine triphosphate) which leads to activation of a second messenger system such as adenylyl cyclase. There are many G proteins, including the product of the *ras* proto-oncogenes. G proteins are self-regulating since the GTP-G protein complex is hydrolysed to inactive GDP-G protein by the GTPase activity of the G protein. Over 100 receptors convey messages through G proteins (see *ras*, signal transduction, Fig. 6.4).

Guthrie spot   Used (incorrectly) to describe the blood spot taken from newborns by heel prick. The blood spot is then used for newborn screening of genetic and metabolic disorders. The name is derived from the newborn screen for phenylketonuria which uses a test called the Guthrie bacterial inhibition assay.

Haematopoietic   Related to the blood; blood forming.

Haemoglobinopathies   Genetic disorders involving globin, the protein component of haemoglobin. Divided into the thalassaemia syndromes, e.g. α or β thalassaemia, and the variant haemoglobins, e.g. sickle-cell anaemia (HbS).

Haploid   The chromosome number found in gametes. In humans this will be 23, i.e. one member of each chromosome pair (see diploid).

Haplotype   A set of closely linked DNA markers at one locus which is inherited as a unit (see Fig. 3.25).

Hemizygous   Having only one copy of a given genetic locus, e.g. a male is hemizygous for DNA markers on the X chromosome (see Fig. 3.14).

Heterochromatin   Condensed, dark-coloured bands following G (Giemsa)-banding of chromosomes. Contains predominantly repetitive DNA (see euchromatin, centromere, Fig. 2.21).

Heteroduplex   Hybrid DNA involving two strands which are different, e.g. there may be a base mismatch (see homoduplex, Fig. 2.17).

Heterologous   Belonging to another species, e.g. the use of salmon sperm DNA to block non-specific hybridisation by human DNA (see allogenic, autologous).

Heteroplasmy   The presence of more than one type of mitochondrial DNA in a cell. There are thousands of molecules of mitochondrial DNA per cell. If there is mutant mitochondrial DNA it can be present in varying amounts. Some cells might have predominantly wild-type DNA, others predominantly mutant DNA (called homoplasmy) and others are said to be heteroplasmic because there is a mixture of both. Thus, phenotypic variation between cells is possible.

Heterozygote   An individual with two different alleles (e.g. gene, polymorphic marker) at a single locus (see homozygote, Figs 3.3, 3.5).

HLA   Abbreviation for *human leukocyte antigen*. HLA is encoded for by a multigene complex occupying ~3500 kb of DNA on the short arm of chromosome 6. Antigens belonging to the HLA system are found on the surface of all cells except the red blood cells. HLA is concerned with normal immunological responses and plays a vital role in graft rejection or acceptance following transplantation. Also known as MHC (*m*ajor *h*istocompatibility *c*omplex) (Fig. 8.5).

Homeobox   A sequence of about 180 bp near the 3′ end of some homeotic genes. The 60 amino acid peptide encoded by the homeobox is a DNA-binding protein (*homeo* – Greek for 'alike').

Homeotic genes   Genes which determine the shape of the body along the antero-posterior axis of the embryo. Mutations in homeotic genes cause a part of the body to be replaced by a structure normally found elsewhere. Conserved DNA sequences within homeotic genes are called homeoboxes. All vertebrates including humans have four homeotic gene complexes located on different chromosomes. In mammals homeotic genes are called *hox* genes (from *homeobox*). Another gene family involved in development is the *pax* genes, the conserved sequence for which is called the *paired box* (see *pax* genes).

Homoduplex   Hybrid DNA involving two strands which are identical (see heteroduplex).

Homologous recombination   A form of gene targeting on the basis of recombination between DNA sequences in the chromosome and newly introduced identical DNA sequences (see homology, Fig. 7.8).

Homology   Fundamental similarity, matched, e.g. homologous (the same) chromosomes pair at meiosis. Homology between DNA sequences means close similarity.

Homozygote   An individual with two identical alleles (e.g. gene, polymorphism) at a single locus (see heterozygote, Figs 3.3, 3.5).

Hot spots   Regions in genes or DNA where mutations occur with unusually high frequency.

Housekeeping (genes)   Genes that are expressed in virtually all cells since they are fundamental to the cell's functions.

Human Genome Project   Multicentred,   multinational,

multibillion dollar project launched in 1988 and estimated to be completed in 2005. The project's aims include construction of a genetic and physical map of the human genome and a number of model organisms. Ultimately there will be a complete DNA sequence for the entire human genome. At the same time technology development and training in human gene mapping will be undertaken.

Hybridisation   The pairing, through complementary nucleotide bases (A with T and G with C), of RNA/DNA strands to produce an RNA/RNA or RNA/DNA or DNA/DNA hybrid (see Fig. 2.3).

Illegitimate transcription   Low transcription of a tissue-specific transcribing gene in non-specific cells, e.g. the detection of mRNA for the β myosin heavy chain gene (a muscle-specific gene) in peripheral blood lymphocytes. Also called 'ectopic' or 'leaky' RNA (see Box 3.7).

Immunoblot   See western blot.

Immunoglobulin   See antibody.

Immunophenotyping   Typing of cells with immunological markers such as monoclonal antibodies (also called cell marker analysis).

Imprinting   Reversible modification of DNA that leads to differential expression of maternally and paternally inherited DNA or homologous chromosomes (see uniparental disomy, Figs 3.20, 10.1).

Informatics   The application of computer and statistical techniques to the management of information (also known as bioinformatics when it relates specifically to biology) (see Human Genome Project, Ch. 10).

Informative (polymorphism)   Means a polymorphism is heterozygous and so able to distinguish two alleles. In a parental mating, at least one parent must be heterozygous for a polymorphism to be potentially informative. If both parents are heterozygous, the polymorphism is fully informative (if there is a key individual to help assign which marker co-segregates with disease, etc.) (see Fig. 3.3).

In situ hybridisation   Hybridisation of a DNA probe (labelled with ^{3}H, fluorescein or a chemical such as biotin) to a metaphase chromosome spread or a tissue section on a slide.

Interleukins   Proteins secreted by mononuclear leukocytes which induce the growth and differentiation of other haematopoietic cells.

Interstitial deletion   Loss of DNA or part of a chromosome which does not occupy a terminal position.

Intron   Segment of DNA which is transcribed but does not contain coding information for a polypeptide (also called *intervening sequence* or *IVS*). It is spliced out of the transcript before mature mRNA is formed (see exon).

Isozymes (isoenzymes)   Different forms of an enzyme.

Karyotype   An individual's or a cell's chromosomal constitution (number, size and morphology). Determined by examination of chromosomes with light microscopy and the use of stains (see Fig. 2.21).

Kilobase (kb)   One thousand base pairs in a sequence of DNA.

Kilodalton (kDa)   A unit which measures the molecular weight of proteins (= 1000 daltons). One dalton approximates to the molecular weight of a hydrogen atom. The molecular weight of a protein will be based on the sum of the atomic weights of the elements contained in it.

Library   A large number of recombinant DNA clones which have been inserted into a vector for the purpose of cloning a segment of DNA (see Fig. 2.9).

Ligand   A molecule which binds to a complementary site on a cell or other molecule.

Linkage   The tendency to inherit together two or more non-alleleic genes or DNA markers than is to be expected by independent assortment. Genes/DNA markers are linked because they are sufficiently close to each other on the same chromosome (see Fig. 3.25).

Linkage disequilibrium   Preferential association of linked genes/DNA markers in a population, i.e. the tendency for some alleles at a locus to be found with certain alleles at another locus on the same chromosome with frequencies greater than would be expected by chance alone, e.g. HLA-DQ and HLA-DR alleles (see Fig. 8.5).

Lipofection   An in vivo or in vitro way to transfer DNA into a cell's nucleus. The gene of interest is mixed with a cationic lipid suspension and then mixed with the cell of interest.

Liposomes   Synthetic spherical vesicles with a lipid bilayer. Function as artificial membrane systems to deliver DNA, etc. into cells.

Lod score   Statistical test to determine whether a set of linkage data are linked or unlinked. Lod in an abbreviation of the 'log_{10} of the *odds*' favouring linkage. For genetic disorders which are not X-linked, a lod score of +3 (1000:1 odds of linkage) indicates linkage whilst a score of −2 is odds of 100:1 against linkage.

Lymphoproliferative disorders Lymphomas and leukaemias of lymphocyte origin.

Megabase (Mb) One million base pairs in a sequence of DNA.

Meiosis Process in which diploid germ cells undergo division to form the haploid chromosome number (see mitosis).

Messenger RNA (mRNA) Transfers the genetic information from DNA to the ribosomes. Contains the template for polypeptide production.

Metastasis A secondary tumour arising from cells carried from the primary tumour to a distant locus.

Methylation (of DNA) Vertebrate DNA contains a small proportion of 5-methylcytosine which arises from methylation of cytosine bases where they occur in the sequence CpG. The methylation status of DNA correlates with its functional activity: inactive genes are more heavily methylated and vice versa (see CpG islands).

MHC See HLA.

Microdeletion DNA/chromosomal deletion which is not detectable by conventional techniques such as microscopy (cytogenetics) or Southern blotting (DNA mapping).

Microsatellites As for minisatellites except that the polymorphism allele size is smaller, e.g. <1 kb and the basic core repeat unit involves a two to four nucleotide base pair repeat motif. Also known as SSRs (*simple sequence repeats*). One example is repeats of the motif CA, e.g. CACACACACA, etc. which is also written as AC (ACACA-CACAC, etc.). There is confusion with terminology since the above are identical. To avoid this problem it has been recommended that the microsatellites are written in alphabetical order (the above would be $(AC)_n$) (see Figs 3.21, 8.4).

Minisatellites Repeat DNA segments which comprise short head-to-tail tandem repeats giving the variable *number* of *tandem* *repeat* (VNTR) type polymorphisms with approximate size of 1–30 kb. VNTRs can be of two types: single locus or multilocus. The latter are utilised in constructing a DNA 'fingerprint' of an individual (see microsatellites, satellite DNA, Figs 8.2, 8.3, 8.6).

Missense mutation A single DNA base change which leads to a codon specifying a different amino acid, e.g. the base change of GGT (glycine) to GTT (valine).

Mitosis Somatic cell division. The process in which chromosomes duplicate and segregate during cell division (see meiosis).

Monoclonal Derived from a single clone, i.e. monoclonal antibody, monoclonal lymphocyte population (see polyclonal, Figs 6.18, 7.11).

Monozygotic twins Genetically identical twins formed by the division into two at an early stage in development of an embryo derived from a single fertilised egg (see dizygotic twins, Box 3.9).

Mosaicism A condition in which an individual or tissue has two or more cell lines of different genetic or chromosomal constitution. In contrast to a chimaera, both cell lines in a mosaic are derived from the same zygote.

Multidrug resistance (MDR) Development of simultaneous resistance to multiple structurally unrelated chemotherapeutic agents (see P-glycoprotein).

Multifactorial disorders Diseases which result from an interaction of environmental factors with multiple genes at different loci (see polygenic, which is sometimes used in the same sense as multifactorial).

Murine Of the mouse (Latin, *mus*).

Mutation An alteration in genetic material. This could be a single base change (point mutation) or more extensive losses of DNA (deletions) (see frameshift mutation, missense mutation, nonsense mutation, Figs 2.6, 2.19).

Nonsense mutation A single DNA base change resulting in a premature stop codon (TAA, TGA, TAG), e.g. TCG (serine) to TAG (stop).

Northern blotting Procedure to transfer RNA from an agarose gel to a nylon membrane (see Southern blotting, western blotting).

Nosocomial Hospital-acquired.

Nucleases Enzymes which breaks down nucleic acid. There are DNase (DNAase) and RNase (RNAase) enzymes. RNA in particular is susceptible to RNases so that preparation of RNA requires a lot more care compared to the robust DNA.

Nucleotide The monomeric component of DNA/RNA comprising a base (A, adenine; T, thymine; U, uracil; G, guanine or C, cytosine), a pentose sugar (deoxyribose or ribose) and a phosphate group (see Fig. 2.1).

Oligonucleotides Small single-stranded segments of DNA typically 20–30 nucleotide bases in size which are synthesised in vitro. Uses include DNA sequencing, DNA amplification and DNA probes (see primer, allele-specific oligonucleotides).

Oncogenes   Genes associated with neoplastic proliferation following a mutation or perturbation in their expression (see proto-oncogenes, *ras*).

***Online Mendelian Inheritance in Man* (OMIM)**   An encyclopaedia of phenotypes for genetic traits, disorders and gene loci established by Victor McKusick. Available on hard copy (MIM – *Mendelian Inheritance in Man*) or via the http://www3.ncbi.nlm.nih.gov/omim/ Internet (OMIM)

P53   A tumour suppressor gene, mutations of which are frequently found in human cancers (see tumour suppressor genes).

Palindrome   Sequence of DNA which is identical in either direction.

e.g.   5'-GTCGAC-3'
       3'-CAGCTG-5'

This is the recognition sequence for the restriction enzyme *Sal*I. Further examples of palindromic sequences are given in Table 2.2. Palindromes involving small to large segments of DNA are found throughout the genome and need not necessarily be sites recognised by restriction enzymes.

Parthenogenesis   The development of an egg that has been activated in the absence of sperm.

Pathogenesis   The steps involved in development of a disease.

***Pax* genes**   Abbreviation of *paired* box. The paired box is a conserved DNA sequence which plays a role in development of the neural crest (see also homeotic genes, Table 10.3).

Penetrance   All or nothing phenomenon relating to the expression of a gene.

P-glycoprotein   A glycoprotein associated with multidrug resistance. A member of the ATP-binding cassette transporter proteins. P-glycoprotein allows the active extrusion of a variety of compounds out of cells. The gene for P-glycoprotein is *MDR1* (see multidrug resistance).

Phase   A term to describe the combination in which polymorphic markers have been inherited within the context of a family study.

Phenotype   The observed appearance of a gene or an organism which is determined by the genotype and its interaction with the environment.

Physical map   Can be constructed in different ways but in contrast to genetic maps it represents measurements of physical length (bp, kb, Mb). Types of physical maps include: cytogenetic, pulsed field gel electrophoresis, fluorescence in situ hybridisation, contigs, e.g. cosmid or YAC (see genetic map).

Plasmid   Cytoplasmic, autonomously replicating extrachromosomal circular DNA molecule. Used as vectors for cloning. In vivo, plasmids are found in bacteria where they can code for antibiotic resistance factors (see episome, vector).

Pleiotropy   Different effects of a gene on apparently unrelated characteristics such as the phenotype, organ systems or functions.

Ploidy   The number of chromosomes in a cell. Euploid, the correct number; aneuploid, an abnormally high or low number; polyploid, a multiple of the euploid number.

***Pol* gene**   Encodes reverse transcriptase enzyme of a retrovirus (see Fig. 5.2).

Polyclonal   Derived from more than one cell (see monoclonal).

Polygenic inheritance   Trait which results from an interaction of multiple genes at different loci (see multifactorial disorders).

Polymerase   RNA polymerases are enzymes which catalyse the formation of RNA using DNA as a template. DNA polymerases are enzymes which can synthesise DNA from four nucleotide precursors (dATP, dTTP, dCTP and dGTP) provided a template or primer is available to start off the process. Functions of the DNA polymerases include DNA repair and DNA replication. Reverse transcriptase is also a DNA polymerase (see Figs 2.11, 7.6).

Polymerase chain reaction (PCR)   DNA method which allows amplification of a targeted DNA sequence (see Fig. 2.7).

Polymorphisms (DNA)   A part of the DNA sequence that can occur in two or more forms which can be detected on the basis of variations in the sizes of DNA fragments produced following digestion with restriction enzymes. Polymorphic variations result from point mutations (see RFLP) or insertions of repetitive DNA sequences (see VNTR). In terms of human genetics, polymorphisms are inherited along Mendelian lines in a family and by definition should occur at a 1% or more frequency within a population (see RFLP, VNTR, Figs 2.15, 3.3, 8.2).

Positional cloning   Cloning of a gene on the basis of its chromosomal position rather than its functional properties. Also called 'reverse genetics' (see clone, functional cloning, Fig. 2.14).

Primer   A short oligonucleotide segment which pairs with a complementary single-stranded DNA sequence. The double-stranded segment formed has a free 3′ terminus which provides the template for extension into a second strand (see oligonucleotides, Fig. 2.11).

Proband (or propositus or index case)   The affected individual from whom a pedigree is constructed (see consultand, Fig. 3.14).

Probe   A single-stranded segment of DNA/RNA which is labelled with a radioactive substance or chemical. The probe will bind to its complementary single-stranded target sequence. Hybrids formed are detectable by autoradiography or by chemical changes. There are a number of different probes: genomic, cDNA, RNA, oligonucleotide. The naming of probes has led to confusion. Therefore, an attempt to induce uniformity has been made by naming loci to which probes will hybridise, e.g. D15S10 indicates human chromosome 15 DNA segment 10. A number of DNA probes could hybridise to this locus (see Figs 2.3, 2.4).

Prokaryotes   Bacteria and certain algae with cells that are not nucleated.

Promotor   DNA sequence located 5′ to a gene which indicates the site for transcription initiation. May influence the amount of mRNA produced and the tissue specificity. Examples of promotors are the TATA, CCAAT boxes (see Cap, Fig. 3.4).

Proto-oncogenes   Normal genes comprising a number of functionally different classes which are involved in cellular growth control. Altered forms of the proto-oncogenes are called oncogenes.

Provirus   Virus that is integrated into the chromosome of its host cell and can be transmitted from one generation to another without causing lysis of the host (see retrovirus, reverse transcriptase).

Pulsed field gel electrophoresis (PFGE)   A type of gel electrophoresis in which large fragments of DNA can be separated by altering the angle at which the electric current is applied (see Fig. 2.16).

Ras   A family of proto-oncogenes (H-*ras*-1, K-*ras*-2 and N-*ras*) which encode for a protein called p21. p21 binds to GTP/GDP and has GTPase activity. *Ras*-derived proteins play a physiological role in regulation of cellular prolifera-

tion. Mutations in *ras* are found in a number of cancers (see G proteins, Fig. 6.4).

Recessive   The products of both normal (wild-type) alleles at a particular locus are non-functional in a recessive disorder (see dominant, Fig. 3.5).

Recombination   Crossing over (breakage and rejoining) between two loci which results in new combinations of genetic markers/traits at those loci, e.g. imagine that one locus has four genetic markers linearly arranged; a–b–c–d, and the second locus is: b–b–c–a. Recombination involving these two regions between the b–c markers would give new genetic combinations, i.e. a–b–c–a and b–b–c–d (see Fig. 3.25).

Repair genes   A group of genes which monitors and repairs DNA errors. There are two major repair pathways: (i) mismatch repair and (ii) nucleotide excision repair (see Fig. 6.11).

Restriction endonucleases (enzymes)   Enzymes which recognise specific short DNA sequences and cleave DNA at these sites (see Table 2.2).

Restriction fragment length polymorphism (RFLP)   Biallelic DNA polymorphism which results from the presence or absence of a restriction endonuclease site (see polymorphisms, Figs 2.15, 3.3, 4.8).

Restriction map   A series of restriction endonuclease recognition sites associated with a DNA locus or gene (see Fig. 2.5).

Retrovirus   RNA virus that utilises reverse transcriptase during its life cycle. After infecting the host cell, the retroviral (RNA) genome is transcribed into DNA which is then integrated into host DNA. In this way the retrovirus can replicate (see provirus, reverse transcriptase, Figs 6.1, 6.2).

Reverse genetics   A name for the recombinant DNA strategy which attempts to clone a gene on the basis of its position on the chromosome rather than its functional properties. The name is now replaced with the more descriptive term of positional cloning (see cloning, functional cloning, Fig. 2.14).

Reverse transcriptase   Enzyme which enables synthesis of single-stranded DNA (called cDNA) from an RNA template (see polymerase, Fig. 1.1).

Ribosomal RNA (rRNA)   The nucleic acid content of ribosomes. The latter are small cellular particles which are the site of protein synthesis in the cytoplasm.

Ribotyping The use of rRNA-specific DNA probes in Southern analysis to distinguish bacteria on the basis of their rRNA patterns on chromosomes. Polymorphic bands so produced facilitate typing of bacterial strains.

Satellite DNA Short head-to-tail tandem repeats which incorporate specific DNA motifs (see microsatellites, minisatellites, Figs 8.1, 8.2, 8.4, 8.6).

Screening (genetic) Testing individuals on a population basis to identify those who would be at risk for disease or transmission of a genetic disorder.

Sequence tagged site (STS) A way to provide unambiguous identification of DNA markers generated by the Human Genome Project. STSs comprise short, single-copy DNA sequences that characterise mapping landmarks on the genome.

Sequencing (DNA) Establishing the identity and order of nucleotides in a segment of DNA. The 'gold standard' in characterising a mutation (see Figs 2.11, 2.12).

Sibship A group comprising the brothers and sisters (siblings) in a family.

Signal transduction Transfer of signals from extracellular factors and their surface receptors by cytoplasmic messengers to modulate events in the nucleus (see G proteins, Fig. 6.4).

Simple sequence repeat (SSR) See microsatellites.

Somatic cells Any cell in an organism which is not a germ cell (sperm or egg).

Somatic cell genetic disorders One of the five groups of genetic disorders. Defects in DNA are found in specific somatic cells. An example of this type of disorder is cancer. By comparison the four other categories (single-gene, polygenetic, multifactorial and chromosomal disorders) have the genetic abnormality present in all cells including the germ cells.

Somatic cell hybrid A hybrid formed from the fusion of different cells. These usually come from different species, e.g. human and rodent hybrids are frequently used for human gene mapping.

Somatic mutation A mutation which occurs in any cell that will not become a germ cell.

Southern blotting Named after E Southern. Describes the procedure for transferring denatured (i.e. single-stranded) DNA from an agarose gel to a solid support membrane such as nylon (see northern blotting, western blotting, Fig. 2.6).

Splicing The removal of introns to produce mature mRNA.

Sporadic No obvious genetic cause.

Start codon Nucleotide codon (ATG) which is positioned at the beginning of a gene sequence in eukaryotes. Prokaryotes do not have such a start codon and so ATG is translated into the amino acid methionine.

Sticky ends Fragments of double-stranded DNA with a few end bases not paired, i.e. they anneal with greater efficiency than blunt-end fragments.

Stop codons Nucleotide codons (TAA, TGA and TAG) are positioned at the 3' end of a gene sequence and indicate the termination of a polypeptide.

Syntenic genes Genetic loci or genes which lie on the same chromosome or same DNA strand.

Tandem repeats Small sections of repetitive DNA in the genome, arranged in head-to-tail formation.

Telomeres The two ends of a chromosome (see centromere, Fig. 2.22).

Transcription Synthesis of a single-stranded RNA molecule from a double-stranded DNA template in the nucleus (see polymerase, translation).

Transduction (gene) Transmission of genetic material from one cell to another by viral infection.

Transduction (signal) See signal transduction.

Transfection Acquisition of new genetic markers by incorporation of added DNA into eukaryotic cells by physical or viral-dependent means (see Fig. 7.2).

Transfer RNA (tRNA) Provides the link between mRNA and rRNA. Each tRNA can combine with a specific amino acid and also bind to the relevant mRNA codon (see codon, mRNA, rRNA, translation).

Transformation (of bacteria) Acquisition of new genetic markers by incorporation of added DNA into bacteria.

Transformation (of cells) Sudden change in a cell's normal growth properties into those found in a tumour cell (see Fig. 6.3).

Transgenic The presence of foreign DNA in the germline.

Transgenic animals are produced by experimental insertion of cloned genetic material into the animal's genome. This can be done by microinjection of DNA into the pronucleus of a fertilised egg or through utilisation of embryonic stem cells. A proportion of transgenic animals will express the foreign gene and transmit it to their progeny (see embryonic stem cells).

Transition   Change of a purine (i.e. adenine or guanine) to a purine or a pyrimidine (i.e cytosine or thymine) to a pyrimidine (see transversion).

Translation   Cytoplasmic production of a polypeptide from the triplet codon information on mRNA (see transcription).

Translocation   The presence of a segment of one chromosome on another chromosome (see Fig. 6.5).

Transposons   DNA sequences which can replicate and insert a copy at a new location in the genome.

Transversion   Change of a purine to a pyrimidine or vice versa (see transition).

Tumour suppressor genes   (Also called recessive oncogenes, anti-oncogenes, growth suppressor genes). These are normal genes which have the suppression of tumourigenesis as one component of their function (see *P53*, Figs 6.6, 6.7).

Uniparental disomy   The inheritance of two copies of a chromosome from the one parent. This can be isodisomy (both chromosomes from the one parent are identical copies) or heterodisomy (the two chromosomes are different copies of the same chromosome). Described with a number of human chromosomes, e.g. 7, 11, 15, 16 (see also imprinting).

Variable *number* of *tandem repeat* (VNTR)   A mutiallelic DNA polymorphism which results from insertions or deletions of DNA between two restriction sites (see polymorphisms, Figs 2.15, 8.1, 8.2).

Vector   Cloning vehicle, i.e. plasmid, phage, cosmid or YAC, into which DNA to be cloned can be inserted (see Fig. 2.9).

Western blotting (immunoblot)   A technique used to separate and identify proteins (see northern blotting, Southern blotting).

Wild-type (gene)   The form of the gene normally present in nature.

X chromosome inactivation   Random inactivation of one of the two female X chromosomes during early embryonic development. Thus, cells in a female are mosaic in respect to which of the X chromosomes is functional.

Yeast artificial chromosome **(YAC)**   A cloning vector which allows large segments of DNA, e.g. 300 kb in size, to be cloned.

Zoo blot   A way to detect conservation of DNA sequence during evolution. In a zoo blot, a segment of DNA being investigated is used as a probe to hybridise against a series of DNA samples from various species, e.g. human, mouse, yeast. If the probe can detect unique DNA sequences in the above, it provides indirect evidence that the DNA sequence has been conserved during evolution, i.e. it has functional significance which might mean it is a gene.

Zoonoses   Infections transmitted from animals to humans.

Zygote   The diploid cell resulting from union of the haploid male and haploid female gametes, i.e. fertilised ovum.

ABBREVIATIONS

A	Adenine nucleotide base
ADA	Adenosine deaminase
AIDS	Acquired immunodeficiency syndrome
APC	Adenomatous polyposis coli
ASO	Allele-specific oligonucleotide
ATM	Ataxia telangiectasia mutated
bp	Base pair
BRCA1	Breast cancer 1
C	Cytosine nucleotide base
CCM	Chemical cleavage of mismatch

CDK	Cyclin-dependent kinase		**HbF**	Haemoglobin F (fetal haemoglobin)
cDNA	Complementary or copy DNA		**HbS**	Haemoglobin S (sickle-cell haemoglobin)
CEPH	Centre d'Etude du Polymorphisme Humain		**HBsAg**	Hepatitis B virus surface antigen
CFTR	Cystic fibrosis transmembrane conductance regulator, i.e. cystic fibrosis gene		**HIV**	Human immunodeficiency virus
CHO	Chinese hamster ovary (cell line)		**HLA**	Human leukocyte antigen
cM	CentiMorgan		**HNPCC**	Hereditary non-polyposis colon cancer
CMV	Cytomegalovirus		**HPFH**	Hereditary persistence of fetal haemoglobin
CSF	Colony stimulating factor		**HSV**	Herpes simplex virus
CVS	Chorionic villus sample (sampling)		**IgG**	Immunoglobulin G
DGGE	Denaturing gradient gel electrophoresis		**IgM**	Immunoglobulin M
DHFR	Dihydrofolate reductase		**IVS**	Intervening sequence (= intron)
DNA	Deoxyribonucleic acid		**kb**	Kilobase
dsDNA	Double-stranded DNA		**kDa**	Kilodalton
DZ	Dizygotic (twin)		**LDL**	Low density lipoprotein
EPO	Erythropoietin		**LTR**	Long terminal repeat (of a retrovirus)
ES cells	Embryonic stem cells		**Mb**	Megabase
EST	Expressed sequence tag		**MDR**	Multidrug resistance
FAP	Familial adenomatous polyposis		*MDR1*	Gene for P-glycoprotein
FDA	Federal Drug Administration		**MHC**	Major histocompatibility complex
FHC	Familial hypertrophic cardiomyopathy		**MoAb**	Monoclonal antibody
FISH	Fluorescence in situ hybridisation		**mRNA**	Messenger ribonucleic acid (RNA)
5′ → 3′	Direction of transcription		**MRSA**	Methicillin-resistant *Staphylococcus aureus*
G	Guanine nucleotide base		**MZ**	Monozygotic (twin)
G-CSF	Granulocyte colony stimulating factor		**neo**	Neomycin
GDP	Guanosine diphosphate		**NIH**	National Institutes of Health
GTP	Guanosine triphosphate		**OMIM**	*Online Mendelian Inheritance in Man*
H-2	mouse equivalent of MHC/HLA		^{32}**P**	Radioactive phosphorus
			PCR	Polymerase chain reaction

PFGE	Pulsed field gel electrophoresis	**SIV**	Simian immunodeficiency virus
PTT	Protein truncation test	**SSCP**	Single-stranded conformation polymorphism
RAC	Recombinant DNA Advisory Committee	**ss-DNA**	Single-stranded DNA
RB1	Retinoblastoma gene	**SSR**	Simple sequence repeat (microsatellite)
rDNA	Recombinant DNA	**T**	Thymine nucleotide base
riDNA	Ribosomal DNA	**tk**	Thymidine kinase
RFLP(s)	Restriction fragment length polymorphism(s)	**TSG**	Tumour suppressor gene
rh	Recombinant human	**VNTR(s)**	Variable number of tandem repeat(s)
rRNA	Ribosomal RNA	*WT1*	Wilms tumour gene 1
RT-PCR	Reverse transcriptase PCR	**YAC(s)**	Yeast artificial chromosome(s)

INDEX

Note: page numbers in *italics* refer to figures and tables.